FROZEN FAUNA OF THE MAMMOTH STEPPE

FROZEN FAUNA OF THE MAMMOTH STEPPE

The Story of Blue Babe

R. DALE GUTHRIE

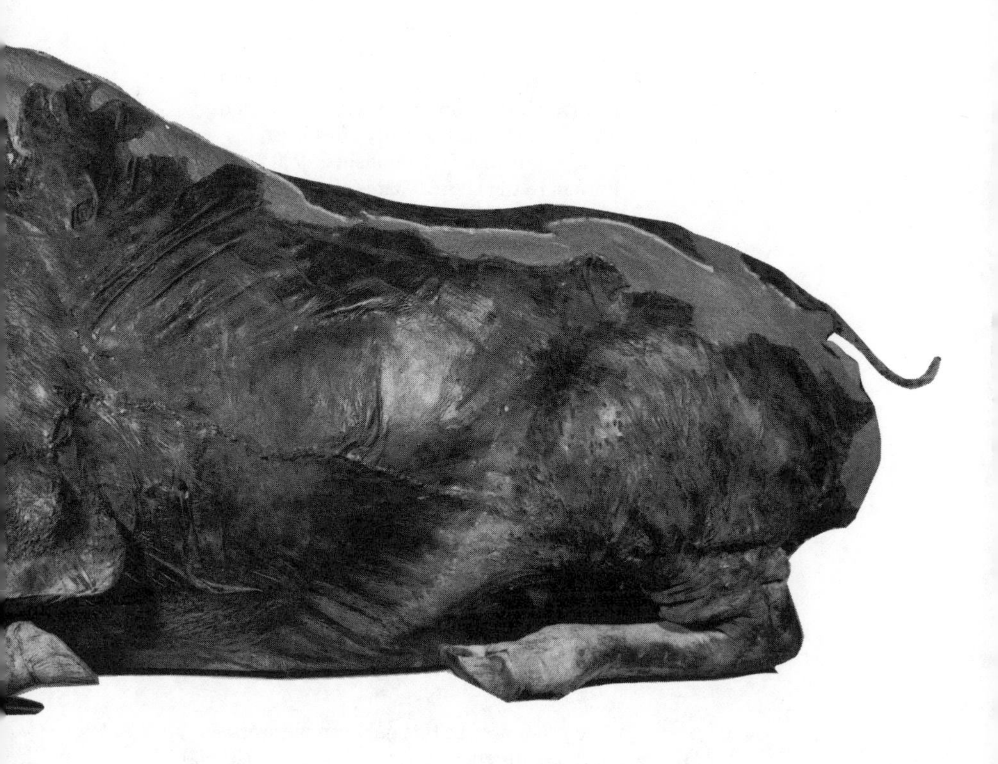

The University of Chicago Press
Chicago and London

The University of Chicago Press, Chicago, 60637
The University of Chicago Press, Ltd., London

Printed in the United States of America
99 98 97 96 95 94 93 92 91 5432

Library of Congress Cataloging in Publication Data

Guthrie, R. Dale (Russell Dale), 1936–
Frozen fauna of the mammoth steppe: the story of Blue Babe / R. Dale Guthrie.
p. cm.
Bibliography: p.
Includes index.
ISBN 0-226-31122-8 (alk. paper).—ISBN 0-226-31123-6 (pbk. : alk. paper)
1. Bison priscus—Alaska—Fairbanks Region. 2. Paleontology—Alaska—Fairbanks Region. 3. Paleobiology. I. Title.
QE882.U3G88 1990
569′.73—dc20 89-4896
CIP

This book is dedicated to my closest comrades and colleagues in pursuit of the Arctic's frozen fauna: Bjorn Kurtén, Eirik Granqvist, Andrei Sher, and N. K. Vereshchagin. Bjorn departed before I could thank him in this formal way. We have his books; we miss the man.

CONTENTS

PREFACE

There are stories throughout the north woods about a giant of a man, Paul Bunyan, who roamed the forests and accomplished heroic feats with his outsized broadax. Paul's companion was an immense blue ox he called Blue Babe. The unearthing in 1979 of a giant Pleistocene bovine carcass, coated with blue vivianite crystals, near Fairbanks, Alaska, recalled the image of Babe buried somewhere in the northern forests, and so from the first we called the mummy Blue Babe.

Were the Bunyan tales a bit more credible, were we living in a time when stories told to children about giants were not so clearly fanciful, the discovery of Blue Babe would have been sure proof of Bunyan and his exploits. Occasional finds of frozen mummies probably substantiated beliefs held by earlier peoples: that a strange community of large creatures existed beneath the northern forests, living entirely underground except when they mistakenly surfaced along a stream bank, for that was where their bodies and bones were found emerging from the mud.

The context of our modern explanations, the story we tell our children of woolly mammoths and great ice ages, is fairly recent. In fact we are still in the formative stages of developing our understanding about the Pleistocene animals, vegetation, landforms and climate that once occurred in the north. While Paul and Babe may be myths, there was a time when real giants did roam the earth. When people first developed the technology to move into the north, as ice of the last glaciation was retreating, they hunted such giants: woolly mammoths and rhinoceroses, and a big woolly bison like the one described here.

We have solid evidence, however, that this bison lived before people colonized North America. Blue Babe walked the Fairbanks hills about 36,000 years ago. In Europe Neanderthal families were still lounging on bison robes beside their fires, eating bison that closely resembled Blue Babe.

Although as a paleontologist I think and talk about things happening tens of thousands of years ago, it is hard for me to imagine the depth of time we measure by 36,000 years. If we were to walk through a geological time scale back to when Blue Babe lived, we would pass the peak of the Roman Empire at 2,000 years ago in the first few strides. At 3,000 years the first pottery occurs in the New World and Tutankhamen has been buried for two hundred years. At 6,000 years ago there are no real urban centers, only tribal villages. Around 9,000 years ago the first domestic plants and livestock occur; prior to that, people all over the world were hunters and gatherers. At 11,000 years ago we find that Pleistocene species of large mammals, like ground sloths and mammoths, still lived in North America. At 13,000 years ago large ice masses remained from the last glaciation. At 18,000 years the world was locked in the midst of a full glacial climate; Hudson Bay was the center of a giant ice mass, over a mile thick, which extended south to central Illinois. Land that is now the American Great Lakes warped downward under the enormous weight of continental ice. This tremendous ice mass flowed imperceptibly southward, grinding the landscape into a flat plane as far south as Missouri. The earth's crust under those areas is still slowly rebounding upward, recovering from its heavy glacial load.

The vegetation of the north was quite different at this time. There were no north woods; trees could not grow in the cold, dry glacial climate. London to Moscow to Irkutsk, Irkutsk to Fairbanks, and Fairbanks to Whitehorse was a windswept, arid, treeless landscape, a grassy collar over halfway around the globe. This landscape was not inhabited by people; they could live only to the south in the woodland borders of the French Perigord and the Russian Plain where there was wood for fuel and warm south-facing slopes even in winter. The most obvious inhabitants of this northern no-man's-land were cold-hardy species: horses, reindeer, mammoths, and bison, which did not depend on woody plants for food or shelter but could survive out in the open, eating grasses and grasslike plants.

By 25,000 years we are closer to a major warming episode in the middle of the glacial, and at 30,000 in our backward-running time scale, trees again return to the north. By 36,000 years the climate is a little more moist and no so extreme as during the full glacial. There is our bison, walking up a creek bottom early one winter morning when the low sun just clips the ridge crests with pink-gold light. Here begin the first steps in a series of unique events that even-

An early lithograph, made from a sketch by T. Woodward, of the "ice cliffs" of Kotzebue Sound, Alaska, where fossil bones of the mammoth fauna commonly occur and where frozen mummies have also been found. The misconception that mummies occur in clear ice is shown in this drawing. The cliffs of Kotzebue Sound are indeed frozen, but they consist of frozen silt penetrated by ice veins and wedges. Unlike those shown here the real cliffs do not look like buried glaciers. Mummies occur in the frozen silt. (Courtesy of the Anchorage Museum of History and Art.)

tually bring this Pleistocene bull bison and his story to us 36,000 years later.

Because Blue Babe was buried and deep-frozen, his carcass is like a book thrown through time—bypassing the first colonists in North America, the first urbanities, and early agriculturists—a book unknown to early Romans or to Charles Darwin. But this story unfolds erratically. We view it from torn fragments teased bit by bit from muddy hair and patches of skin, woven from apparent digressions including Montana bison, African lions, and Iberian cave artists—the story of a real Blue Babe.

Prior to this study I worked with bones of hundreds of Alaskan steppe bison who, in life, were neither more nor less important than Blue Babe. In much the same way, the thousands of boys who lived at the time of young Tutankhamen may have also had uniquely interesting lives. The bodies of Blue Babe and Tutankhamen are special because they come with a story—stories decipherable from strange marks and once-exquisite robes.

ACKNOWLEDGMENTS

Walter Roman and his family, who operate the mine where Blue Babe was found, are at the top of my list of acknowledgments. Thanks also are due to Dan Eagan, then president of the Alaska Gold Company, which owns the mine property. Their interest and cooperation enabled us to salvage enough information to get the whole Blue Babe story. Their generous donation of Blue Babe to the University of Alaska Museum lets us all enjoy him.

My most grateful thanks goes to Mary Lee, my wife, who was a colleague on the excavation. She also played such a large role as editor of the manuscript that her voice is as much a part of the book as mine is. My son, Owen, also was collared into being a field assistant. John Bligh, then director of the Institute of Arctic Biology, realized the importance of this mummy when it was still a ball of stinking mud, and he provided critical support when it was most needed.

David Norton's initial suggestion that I write the report on Blue Babe in book form and his enthusiastic support of the project has meant a lot. Tina Picolo, the departmental secretary, has labored through the rewrites and additions with her usual pleasantness. Many people have added to the book: museum and institute staff, students, and proofreaders. The National Science Foundation provided a small grant to pay for two part-time student assistants for one semester. Petrie Viljoen graciously allowed me to accompany him in his studies of lions near Savuti, Botswana.

Traveling to the opposite end of the earth in hopes of learning something about 36,000-year-old Alaskan lions may seem like a long shot, but the firsthand exposure gave me insights that books simply cannot supply. Likewise, the opportunity to meet and study with Soviet colleagues during an earlier trip was essential to the quality of my work with Blue Babe. These and so many other occasions provided by the University of Alaska sabbatical system have truly borne fruit for me. I am especially grateful to the Institute of Arctic Biology, which was the chief sponsor of the Blue Babe project, as it is of the rest of my Alaskan research.

1
THE CURSE OF THE FROZEN MAMMOTHS

Woolly mammoths, woolly rhinoceroses, steppe bison, and other mummies found in the frozen soil of the far north have fascinated us in both fiction and fact. Yet much about these animals—the environment in which they lived, and how they died and were preserved—is unknown or controversial.

In the summer of 1979, the frozen mummy of a steppe bison was unearthed at a placer gold mine just north of Fairbanks, Alaska. As the local specialist in Pleistocene mammals, I was asked to visit the site, and so began my involvement with the remains of the animal I came to call Blue Babe. This book is the story of Blue Babe and others of his kind. It is also an account of how we can look at the remains of an animal such as Blue Babe and reconstruct his history and his life, and a good deal of the world in which he lived and died. In this sense, this book is also a detective story, one that begins with the first studies of frozen mummies more than a hundred years ago.

Floaters or Sinkers

Modern views of the origin and history of the earth and the evolution of life itself had their origins in the first six decades of the nineteenth century, and naturalists in Europe were stimulated to develop new ideas and hypotheses in part by reports of the giant woolly mammoth carcasses found frozen in the ground in northeastern Asia. Parts of woolly mammoth and rhinoceros mummies were actually sent to Paris and London to satisfy the curiosity of naturalists and the public, which had become increasingly fascinated by the reports and finds of explorers and travelers who brought back tales and specimens of exotic plants and animals from around the globe.

The great geologist Charles Lyell supported the "floater" theory, arguing that the woolly mammoths and rhinoceroses, so much like the elephants and rhinos recently discovered in Africa, had lived in warmer climates in central Asia, had died in floods, and had subsequently been carried far northward in floodwaters (the major rivers in northern Asia really do flow north), where they had been frozen and buried. Georges Cuvier, the famous French anatomist, wrote a lengthy challenge to this idea in 1825. He observed that these animals were adapted to cold, with long, dense pelage and thick subcutaneous fat, and that they differed from living elephants and rhinos. Cuvier proposed that they were natives of the country where they were found. Hedenstrom (1830) joined Cuvier and showed that the carcasses of lower-latitude flood victims would have been destroyed long before they reached northern latitudes, and that the frozen carcasses and bones showed no signs of alluvial transport.

The Arctic explorer Middendorff (1848) continued to argue for river transport from milder, southern climates. An attempt at reconciliation was made by Howorth (1887), who proposed that these mammoths had lived in the north in a very warm climate that existed prior to the biblical flood and that they had been buried by silt when the waters receded. Since warmer climates did not return after the deluge, we find the presence of proscidians incongruent with northern cold. Lapparent still (1906) embraced Howorth's compromise in his famous *Traité de géologie;* however, in later editions of his book, Lapparent attributes the extinction of the mammoth to a gradual increase in cold and a decrease in the supply of food, rather than to a cataclysmic flood.

These controversies indirectly helped fund a number of successful expeditions commissioned by the Russian Academy of Sciences and an American Museum of Natural History expedition that was funded by J. Pierpont Morgan. After that later expedition, Quackenbush (1909) concluded that the partial mammoth mummy from Eschscholtz Bay, Alaska, was so deteriorated as to exclude "sudden fall in temperature" theories and that the mammoth had not been retransported after burial. The Russian and American expeditions obtained enough evidence to show that mammoths had indeed lived in the same areas in which their remains were found and that the former climate was as cold as now.

Once the issue was settled to the satisfaction of all, or nearly all, the next question was how the mummies had been buried. There were two main views. One was that the animals had fallen into a glacier crevasse or similar hole, often snow covered, but sometimes mud walled. Geikie (1881), in his major work on Pleistocene geology, championed the theory of Schrenk and Nehring that these creatures had been buried in snowstorms. The alternate view argued for some kind of entrapment in mud. These debates still continue.

Brandt (1866) was an adherent of the mud trap theory, as was Tolmachoff (1929) much later. Vollosovich (1909) also supported this idea and backed it with numerous anecdotes. He himself had been caught in such a mud trap and extricated only with the help of his guides. Vollosovich proposed that a trapped mammoth would effectively obstruct a small drainage, damming up mud and creating its own depositional environment.

There were many adherents to the snow-covered crevasse theory or some version of it, but the most influential of all was Digby (1926), who wrote a popular book titled *The Mammoth and Mammoth Hunting in North East Siberia*. Nevertheless, data from people who had seen these carcasses excavated indicated that the mammoths had been buried in mud rather than ice, although ice lenses were present around their carcasses. As I show, there are several interpretations for these ice lenses, and they form the basis of present-day theories about frozen mummies.

The Berezovka Mammoth

Until a few years ago, the Berezovka mammoth, excavated by Herz at the turn of the century, was the centerpiece of information about frozen mammoths. Unfortunately, this mummy was found before the days of rapid transportation, when it took a year for word to trickle back from Siberia to the Russian capital, Petrograd, and for an expedition to reach the site. In the winter of 1900, a cossack dealer in ivory mammoth tusks, named Yavlovski, bought a number of tusks from a Lamut tribesman on the Kolyma River. The tribesman, named Tarbykin, said he had chopped a pair of ivory tusks from one of the hairy beasts (Digby 1926). Yavlovski reported the statement to a local police official named Horn, who forwarded word to the governor general at Yakutsk, who in turn telegraphed the Imperial Academy of Sciences in Petrograd. This relay took several months, and it was not until May 1901 that an expedition set out to investigate Tarbykin's find. The expedition was led by Otto F. Herz (sometimes spelled Hertz), a zoologist on the Academy staff. He was accompanied by a geologist, M. D. P. Sevastianov, and a zoological preparator, M. E. V. Pfitzenmayer. It took them all summer to reach the site, where, hundreds of miles from the nearest source of supplies, they immediately began work on the mammoth. It was late

September, and soon snow and frost began to hamper their work. They had to construct a log and canvas structure heated by a stove so that thawing and excavation could continue. According to the Lamut tribesmen, the head of the mammoth had been exposed two years before and many soft parts were already missing.

The mammoth was surprisingly well preserved but had undergone decomposition (fig. 1.1). In addition to its skin parts, some internal tissues such as the tongue were also well preserved (fig. 1.2). The mammoth even had food between its teeth, including flowers distinguished as buttercups. Herz described many other strange phenomena. In the early 1900s, the characteristics of Arctic frozen ground were strange to Europeans. Ice wedge features were wondrous things—"massive walls of ice," interpreted as underlying glaciers. Arctic explorers such as Kotzebue, Beechy, Stehanson, and Nelson (in Quackenbush 1909) had strange interpretations of this ice. They might indeed, for the origins of ice wedges are not intuitively obvious. These large ground ice formations were responsible for Herz's idea that the mammoth must have fallen into a snow-covered glacial crevasse, an idea that still survives. However, some of the ice lenses included parts of the mammoth, and much of its hair was embedded in ice.

Tales of earlier explorers suggested that mammoths ate pinecones. Herz was disappointed to find no pinecones or larch needles, but only bits of various grasses, in the large bolus of food frozen between the mammoth's teeth. Much of the animal's head had been eaten away by carnivores during the two years of exposure, but enough remained for Herz to find real information about mammoths. He and his associates were the first professionally trained biologists to see a mammoth mummy "in the flesh." Herz described hair color and length in detail, as well as other parts of the anatomy. The bare tail, for example, was short, only 14 inches (355 mm) long, with 10-inch (255 mm) hairs (fig. 1.3).

The death of the Berezovka mammoth seems to have been almost instantaneous. It had several broken bones—ribs, shoulder blade, and pelvis—and vertebrae were wrenched to one side. There were large blood clots in the viscera. It is possible, however, that some bones were broken diagenetically by the movement of sediments after burial. Despite a good description of the carcass, it is still unclear how the Berezovka mammoth died.

How the Berezovka mammoth was buried is likewise unknown. It may have been trapped in a cavity or mud sink as Herz suggests, but how it was buried remains a problem and probably does

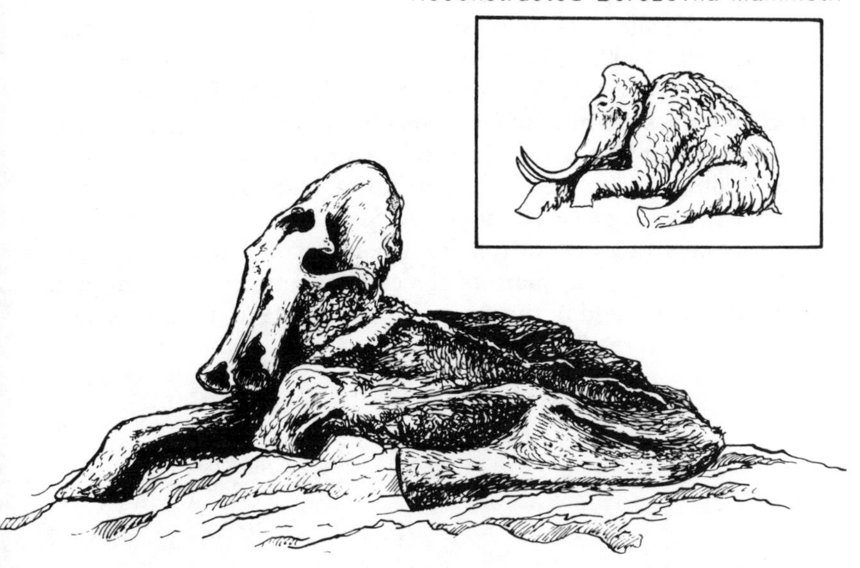

Fig. 1.1. The Berezovka mammoth. (Drawn from photo in Pfitzenmayer 1926.) Reconstruction of the mammoth's appearance at time of death is shown in box.

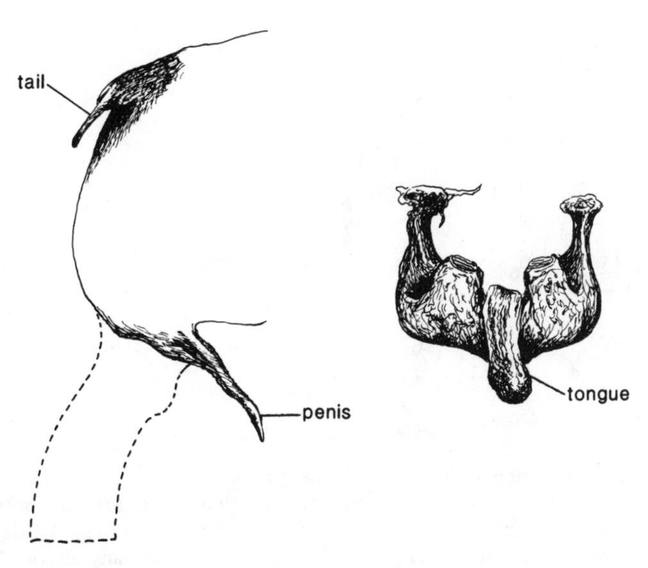

Fig. 1.2. Tongue, penis, and tail of the Berezovka mammoth. (Drawn from photos in Pfitzenmayer 1926)

not relate directly to how and only indirectly to where it died. The geology of the site is sketchy, as one might assume that geologists from the early 1900s did not have a sophisticated understanding of the sedimentary features of reworked loess (fig. 1.4). The animal had about 3.6 inches (90 mm) of fat on its torso, suggesting that it was in good health and that it died in the late autumn, after a good summer's feeding. Fortunately, the Berezovka mammoth was brought back to Petrograd and saved for posterity.

This was the sum of our knowledge of frozen woolly mammoths for almost three-quarters of a century. Some very incomplete mummies were found in the interim, but until 1977, none was more informative than the Berezovka mammoth.

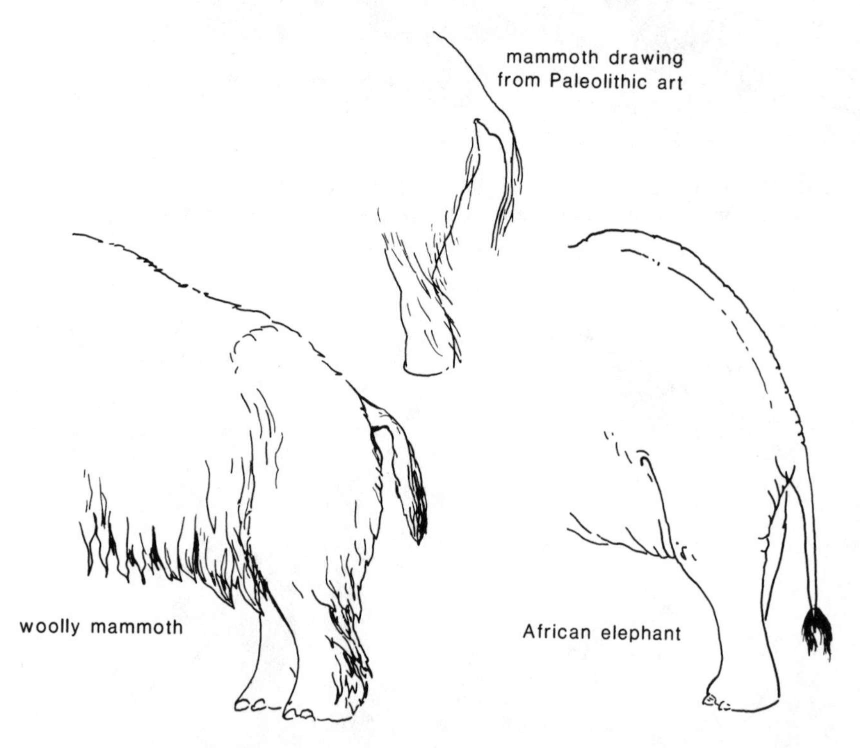

Fig. 1.3. Comparison of mammoth and elephant tails. Mammoth tails were short and hairy compared to those of elephants. The total length of the Berezovka mammoth's tail was about 60 cm, whereas, elephants have tails over 160 cm long. A Paleolithic drawing of a mammoth tail from Gönnersdorf in northern Germany, above, shows a relatively short structure. In this drawing one can see the anal flap characteristic of proboscidians.

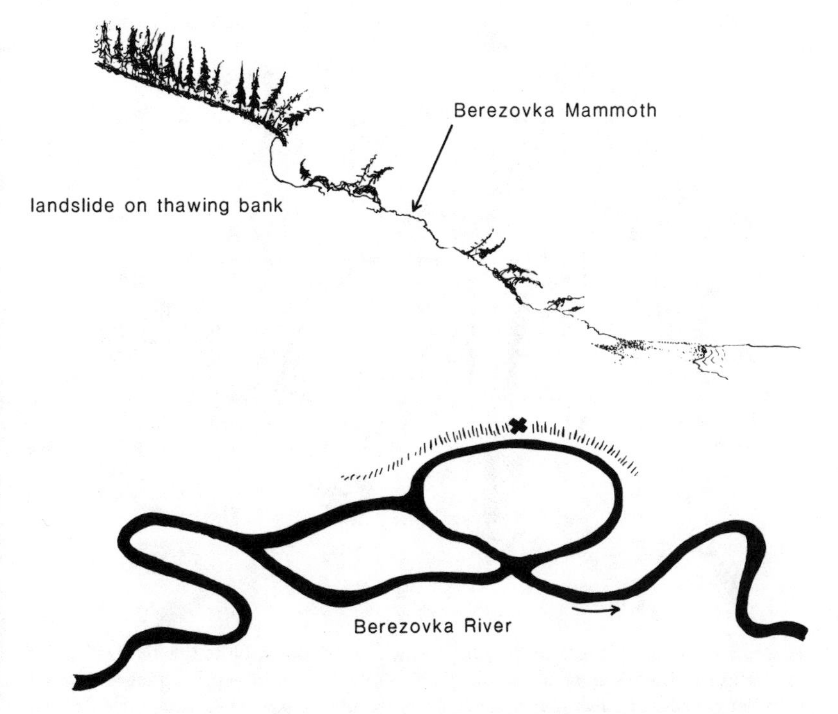

Fig. 1.4. Berezovka mammoth site. Like other early finds of Siberian mammoth mummies, the Berezovka mammoth was discovered in a naturally eroding stream bank. The mummy was exposed by slumping of the hillside above a loop of the Berezovka River. Such erosion is common in Alaska as well, usually triggered by melting of underlying ice wedges. The weight of the water-saturated soil carries it downslope. (*Above*) A diagrammatic portrayal of how the mammoth was exposed (taken from the photos of the Herz expedition). (*Below*) Aerial view of Berezovka River channels showing the eroding bank where the mummy was found.

Dima

Perhaps the most famous frozen mummy from the far north is Dima, a baby mammoth found in the USSR in 1977 by a placer gold miner who was using a bulldozer to strip the thawed soils from underlying frozen ground. Dima was uncovered on a terrace of the Kirgiliakh River, a tributary of the Kolyma, north of Magadan in the Soviet Far East. Dima is perhaps the best-preserved specimen of all of the frozen large-mammal carcasses found in the far north. It had not been scavenged, so the carcass was intact (fig. 1.5), and very little decomposition had taken place before the body was frozen. A team

Fig. 1.5. Dima, a baby mammoth found in the Soviet Union. Note the "wings" of the trunk, the small ears and tail. Its relatively empty gastrointestinal tract and small size must have allowed the body to cool rapidly after death. There was little decomposition before the body was completely frozen, and even the viscera were preserved. Dima is displayed at the Zoological Museum in Leningrad.

of Soviet scientists was assembled to work with the mummy; they were able to conserve and analyze the specimen soon after discovery, and their analysis was quite thorough. The team included experts in paleontology, geology, geography, botany, palynology, entomology, histology, and anatomy. Each contributed a chapter to a monograph about Dima. Unfortunately, their book (Vereshchagin and Mikhel'son 1981) has not yet been published in English.

This team of experts generally agrees on basic data pertaining to Dima; their interpretations, however, vary widely. Each of the scientists who worked with Dima provides a scenario of his death and burial. Their report brings new life to the parable of the blind men and the elephant; each man, touching only part of the animal, describes a quite different beast.

Dima was truly a baby mammoth, just slightly over one meter high at the shoulder. Most hair had slipped, but much was found embedded in the mud where the animal laid. In fact, the hair actually helped the scientists, most of whom arrived at the site a little over a week after discovery, to relocate the exact spot where the carcass was found, because immediately after the discovery Dima had been moved to Magadan and refrozen. Dima's small size, lack of fat, and empty gut probably helped the carcass cool rapidly at the time of death, reducing the rate of decay so that even his viscera were well preserved. The carcass was virtually complete; only one side had been torn loose by the blade of the bulldozer at the time of discovery.

The Soviet team found a number of points of information pertinent to a reconstruction of Dima's death and preservation. The high quality of tissue preservation is itself unusual and indicates that the body cooled rapidly after death and was not subsequently exposed to long episodes of warm temperatures. The degree of completeness of the carcass is also unusual. Because there were no signs of scavenging by avian or mammalian carnivores, the carcass probably was not exposed to scavengers for any length of time.

Dima was a very young animal; judging from his size and the degree of tooth eruption and wear, his age was assessed at 7 to 8 months. His gastrointestinal tract was empty of food but contained plant detritus (unetched by stomach acids) and considerable amounts of mineral particles, silt, clay, and gravel. Most of the gut material (3.5 kg) was in his colon. "Many" hairs from the animal's own body were found in its gut, along with a few insect (Coleopteran) parts. The botanist identifying seeds from the gastrointestinal tract found them to be late summer to early winter in their degree of maturity; however, the team's palynologist found many pollen grains in the gastrointestinal tract that had not yet reached maturity. Mineral particles were not only found in the gastrointestinal tract, but also throughout the trachea, bronchi, and alveoli of the lungs. Dima was emaciated; no fat of any kind was present.

The animal lay on its left side, with the head pointing downslope and dipping down below the body. This position was fortunate, for it allowed the bulldozer blade to miss the head entirely. The animal was found about 2 m below the surface, where the zone of annual thaw is about 1.2 m. There was some ice around the carcass. Portions of ice were clear and others quite brownish yellow with mineral and organic particles. A very small wound was found on the lateral or outer side of Dima's right wrist, which showed some bruised tissue but no inflammation.

Radiocarbon dating showed that wood found immediately around the carcass had been buried from 9,000 to 10,000 yr B.P. However, a number of Soviet laboratory tests on the mummy's tissue gave dates in the range of 40,000 yr B.P. One tissue sample submitted to a lab in Pittsburgh, Pennsylvania, gave a date of 26,000 yr B.P. Histological evidence suggested that Dima contained a considerable number of helminth eggs. Other histological evidence showed that he had been under considerable physical exertion before death.

A few days prior to finding Dima, the same miner had unearthed a crushed horse head and legs with hooves. Because the miner thought this animal was a moose (*Alces*), he had not saved it, and the specimen was lost and could not be found. His descriptions of the hoof as "not split" helped the investigators conclude it had

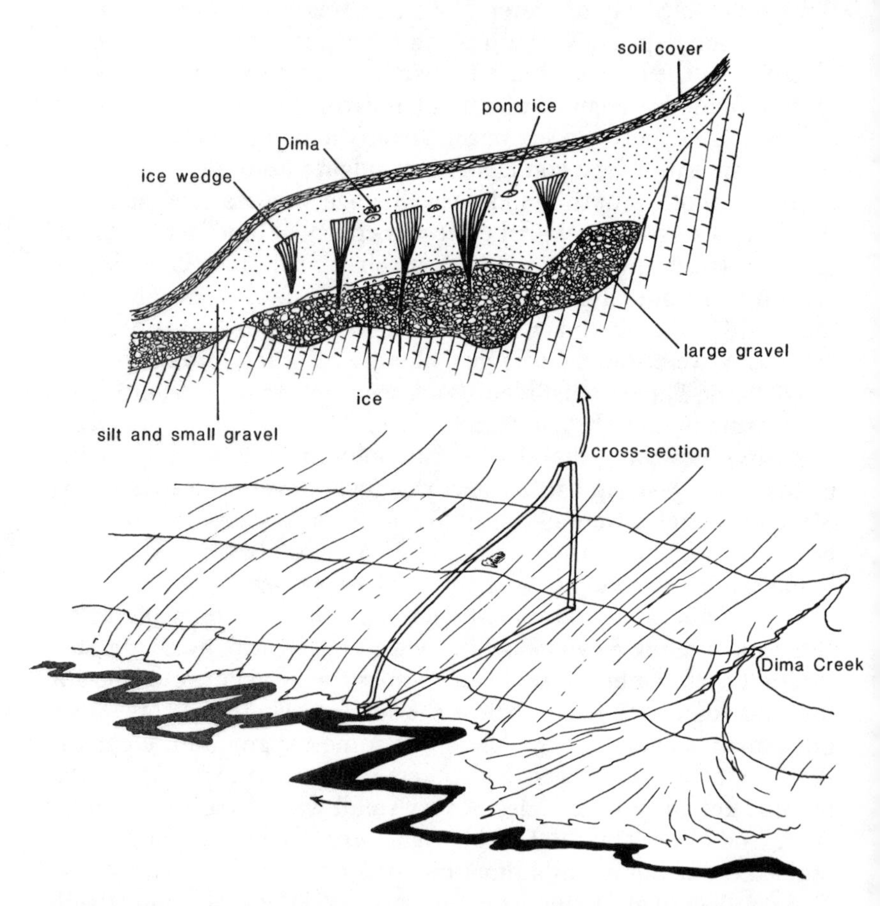

Fig. 1.6. The Dima site. The baby mammoth was found by a placer gold miner while removing silt overburden near a small creek called Dima, shown to the right. Both the cross section and the three-dimensional diagram are reconstructed from two-dimensional illustrations published by Soviet researchers.

been a horse. Other horse bones and the bones of bison (*Bison priscus*) and woolly rhinos (*Coleodonta antiquatus*) were found in nearby sediments.

The heads of large vertical polygonal ice wedges surrounded Dima (fig. 1.6). These reached 7 to 8 m in depth and were about 1.3 to 1.4 m across, creating a polygonal core about 15 to 20 m across. Dima lay within one of these cores, yet he was stratigraphically well above the tops (dorsal surfaces) of the ice wedges. He lay on top of a 10 m terrace of the Kirgiliakh River, not in the river's floodplain. The material covering him was unsorted clay, silt, sand, and gravel from the side slope of this Kirgiliakh River, near a small side drainage called Dima. From pollen found in and around Dima, the team's botanists were able to describe the vegetation as mostly herbaceous, a dry tundra-steppe with some trees in the valleys.

Interpretations of the Dima Mummy

N. K. Vereshchagin, a famous figure in Soviet paleozoology who has conducted the bulk of recent research on Soviet frozen mummies, was the senior scientist on the Dima project. He developed the most complete theory about Dima's death and preservation, and his interpretations are a baseline from which to compare the ideas of other researchers.

Vereshchagin argues that early hunters killed Dima's mother and wounded Dima in the leg. Dima must have escaped and hid. Without his mother he was unable to get milk and began to starve. Not knowing what to eat, he consumed silt and plant detritus. Exhausted, Dima came to a pool to drink, collapsed into the pool, and drowned, inhaling the silt in the water in the process. The cold water preserved the body, but the carcass still underwent some maceration, causing the hair to slip. It froze in the ice that winter and was thus preserved. New flows of mud and waste rock from upslope buried the animal the next spring. Vereshchagin argues that Dima must have died during late summer, because that would have been the time of maximum thaw. He thinks Dima is of a very late age, nearer the time of the surrounding wood (9,000–10,000 yr B.P.) that the skin dates (40,000 yr B.P.), but he recognizes the problematic incongruity of dates.

To account for these differences in dates, several of the Soviet scientists who studied Dima proposed or implied the retransporation of a 40,000-year-old carcass into later sediments; others argue that the ice in which Dima was frozen was never thawed. Supporters of redeposition claim that little hair was associated with the carcass and that the hair was left behind during retransport. N. A. Shilo and E. E. Titov, two geologists, are among those proposing the reburial of Dima during the Holocene, perhaps by an earthquake. However, they propose that the death was accidental, not caused by humans, and they rely on the "lost mother" scenario to explain Dima's starved condition. They suggest the animal fell into a polygon pond during early winter, froze, and was covered by a seismic landslide the next spring.

I. A. Dubrovo, another geologist, finds evidence that Dima did not die from drowning in a polygon pond but rather fell into a thermokarst pit, starved, and was subsequently buried by snow, then covered by solifluction (the slow sag of soil downslope every summer as winter frost thaws) soon thereafter, probably within a month. To answer the question of why Dima was not rescued by his mother, Dubrovo also employs that lost-mother scenario.

S. Timordiaro and V. K. Riabchun, a geographer and a permafrost specialist, respectively, contend that mummification requires surface exposure for an extended period before burial. They realize that scavengers would have eaten an exposed animal and proposed that Dima's body was guarded by his mother—the opposite of Vereshchagin's lost-mother scenario. They propose that following a period of desiccation, the baby mammoth was buried by a mud flow of at least 1.5 m, providing a cover that kept it from decomposing. They identified the ice enclosing Dima as neither vein ice (polygonal ice-wedge ice) nor frozen pond ice, but as segregated ice, ice withdrawn when water-saturated soil freezes. They also argue that a water-saturated mudflow retransported Dima, removing its hair in the process.

V. V. Ukraintseva's study of pollen percentages shows a surrounding vegetation dominated by woodless communities (67% grass pollen), similar to what we see today in cold mountain steppes. Pine, birch, and willow were distributed along the river valleys. However, V. P. Nikitin, examining the macrofossils in the gastrointestinal tract, concludes Dima lived in a swampy tundra. Ukraintseva found many immature pollen grains, which suggest the animal died in the spring. On the other hand, the mature seeds Nikitin found in the gut indicate to him a death in late summer or early fall. Nikitin also subscribes to the lost-mother scenario and frozen thermokarst trap. He thinks Dima probably used his trunk to pull at the edges of the pit, getting soil and plant detritus on his trunk and carrying this material to his mouth, accounting for the strange gut contents. Ukraintseva interprets the almost empty stomach as indicative of an animal trapped in sticky mud, unable to work its way out, starving and desperate, finally eating plant detritus and mineral particles.

S. P. Il'inskii found mineral particles throughout the respiratory system and proposed that Dima died either of suffocation under a large block of dusty silt or from drowning in silty water. Widening of the pulmonary alveoli suggested death by asphixia. Heart muscle hemorrhaging and other damage were found, suggesting great exertion just before death, which Il'inskii interpreted as running. E. I. Ivanova found many cysts and eggs of helminths, which she contended could have contributed to the animal's death.

Dima's Case and Others: Varying Views

This brief summary does not do justice to the detailed discussions by Dima's investigators, but it does indicate the great diversity of interpretations. Several issues still warrant further discussion and analysis, particularly the carbon-14 dates. I do not know what to make of the one Rochester, New York, U.S., date of 26,000 yr B.P. and the Soviet dates, from two different labs, of around 40,000 yr B.P. I suspect the Rochester date may be incorrect, but this disparity in dates should be pursued.

As to the general pattern of surrounding vegetation, Ukraintseva's steppe reconstruction agrees with many subsequent studies. The northern extent of pine and birch also agrees with the general picture of northward extension of tree species during the slightly warmer and wetter period of the interstade (Hopkin's Beringian Boutellier Interval or marine isotope stage 3). The pollen and macroplant fossils indicate a dry, grassy landscape with some tree species growing along valley bottoms.

Dima's ontongenetic age is an important factor in determining the season in which he died. Among northern ungulates, timing of parturition has evolved to coincide with the first growth of plants in spring. I think woolly mammoths would also have had their young in the spring, for two reasons. The first concerns the mother: lactation is a costly endeavor, and the protein-rich greenery of spring helps the mother recover from winter and nurture her young. The second reason involves the young: a calf must maximize growth during its first summer or it stands no chance of surviving the lean times of its first winter.

Winter forage is so poor in available nutrients that northern ungulates do not grow and even fail to maintain body condition over the winter. In less than five months, they must recover winter energy debts and build reserves for the next year. Young of the year face the same rigors as the adults. Unlike Asian and African elephants, which may be born any time of the year, baby mammoths must have arrived synchronously, in the early spring. Therefore, if the interpretations of autumn death are correct, it would not seem possible for Dima to be 7 to 8 months old. Nor would a spring death be consistent with that age, for then Dima would have had to have been born in the previous January or February, arriving in −40° F (−40°C) temperatures and winds. Wet fur and amnionic membranes would have frozen immediately. Determining Dima's age depends on a clear assessment of the season of death.

The presence of mature seeds and immature pollen in the gut indicates, to me, that the young mammoth ingested *seasonally mixed* plant detritus and that plant studies alone cannot specify season of death. One reason an autumn death seems most logical is that the carcass would not be subject to lengthy decay or scavenging. If Dima did drown, the body would have bloated with enzymatic activity from its own heat after death and floated to the surface. Scavengers could have penetrated Dima's relatively thin hide. The quality of preservation and lack of scavenging suggest fairly rapid cooling and freezing, but all evidence argues against a winter death. Mud in the gastrointestinal tract, silt in the respiratory system, and skeletal parts of Coleopteran beetles are incongruent with death in winter. An autumn death seems most likely. For this reason, I think the estimate of Dima's age may be incorrect.

Birth in spring and death in autumn would make Dima approximately 4 months, or 1 year and 4 months old. I think the animal's size is consistent with an estimate of 4 months. The very rapid growth rates of mammoths during the short summer season would probably have been slightly higher than those of elephants, for young elephants enjoy a longer growing season and even grow year-round in places where forage permits. Elephants average 85 cm at the shoulder at birth and weigh about 120 kg, sometimes even as much as 160 kg (Hanks 1979). Dima was slightly over 100 cm tall and weighed an estimated 120 kg in life. This amount of growth, a 15% to 20% increase in shoulder height, can be expected during the first year of life. Northern ungulates, remember, complete a year's growth during their first four months because resources to support growth are simply unavailable all winter. Dima would have been about four months old in September.

Determining Dima's age from tooth eruption pattern is more ambiguous, but Dima's teeth do not contradict an estimated age of about four months. We do not know the age of dental eruption of mammoths; we can only assume it was roughly analogous to that of elephants. The second deciduous premolar is the first tooth to erupt, as the first deciduous premolar has been lost in proboscian evolution. In Dima's case this second deciduous premolar was in use, and the third deciduous premolar was beginning to show wear. These two teeth provided a modest-sized occlusal surface and undoubtedly prompted the original age estimate of 7 to 8 months. Several researchers in this study assumed that a mammoth in its first year does not find much of its own food, but rather lives mainly on milk. This is the case for elephants, but it may not have been true for young mammoths. Both the first and second teeth are high crowned and have complex occlusal surfaces. These teeth are worn out in the first two or three years of life. Most high-quality herbage available during the growing season would have been digestible by a young mammoth. Young mammoths nearing the end of their first growing season may necessarily have been eating quite a bit on their own, and one might expect to find some food throughout Dima's gastrointestinal tract.

That Dima's gut was empty of suitable food is, I think, a clear indication that death was not instantaneous. Autumn, however, is not a season of starvation; forage would have been at its maximum volume or biomass. Also, how do we explain the mud, plant detritus, and hair found in Dima's gastrointestinal tract? I agree with Ukraintseva that these features suggest a trapped animal. Studies of animals trapped in tar pits and other sticky substrate have shown that adults of larger species, such as bison and mammoths, are seldom entrapped. Adults are strong enough to pull themselves out (Akersten, pers. comm.). Young animals are apparently the most common victims. If a young mammoth were trapped in organically rich, water-saturated silt, its struggles may have pulled it even deeper. Comparable struggles have been seen and filmed in African springs and witnessed by various researchers (e.g., Gary Haynes, pers. comm.). Mothers do try to help their young and sometimes succeed, but not always. A mother's presence, however, prevents predation.

A lost-mother scenario seems unwarranted in Dima's case. The end of the growing season would have seen mammoth cows in their best condition, an unlikely time to die. We know indirectly that Dima's mother was probably not in poor shape because, among elephant cows, only those in relatively good condition can become pregnant and have young. Among proboscidians, it is the young which have the highest mortality; Laws, Parker, and Johnstone

(1975) estimate about a 30% mortality during the first year among African elephants. The death of a mammoth in its first year would probably not have been unusual either.

We can imagine that it was not uncommon for a young mammoth to die, while the mother lived on to old age. Cow mammoths had few predators, and surely any predators capable of taking an adult mammoth would have killed this tiny one-meter-tall baby as well. Vereshchagin's proposal that the predators could have been human hunters, although possible, is not very convincing. Babies of Dima's size do not run away from their dead mothers when hunters cull elephant cows today. The social bond to the mother elephant is strong (Laws, Parker, and Johnstone 1975). Also, I know of no evidence that human hunters were above 60° latitude north 40,000 years ago, if indeed Dima dates from that time.

It is not Dima's death which we have to explain; there were probably hundreds of thousands of woolly mammoth calves dying every year across the Pleistocene northern steppes. It is Dima's preservation which is unique. The reflex reaction of most people to fossilized animals, especially Pleistocene mummies, is to try to account for their death. The more appropriate issue is usually one of preservation. An example is Ivan Sanderson's (1960) article about mammoths frozen with buttercups in their mouths. Sanderson's emphasis was on death, focusing on change in climate, but he started at the wrong end. The interesting point is not that a mammoth died midway through its last bite. That, no doubt, happened to millions of unfortunate mammoths. The real puzzle is how a few mammoths became permanently preserved.

Even Vereshchagin, in his discussion of the Selerikan horse mummy (1977), conflates death and preservation in discussing the two main episodes of mammoth mummification: 10,000–12,000 yr. B.P. and 33,000–45,000 yr B.P. Instead of seeing mummies from these periods as products of unique depositional circumstances, he comments that these are two main times of die-off due to climatic change.

Most of Dima's body might have sunk rather quickly as he struggled in deep mud. The vacuum created by attempts to pull a large object from silty-clay mud is quite strong; I have twice seen horses stuck in organic muds and had to work most of the day to pull each out, even with the help of other horses. During a recent hunting season, my partner sank a meter deep in soft organic mud while he was backpacking out a heavy moose quarter. Even with the pack removed he could not get out of the mud by himself, nor could I pull him loose with a rope. We had to lay a scaffolding of poles around him to pull him free. Such deep, organic mud is not everywhere common, but it does occur.

Vereshchagin rejected entrapment because the region today lacks deep thermokarst cracks like those proposed by Dubrovo. Water-saturated soils in that area also do not now provide conditions for mud entrapment. However, as Ukraintseva and others argued strongly, today's conditions are quite different from those in which Dima lived. I have proposed that the complete ground cover and stable soils we know today were not characteristic of the Pleistocene landscape (Guthrie 1982). We can draw this conclusion from the mass transport of the sediments in which Dima was found. There seems to have been a basically dry landscape, with bare soil commonly showing through the vegetation cover, and a few wet spots in polygonal ground where unsorted sediments with a high organic content occurred along poorly drained benches.

In my opinion, the evidence surrounding Dima points to mud entrapment, but we have to explain why a trapped animal survived several days without being molested by predators. As long as Dima was alive, his mother probably would not have left (fig. 1.7). By the time Dima died, he may have been so deeply mired in the mud that predators could not see or smell him. Mud in the respiratory system and distended bronchi indicate Dima died when he no longer had the strength to keep his head above the surface. Once under the mud, the body would have been protected from scavengers. An empty gastrointestinal tract would produce few visceral gases; however, enzyme activity would have been sufficient to slightly rotate they body and swing the abdomen up and the heavy head down—the position in which the frozen mammoth was found.

An animal exerting itself in cold mud would have depleted its fat stores rapidly. And Dima may have been a rather sickly young animal in the first place, as the parasite load suggests. Perhaps Dima lacked the normal strength required to overcome the tenacity of soggy, deep mud.

If Dima had fallen into a small pond and drowned, visceral gases, even if Dima had had an empty stomach, would have bloated the body and floated it to the surface where it could be smelled and dragged to shore by predator-scavengers. Nor is snow a sufficient insulator to serve as a protective cover for the odor of a dead mammoth. Many mammals are good at scavenging under snow, but, as stated earlier, Dima's gut contents allow us to rule out winter as a time of death.

Fig. 1.7. Postulated burial and preservation. Dima seems to have been a young animal, a baby of the year. Apparently it got stuck in wet silt late in summer or early autumn. Struggling to get out, it sank farther, probably over a period of several days (its body fat was depleted). Under the stress of entrapment, it ate plant detritus and silt. Finally, it sank deeply enough to die from inhaling the mud (silt particles in the trachea and lung alveoli). It rapidly sank beneath the surface (no sign of scavenging). The head being slightly heavier, it sank deeper than the torso and was frozen beneath the surface that winter. Segregation ice formed in the water-saturated soil surrounding the mummy. Dima then was covered by mass wastage of soil the following spring.

The presence of hair on Dima's trunk, tail, and feet, and the better histological detail preserved in these structures, can be explained if we remember that the peripheral appendages cool faster than the body core. Cool mud would have quickly removed heat from the appendages but not from the more massive abdomen, where heat from enzyme action would have maintained body temperatures longer. This in turn would cause hairs on the torso to come loose from their roots, but mud would have held the hair in place. Movement by ice segregation or by the miner would leave that hair in the mud. Such hair is not as readily apparent as one might suppose. I think this phenomenon of hair actually present near the carcass but lost in the mud can explain Riabchun's quandry—that Dima must have decomposed before burial to explain the absence of hair.

Mud entrapment would not only account for Dima's debilitated condition and features of his preservation but also for the strange gastrointestinal tract contents. Trapped animals usually die from the stresses of being trapped and from starvation over a period that may last for days. In their delirium of anxiety, trapped animals exhibit strange displacement behavior.

As part of other natural history studies, I have necropsied several thousand carcasses of mammals caught in steel traps. These carcasses often had dirt in the gastrointestinal tract. Trapped and starving animals desperately eat or chew on anything in reach, often their own hair. I originally found the presence of hair perplexing, but realized that the trapped animal often pulls and chews at its own body when it is in a great deal of pain. All of the hair in Dima's gastrointestinal tract had been broken, not shed like hair combed out during grooming. A scenario of entrapment in cold autumn mud can thus explain several things: (1) why the carcass was not scavenged (it was totally buried in wet mud); (2) why the carcass was preserved so completely (the cold mud curtailed maceration until the carcass was frozen in winter); (3) why the gastrointestinal tract was empty of suitable food (being trapped, Dima did not have access to normal diet); (4) why both silt and Dima's own hair were present in the gastrointestinal tract (delirious displacement feeding and self-mutilation); and (5) why there were characteristic horizontal ice lenses in the silt surrounding the carcass.

Whatever the proximate cause of death, it is apparent Dima was embedded in water-saturated soils. The first freeze would have segregated ice from the silt. Fairbanks potters use this freezing action to dehydrate their clay slurry every autumn. The creamy mix is placed outdoors, and, as it slowly freezes, the water segregates into an icy block that the potter can then chip from the clay. Damp clay ready to work into ceramic ware remains. The ice lenses described surrounding Dima are of similar origin. The permafrost specialist, Riabchun, recognized these characteristic long, lateral lenses as segregated ice.

Unfortunately, different kinds of ground ice are not always easy to distinguish, but their identification is critical in the case of mummies because it can determine how the animal was preserved. Geologists (Watanabe 1969; Péwé 1975a) distinguish a number of categories (fig. 1.8) of ground ice:

Glaciers are sometimes buried by sediments and become "fossilized." (Herz, remember, thought the ice around the Berezovka mammoth was glacier ice.)

Aufeis commonly occurs in small northern streams that freeze to the bottom early in winter. The stream continues to flow, build-

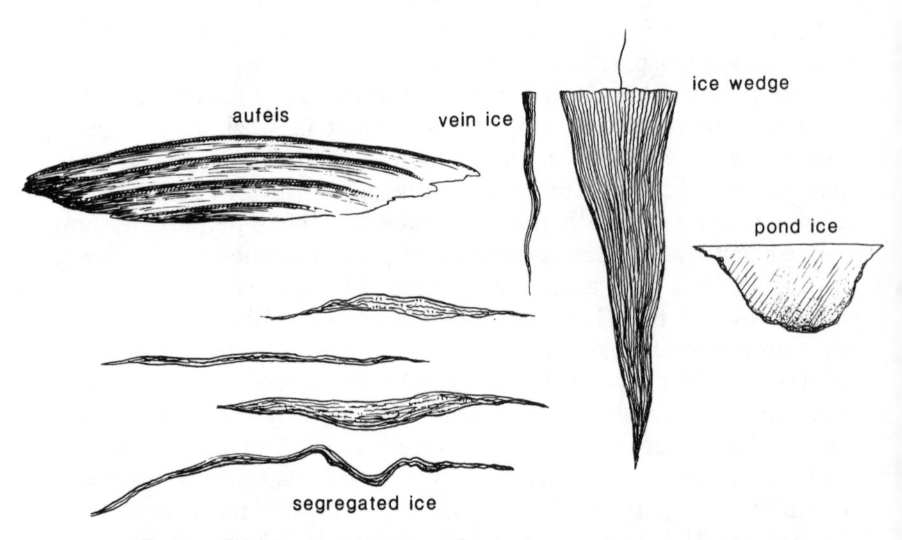

Fig. 1.8. Quaternary ground ice. Ice found in frozen silts surrounding large mammal mummies can provide important clues as to the agents of burial and preservation. Misidentifications and misinterpretations of this ice have lead in the past to some exotic theories of mummy preservation. Different kinds of ice are difficult to portray in pen and ink, but they resemble these forms: *aufeis*—from stream overflow, which is buried by sediment the following spring; *ice veins*—which includes *ice-wedge* ice from ground contraction due to extreme cold; *pond ice*—which is buried the following spring by sediments; *segregation ice*—which forms in water-saturated sediments as they freeze. These latter are the most common form of ground ice associated with large mammal mummies.

Fig. 1.9. Ice Wedges.

ing up ice on the surface, and by spring aufeis is often thick and massive. It is recognized as very clear, clean ice that is stratified horizontally. Aufeis can be buried by silt in spring and preserved.

Pond ice is found as solid, flat-topped ice that was buried during spring and insulated from summer thaw. It is recognized as a homogenous structure with pond inclusions.

Pingos are the main representative of subsurface ice intrusions. They are formed when liquid moisture is trapped between growing permafrost margins. They usually occur in the aftermath of thaw-lake senescence, when frost invades saturated, thawed ground where a thaw-lake has recently been.

Ice wedges occur when frozen ground contracts, leaving deep cracks that fill with water in spring and subsequently freeze. This process, repeated many times, results in massive, vertical, wedge-shaped curtains of ice in a polygonal pattern. They contain bubbles and dirt aligned vertically. These existed near both Dima and the Berezovka mammoths but not immediately around them. In the far north the tops of these active ice wedges are just under the insulating sod (fig. 1.9).

Ice veins are smaller versions of ice wedges. They are by definition quite thin and usually not wedge-shaped. They are also crack fills but occasionally are of more complex origin than ice wedges.

Segregated ice can occur in two primary forms:

Massive ice beds are horizontally oriented and developed in situ, usually within an open system of underground water movement or within a closed system of saturated or supersaturated soils. Segregation ice which forms these massive structures is rare.

Lenticular segregation is the most common form of segregation ice. It consists of horizontally oriented lenses of ice interspersed throughout the soil. Both of these segregated ice forms are horizontally aligned, elongated, and hold bubbles and dirt particles. My work with Blue Babe and descriptions of Dima and the Berezovka mammoth show that sediments surrounding mummies contain mainly segregated ice.

The physics of water segregation is still debated (Washburn 1980), but water is indeed drawn toward the freezing front. Additionally, contrary to one's intuitive sense, there is still liquid moisture in frozen ground, and this liquid moisture may migrate. Thus, especially in silts, segregation ice lens formation can continue behind the freezing plane (Washburn 1980). For extreme segregation to occur, slow freezing is important. In a closed system, such as that of saturated soils well below surface and subject to slow freezing, the buildup of ice lenses desiccates the soil. This desiccation process is critical in understanding the diagenetic changes that affect frozen carcasses and result in the true frozen mummies.

It is this water withdrawal, in the form of segregated ice, which dehydrates a mummy in frozen ground. This process is misunderstood by Shilo and Titov, who assumed that Dima's carcass would have to lie in the open air to be desiccated.

Is Mummy a Misnomer?

The word *mummy* has long been used to describe carcasses preserved in northern permafrost. Some have objected to this usage on the basis that preservation by freezing is unlike "real" mummification of an embalmed or dried corpse. However, frozen corpses, like Dima and Blue Babe, are indeed desiccated and fully deserve to be called mummies.

The process of frozen-ground desiccation is a fascinating story. Underground frost mummification should not be confused with freeze-drying, which occurs when a body is frozen and moisture is removed by sublimation, a process accelerated by a partial vacuum. The freeze-dried mummy is left in its original form, minus most of its moisture—a common technique used in food preservation and taxidermy studios. I have often freeze-dried items, sometimes inadvertently, during our long Alaskan winters, where the temperature seldom rises above freezing for eight months of the year. A bison or caribou foot left on the barn ledge, near where it was butchered in early winter, is freeze-dried to perfection by the following spring. Winter freeze-dried jerky can be made through this same process, by hanging meat strips up away from carnivores and birds.

However, the desiccation of fossil mummies is quite different than freeze-drying. Moisture contained in a buried carcass is not released to the atmosphere but is crystallized in place, in ice lenses around the mummy. This process is more comparable to tightly wrapped food left too long in a freezer. When a stew is first frozen, it swells to a somewhat larger size, bulging the sealed plastic container. The longer it stays in the freezer, month after month, the more moisture begins to separate, forming ice crystals inside the container. The stew itself shrinks and desiccates. Year follows year, and the stew becomes more and more desiccated, as ice segregates from it. Eventually, the stew has become a shriveled, dehydrated block; unlike freeze-drying, in which the object theoretically retains its original form, the stew is shrunken in size and surrounded by a network of clear ice crystals. This segregated ice also occurs in the silt around frozen mummies. Soft tissue becomes mummified and shrunken down, looking like a desiccated mummy dried in the sun. These two processes of cold mummification and freeze-drying were not distinctly understood by people unfamiliar with long winters and the back corners of deep freezers.

Dima's Burial

Once buried and cooled in the mud, which overlaid permanently frozen ground, Dima could have been preserved at "refrigerator" temperatures for a few weeks until it was completely frozen, without being totally macerated. As winter pulled heat from this mud, moisture separated from silt as segregated ice. The place where Dima lay was apparently covered the following spring by more silt and mud flushed from upslope when the little drainage brought a load of sediment from steeper slopes down to the gradual (7 to 8 degree) slope of the terrace. The flow slowed, and sediment was deposited on the terrace. This is not really alluvial deposition in a conventional sense. Removal and concentration of surface soil downslope during the spring following Dima's death would explain why material around the carcass (and the gastrointestinal tract) contained immature pollen grains as well as mature seeds from autumn dispersal; new spring seeds would have remained attached to the green plants and not dispersed downslope, nor would they have been embedded in the plant detritus.

A thick blanket of material, variously estimated at from 0.4 m to 1.5 m thick, would have given insulation sufficient to keep Dima frozen through the summer. The baby mammoth corpse was buried deep enough to be within the upper margins of permanently frozen ground, until it was uncovered by a bulldozer in 1977. Dima is now mounted and on display beside the famous Berezovka mammoth in the Zoology Museum in Leningrad.

Other Frozen Mammoths

At about the same time Dima was discovered (1977–78), another less complete mammoth mummy was found on a small tributary of the Khatanga River (Vereshchagin and Nikolaev 1982). It was found in 1977 by a reindeer herder, but not until the summer of 1978 could an expedition be fielded. The size of the tusks and the very large size of the animal indicate that it, like most of the other mammoth mummies, was an adult male.

Some skin remained on the head, including medial parts of the trunk and entire left ear. Two radiocarbon dates on the mammoth carcass itself were around 53,000 and 45,000 yr B.P. Two feet were also present, but most of the remainder consisted of a few scattered bones and hair. The carcass had been heavily scavenged and had decomposed. Although the animal was quite incomplete, the trunk tissue was moderately well preserved and consisted of 60% fat.

Like many Soviet mammoth fossils away from the regions of loess deposition, the Khatanga mammoth was preserved by burial in cross-bedded alluvial sands. Vereshchagin interprets these sands as indicating stream deposits. Ripple-marked sand had even been washed in between the rami of the lower jaw. Vereshchagin proposes that the animal lay in the open and was buried in a river channel.

Few other adult mammoth ears have been preserved in museum collections (young Dima's, of course, were present), although several had been described by early explorers who saw mammoth mummies. A complete mammoth ear, well preserved, may sound like the epitome of esoterica to some, but it provides important information for those of us who try to visualize exactly what these creatures looked like and understand the implications of that appearance. For a big male proboscidian, the Khatanga mummy's ear was quite small, only 33 cm high and 16 cm wide, slightly larger than the ear described from a mummy found in the 1700s (fig. 1.10).

This was an elderly mammoth; the third molar was worn halfway through, indicating, as Vereshchagin concludes, that the animal was about 50 years old. Fossil mammoth teeth recovered from late Pleistocene deposits in the Fairbanks area are usually in this stage of wear; apparently fifty years was a common life span for woolly mammoths.

That two of the feet, parts of the trunk, and the left ear were not eaten is curious. They are they most accessible body parts; abdominal skin on mammoths is so thick that few predators or scavengers could open it. I have observed that lions and hyenas in Africa usually go for elephant ears, trunks, and feet soon after the animal's death. We do have one other specimen of an adult mammoth trunk tip collected in 1926 (fig. 1.11). Actually, two feet, the distal third of the trunk, and one ear of the Khatanga mammoth were missing. However, the presence of some of the vulnerable soft parts suggests that these parts were buried in cool soil not long before freeze-up in

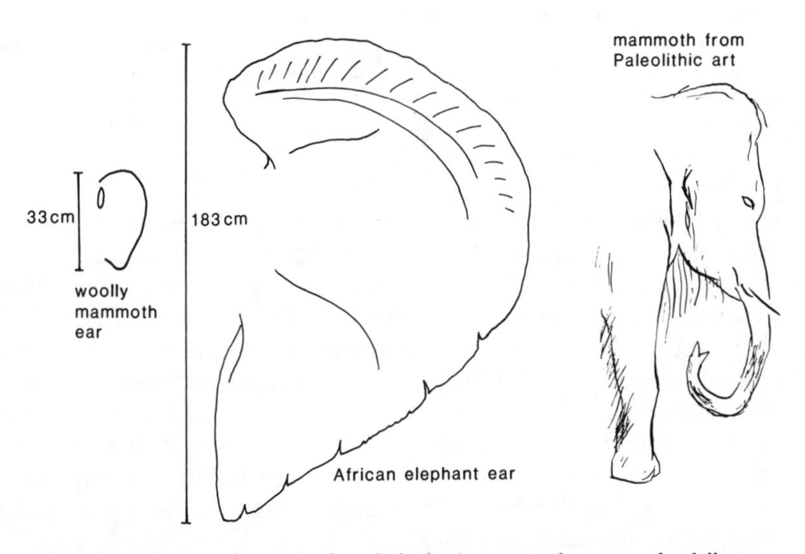

Fig. 1.10. A comparison of mammoth and elephant ears. To dramatize the difference in size I have compared the Khatanga mammoth ear (Vereshchagin and Baryshnikov 1982) with that of an African elephant of similar age and sex (Sikes 1971). The Paleolithic drawing from the Gönnersdorf site in northern Germany shows a woolly mammoth with quite small ears.

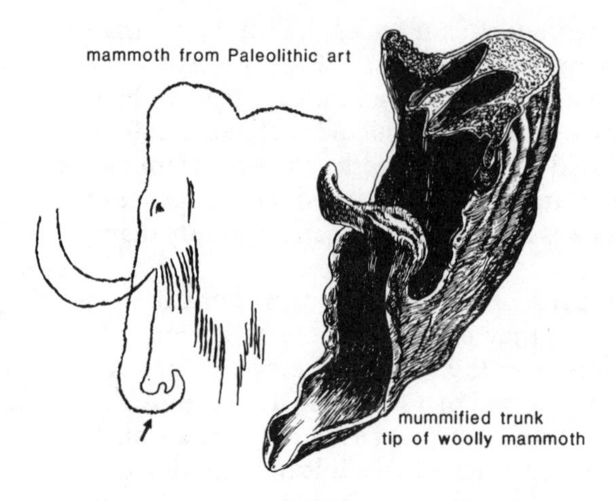

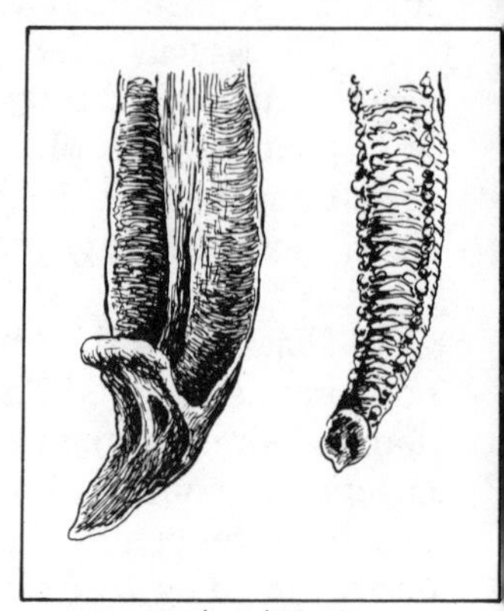

comparison between mammoth and elephant

Fig. 1.11. Unique trunk structure. The trunk tip of the woolly mammoth was unlike that of living elephants. (*Far left*) A mammoth drawn on the walls of Rouffignac Cave in France; (*next right*) a mammoth's trunk tip displayed at the Zoological Museum in Leningrad (the only one in existence from an adult mammoth, to my knowledge). The trunk obtained in 1924 from a Siberian native who had cut it from a frozen mammoth carcass. Note the handlike structures around the nasal opening and also the wide "wings" on each side. I have interpreted the former as a specialized grazing structure for a low-density sward. The latter "wings" could have served to take snow to the mouth. Free water is usually unavailable in the winter, and proboscidians require large volumes of water because they eat dry winter food, have a fast gut transit time, and lose a lot of water in feces.

the late summer or autumn, as they probably did not stay exposed unfrozen in the open for long. Because frozen skin is rather impenetrable to scavengers, these parts may have gone through a winter exposed. The presence of considerable adipose tissue in the trunk also indicates an autumn or early winter death. But it is unlikely the animal died and was buried in winter, because sediments are relatively stationary during that time of year. It is more likely that parts of the mammoth were buried during the autumn it died and that other portions were buried during the ice breakup the next spring in sand reworked by flooded northern steams—a scenario that may also account for the scatter of bones.

Assuming that this mammoth died on the floodplain some time in early autumn, lying partially underwater or on soft, wet sand, we might explain the preservation and lack of scavenging of those parts on the cold surface, like the right ear (if the head was lying on the right side). The remainder of the carcass would have been scavenged all winter. There would have been partial dismemberment by scavengers, both before freeze-up and afterward. Most of the muscle tissue would have been eaten away. During the spring, much of the remaining carcass would have been totally disarticulated and dispersed. Part of the carcass was lost, either not buried or washed farther downstream. Portions of the long bones still may have protruded for a season because they show considerable carnivore gnawing. Likewise, most of the tissue is macerated from these other long bones, again suggesting exposure for at least another summer. These long bones were, however, eventually covered by silt during later alluvial activity. No mention was made in research reports of fly pupae or gastrointestinal tract contents; these could have well been washed away during stream deposition.

In 1971 another frozen mammoth was found on the Shandrin River. This specimen had gastrointestinal tract contents but very little other soft tissue. The carcass dated within the Boutellier Interval: one radiocarbon date was about 42,000 yr B.P. and another 32,000 yr B.P. The Soviet literature portrays the Shandrin mammoth as a skeleton with an enormous mass of frozen viscera (around 300 kg) among its ribs. These gastrointestinal tract contents were originally studied and reported by N. G. Solonevich, B. A. Tikhomirov, and V. V. Ukraintseva (1977). They examined pollen and large, grossly identifiable woody plants and mosses.

This approach can be misleading. Pollen may provide an understanding of general vegetation but not necessarily record what the animal was eating. Likewise, a cursory examination of plant material exaggerates the proportion of easily identifiable specimens such as mosses and woody plant leaves. This approach misled people in the last century to incorrect conclusions about woolly mammoth diets, for they interpreted the presence of mosses and woody plants as evidence of the staple diet, and even contemporary researchers have stated that mammoths could survive eating present-day tundra vegetation (Colinvaux and West 1984; Colinvaux 1986). It is this cursory approach that causes Solonevich, Tikhomirov, and Ukraintseva to repeat those mistakes and conclude, again, that modern northern forests can serve as an analog for vegetation at the time the mammoth perished.

Percentage of pollen was averaged with that of spores, so the abstract indicated pollen from herbaceous plants constituted only 19% (spores constituted 77%). In fact, only one small sample of pollen was studied, and in that sample, grass (Poacea) was the most common, followed by sedge (Cyperaceae) and significant amounts of wormwood or sage (*Artemisia*) and larch (*Larix*). Other genera were found less often. Some woody plants were identified to species, such as cranberry (*Vaccinium vitis-ideae*) and larch (*Larix dahurica*), others only to genera, like willow (*Salix*). Woody plants, however, accounted for a small part of the gastrointestinal tract pollen and total biomass.

The plants from the stomach of this Shandrin mammoth were later studied by R. N. Gorlova (1982). She concentrated primarily on a histological analysis of vegetative parts and seeds, concluding that the habitats in which mammoths once grazed were wet and that present-day vegetation 100 km south of the Shandrin mammoth location corresponds to the remains of vegetation found in the gastrointestinal tract; mammoths had lived in a marshy, tree-scattered tundra. Gorlova, however, finds that of the identifiable macrofossils, 80% were herbs, 15% woody plants, 1% mosses, 1% leaves of heathers, and 5% completely decomposed and unidentifiable plant material. She does not specify the grasses or sedges that were in the stomach but does rank grass at the top among Angiospermae, suggesting that grass was the most dominant group, as indeed it was in the previously mentioned study on pollen from the stomach of this same mammoth.

I suspect it is this careful identification of species of mosses and woody plants and neglect of grasses (which were not even identified to genera) that make the flora appear more analogous to modern communities. The predominance of herb in the Shandrin mummy's gastrointestinal tract and the dominance of grasses are actually quite similar to studies of other mummies, which indicate that while there is continuity of many plant species, many Pleistocene vegetative patterns were not analogous to those dominating the north today (Sher 1974; Yurtsev 1982).

Gorlova found ripe fruits and seeds and examined the annual rings on the twigs of larch, dwarf birch, and willow, concluding that the mammoth died in the second half of summer.

Two mammalian anatomists, Yu. I. Yudichev and A. I. Averikhin (1982), also examined the frozen 300 kg monolith brought back from the site. This mass consisted mainly of gut contents, that is, vegetation, although parts of kidney, spleen, pancreas, and liver could be identified. These anatomists have an odd conclusion—that the animal died in spring from asphyxia. No reasons, however, are given for this assertion other than the mammoth died from bloating due to eating indigestible food. The validity of this assertation is uncertain, but one aspect of their point is well taken. Among herbivores, if death is not related to predation it is usually influenced by food. Moose and caribou not killed by predators usually die of starvation—they run out of digestible food and are forced to eat indigestible items. Thus, if we know enough about a species' food tolerances, we can appraise death by starvation not only in the case of an empty gastrointestinal tract, but also, and much more commonly, when plant materials are present. Most carcasses of starved ungulates, for example, have a rumen filled with vegetation, but it is indigestible vegetation. Since we are just learning what normal food for a mammoth would be, it is difficulty to judge whether the vegetation found in the Shandrin mammoth was normal or indicative of an animal literally starving with a full belly.

The taphonomy of the Shandrin mammoth is unclear. It seems to have died in the late summer or autumn. How it was buried is not mentioned in the literature I reviewed. That most of the carcass was decomposed except for the bones and gut contents is not strange. Coarse plant material is less subject to destruction or decomposition than animal tissue. A number of taphonomic permutations can render a carcass to bones and gut, including differential scavenging on the surface or differential decomposition after burial. From reports listed here, it seems that most bones were present and that the rib cage was complete around the frozen viscera. If that is correct, it would seem the entire animal was buried after some aerial decomposition (the anatomists reported blowfly pupae cases on the outside of the viscera). However, the presence of visceral soft parts (kidney, heart, etc.) is more difficult to explain according to most models of preservation.

Perhaps the plant-filled viscera survived intact because it froze before it could be scavenged and spread apart. Certainly, if it had not frozen it would have been quickly dispersed by animals in search of the most delectable portions. Perhaps this mammoth went down on its ventral abdomen and lay on frozen ground, both cooling the viscera and inadvertently protecting the thin abdominal skin from predators. Once frozen, the fibrous, vegetation-filled mass would have protected other portions of the viscera from scavenging. The differential distribution of soft tissue and preservation of viscera after that point are unclear.

Kubiak (1982) presents an excellent review of mammoth appearances as determined from these mummies and how woolly mammoth morphology functioned as an adaptation to the cold steppe environment. He takes each part of the external anatomy, portion by portion, and discusses the relevance of each feature to life in the far north. As these are not directly related to the preservation processes, nor to the Blue Babe bison mummy, I do not review his conclusions but simply refer the reader to his paper.

The Selerikan Pony

Although virtually unknown compared to Dima, a mummified Pleistocene pony was discovered in Siberia in 1968 (Skarlato 1977). This mummy, a mature stallion, was carefully studied. Like Dima, it dates from marine isotope stage 3, the Boutillier Interval; however, an exact date is still in question. Radiocarbon dates on the carcass range from 35,000 to 39,000 yr B.P. This Selerikan pony is clearly related to the true caballine horses of Europe and central Asia, known variously as *Equus caballus, E. ferus,* and *E. przewalskii,* but all can be considered the same species. Vereshchagin argues in the Skarlato volume for a separate specific status for the northern horses but acknowledges a close similarity to caballids.

Actually, this is not the first horse mummy from Siberia. Lazarev (1977b) notes that in 1878, Bunge reported the carcass of a white horse thawed from frozen ground on the Yana River, 60 km above Kazach'ye, but that horse carcass was not saved. In 1950 the mummy of a mare with a large embryo was found on the upper course of the Indigirka River on Sana Creek when a mine tunnel was excavated; only part of this mummy was saved for study.

An interesting story is associated with the Selerikan find (Lazarev 1977b). The horse was found deep underground by drift gold miners. Two legs and a tail emerged from the ceiling of the mine, 9 m below the surface in frozen ground. The miners used the horse's hind legs to hold cables and hang lanterns, but when the legs got in the way, they were blown out of the frozen ground with blasting powder and thrown away. Several months later, word of the find got back to Yakutsk and the Siberian Academy of Sciences sent a delegation to Selerikan to investigate. The body of the horse still remained, frozen in the ceiling. Using small blasting charges, the remainder of the horse was blown out. Later that summer, the horse's legs and tail were found in the dump. However, the head could not be found, and examination of the neck skin showed that it had not been preserved with the carcass.

The block of earth containing the body was kept frozen and flown to the Zoological Institute in Leningrad, where a team of experts thawed it and examined the carcass. Most of the trunk and the forelegs were intact. Maceration between death and freezing had consumed much of the abdominal skin, but the thoracic contents and some of the abdominal parts still existed, although they were quite decayed by enzyme action. The gastrointestinal tract was full, indicating the horse had died an almost instantaneous death.

When a team returned the following summer to examine the

stratigraphy and context of the mummy, they found part of the tunnel collapsed and full of water. There were other sections in which general stratigraphy could be studied, but the cave-in left details of the pony's burial even more unclear than was the case for Dima. In his article about the Selerikan pony, Vereshchagin provides a number of scenarios that might account for death and burial, but he does not prefer one over another. The position of the carcass suggests that the pony was stuck in some mire. Its hind legs were pointing more or less downward, and its forelegs were more horizontal. The body, although resting on its right side, is generally angled upward, and absence of the head is consistent with this interpretation. While it was still alive, the pony would have held its head up, and after death the head would be the most likely part of the body to be chewed on and dragged away by carnivores. The lack of scavenging on the lower body indicates it was buried and inaccessible.

No fly pupae or cases were found, again showing that the carcass was not exposed for long to summer air. Also, fat deposits surrounding the heart and other viscera suggest it did not die in late winter, early spring, or even early summer. The pony's sex could be determined by the presence of male genitalia and by the characteristic conformity of its pelvic girdle.

Like the Berezovka mammoth, the Selerikan pony has some bones that seem to have been broken after deposition. Its humeri are both broken in midshaft. Some of the ribs are also broken, and Vereshchagin (1977) proposes that this occurred by diagenetic processes, probably cryogenic deformation. To me these data suggest an animal deeply mired and sinking up to its neck. Predators soon killed the pony (recall its full gastrointestinal tract) and dragged away only the head and distal neck parts (fig. 1.12).

Like Dima, the Selerikan pony's carcass lay within organically rich sediments, interlaced with segregation ice (Lazarev 1977a), indicating that the sediments were water saturated at the time of freezing. Although the circumstances of its death and subsequent preservation are uncertain, there is sufficient information to reconstruct the time of the year the pony died. Pollen from the gastrointestinal tract was studied by N. G. Solonevich, B. A. Tikhomirov, and V. V. Ukraintseva (1977) and identified as mature, suggesting a late summer death. This interpretation was supported by T. V. Yegorova (1977), who identified mature seeds in the gastrointestinal tract. Pelage length is characteristic of a full winter coat. These facts point to late autumn as the time of death. Vereshchagin (1977) notes that winter hair is lost quite late in many northern species, but I doubt winter hair would have persisted until late summer. Also, equids

Fig. 1.12. Postulated entrapment and preservation of the Selerikan pony. The body was relatively well preserved and missing only the head and neck. The scenario of death and preservation that best fits the data is that the pony was mired up to its neck; then the head and neck were eaten by predators or scavengers. Later the body was further buried by the flow of silt and gravel downslope. The body was found from below, in the ceiling of a gold mine drift, and its hind leg was used to attach cables and hang lanterns until someone realized it was the mummified leg of a Pleistocene horse.

have both a winter pelt and a summer one, unlike extant northern ungulates.

Moose (*Alces*), caribou (*Rangifer*), mountain sheep (*Ovis*), and musk-oxen (*Ovibos*) shed their coats only once a year, usually in late spring. They use the early, short growth as a summer coat; thus one pelage suffices for the entire year.

The pelage of the Selerikan pony was almost identical to that of the extant central Asian wild horse, Przewalski's horse (Vereshchagin 1977). The hair was a reddish brown overall, with dark mane, tail, and legs. The tail had long hair to the base, unlike that of asses or hemionids. The undersides of the carcass were yellowish to white in color, and a dark dorsal stripe ran from the base of the tail to the mane. Only a short segment of the mane was present on the shoulder. It was quite black and stood erect. Vereshchagin proposed that in overall color the Selerikan pony was slightly darker than Przewalski's horse, but most of his comparisons were made with older museum specimens which tend to lighten with age. I have seen some quite dark individuals of Przewalski's horse, especially older stallions.

No definite age could be obtained because the mummy's head was missing, but epiphyses were all closed; from this Vereshchagin concluded that the animal was probably 7 to 8 years old—not the time of life one would expect stallions to die. I have shown that northern Pleistocene stallions tend to suffer peaks of mortality between 4 to 5 years of age and again between 10 to 13 years of age. These ages probably mark ascension to dominance (gaining a harem) and the loss of that dominance. Alaskan fossil material indicates death was unlikely at 5 to 10 years of age. Thus if Vereshchagin's age estimate is correct, it is consistent with an accidental death rather than mortality from intraspecific combat.

Well-preserved stomach contents included identifiable epidermal fragments, seeds, and pollen, giving us an idea of the horse's diet (in late autumn) the day it died. Epidermal fragments and pollen both showed a predominance of grass. N. G. Solonevich and V. V. Vikhireva-Vasil'kova (1977) found more than 90% of herbaceous material, virtually all of which was festucoid grasses, mainly *Festuca*. They also found a few sedge parts, some dicot herb leaves, and some woody plants, but they concluded: "Woody plants constitute an insignificant percentage of the total biomass" (p. 205). Both birch (*Betula*), and willow (*Salix*), were found. There were traces of moss, mainly *Polytrichum*. Ukraintseva's (1981) pollen studies from this same material showed a similar pattern. It was almost totally graminoids, with grasses outnumbering sedges 2 to 1. There were a small number of pollen grains of spruce, pine, and alder, all species presently found 1,000 km south of where the Selerikan pony was found. It was unclear to me whether these could be long-distance contaminants. Seeds found in the gut and identified by V. I. Yegorova (1977), however, were quite different than the stems and leaves. These seeds were mainly of *Kobresia* and *Carex*, both sedges. She identified seeds from *Kobresia capilliformis*, which today does not live in the Arctic but is a characteristic meadow grass of highlands in central Asia and Mongolia.

Interestingly, John Matthews identified a species of *Kobresia* seeds in the pelt of the Colorado Creek mammoth, found near McGrath, Alaska. This rather xeric sedge also grows in Mongolia. The mat of mammoth hairs were loaded with these small black *Kobresia* seeds, although the stomach contents (epidermal analysis) showed mostly grasses (85%) and very few sedges. Robert Thorson and I are preparing a manuscript on the Colorado Creek mammoth.

Woolly Rhinos

The Churapachi rhino mummy, found in 1972 by a villager digging a cellar, is also described in the Selerikan pony volume (Lazarev 1977c). Churapachi is a small village located on the Lena-Amga interfluve. Lazarev briefly discusses other rhino mummies found in the Soviet Union. In 1771 an entire rhino carcass was found on the Viyuy River, but only the head and two legs were saved. Another carcass was found on the Kahlbuy River, a drainage of the middle Yana River, in 1877. Lazarev does not mention the most famous of all woolly rhino mummies—those found in Starunia (formerly part of Poland and now part of the Soviet Union) in 1907 and 1929. The rhinos at Starunia were not preserved by freezing as were other mummies discussed here. Rather, they were trapped in a petrochemical seep associated with a salt deposit (fig. 1.13); they were pickled by the saline conditions and surrounded by a mineral wax called ozocerite. The original of one rhino is displayed in Krakov, Poland, and a plaster cast is displayed in the British Museum (Natural History).

The Churapachi rhino from Siberia, described by Lazarev, was a female, as deduced from her pelvis and slender horns. And judging from the well-worn molars, she was an old animal. Most of the carcass had rotted away; the lower legs were in fair condition, and yellowish fur was found in mud around the carcass. The skeleton was rather complete, indicating that it had not been heavily scavenged, perhaps not scavenged at all.

Conditions of death and burial of the Churapachi rhino were uncertain. Analysis of gastrointestinal contents in the carcass showed it to be 89% grasses, 4.5% composites, and 2.5% wormwood or sage, with the remainder diverse forb species. Lazarev (1977c) says that this is similar to plant material taken from the teeth of a different woolly rhino fossil by Garut, Metel'tseva, and Tikhomirov (1970). I have also extracted plant fragments from the large infundibula of a Siberian woolly rhino skull in the American Museum of Natural History and found, from analysis of the cuticle fragments, mostly grasses: 96% grasses, 2% moss, and 2% forbes.

Fig. 1.13. One of the Starunia woolly rhinos (*Coelodonta*).

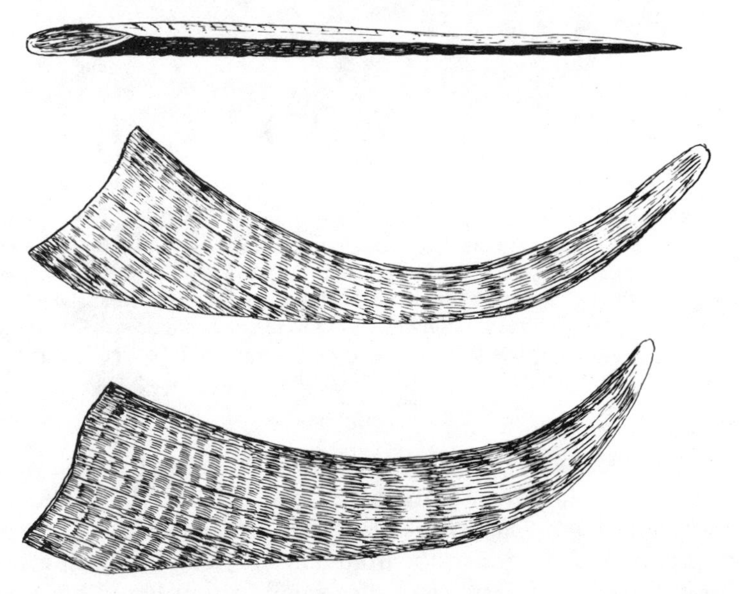

Fig. 1.14. Nasal horns of the woolly rhino (*Coelodonta*). These horns are almost flat in cross section (*top*). If woolly rhinos were like living rhinos, the nasal horn of the female was longer and more slender, as in the above comparison. (I do not believe the actual sexes of these two specimens are known.) Only six or seven of these horns have been found. One is now in the British Museum of Natural History, one in the Zoological Museum in Helsinki, one in the Natural History Museum in Krakov, Poland (the elongated one shown here, with 28 annuli), and several in the Soviet Union, both in the Zoological Museum in Leningrad and the Paleontological Museum in Moscow. The shorter (*bottom*) of the two pictured was illustrated by Pfitzenmayer 1926.

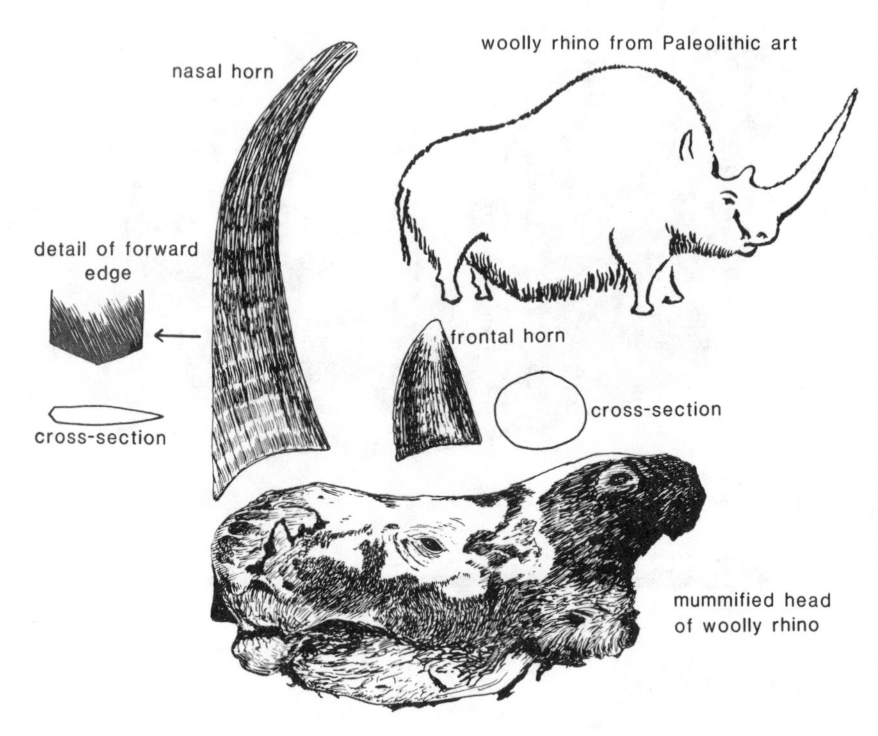

Fig. 1.15. Woolly rhino. This species lived across Eurasia during the latter part of the Pleistocene, but no rhino fossils have been found in Alaska. Woolly rhinos appear in Paleolithic art, and this one is from Rouffignac Cave in France. The anterior edge of the nasal horn was worn into bevels with left-right facets from head-swipe movements on the ground. The function of these movements is unknown. The upper lip was quite wide, indicating a grazing adaptation. The mummified woolly rhino head was found in Siberia and is now at the Zoological Museum in Leningrad.

To my knowledge, the Churapachi specimen is the first woolly rhino mummy in which horns have been found in place. The large anterior nasal horn is flattened laterally and is almost 1,300 mm long (fig. 1.14). In contrast, the frontal horn is short and conical (fig. 1.15). Over half a dozen, long, flat woolly rhino horns have been found, but it was impossible to tell whether these had once been oval and had simply weathered or decomposed on their lateral surfaces or whether they had indeed been rather flat in cross section when the animal was alive. No living rhino has a flattened horn. I think the Churapachi rhino has resolved that issue—the horns were indeed flat-sided.

The woolly rhino species (*Coelodonta antiquitatis*) did not reach Alaska; at least no Alaskan rhino bones have yet been found.

However, they did live in the far north of Eurasia for tens of thousands of years, through the late Pleistocene. Smaller than living rhinos and relatively long legged, the woolly rhino had a long, soft pelt, somewhat like the mammoth. Its upper lip was wide, like the white rhino (*Ceratotherium*) of Africa, and we know, again like the African white rhino, that the woolly rhino was mainly a grazer (fig. 1.15). Enamel patterns on its cheek teeth are very complex, and the teeth are high crowned. Why its range did not extend into Alaska and the Yukon Territory is a puzzle. We know very little about its biology; perhaps some day a well-excavated woolly rhinoceros mummy will supply the critical pieces of information we now lack.

The Alaskan Frozen Mummies

Other Pleistocene mummies have been found in the Soviet Union and Alaska. A large, mature male bison (fig. 1.16) was found in 1952 at Dome Creek, near Fairbanks. A date of 28,000 yr B.P. (L-127) was obtained from this bison. The partial carcass of a female bison (figs. 1.17, 1.18) was found in 1952 on Fairbanks Creek and reported by Flerov (1977). The date on this specimen was 11,950 +/−135 (ST 1633). Both specimens were excavated using a hydraulic mining monitor, with the loss of much associated information. Like Blue Babe, the Dome Creek bison appears to have been incompletely scavenged; it is now a central part of the Smithsonian Institution's Pleistocene Alaska display.

As best I can discern, only one partial Pleistocene bison mummy has been found in Siberia. This is a young (two-and-a-half-year-old) female found on the Indigirka River (Flerov 1977), dated at 29,500 +/−1000 (SOAK1007). Frozen mummies of legs, skins, hair,

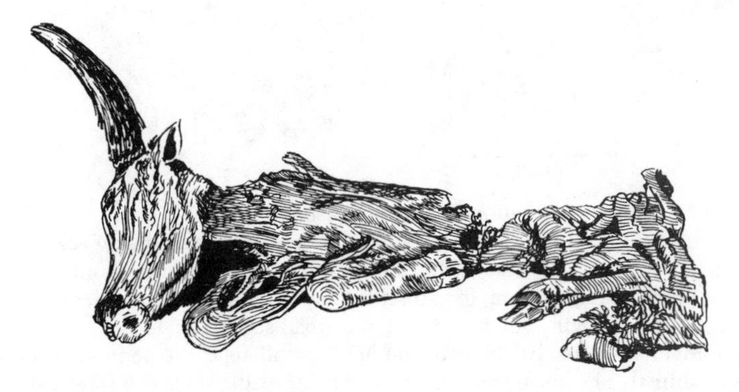

Fig. 1.16. Dome Creek mummy. The only other large bull steppe bison mummy found in Alaska came from Dome Creek in 1952.

and horn sheaths have also been found in both Siberia and Alaska (fig. 1.19).

In general, the number of frozen mummies discovered and the quality of scientific work done on those discoveries have been much greater in the Soviet Union than in Alaska. Of course, the Soviet Union simply has a much larger area of unglaciated northern Pleistocene sediments (fig. 1.20), but that is not the whole story. Several mummies were discovered in Alaska (fig. 1.21) during the first half of this century: parts of a baby mammoth named Effie were found in the Fairbanks mining distinct and dated 21,300 ± 1300 (fig. 1.22).

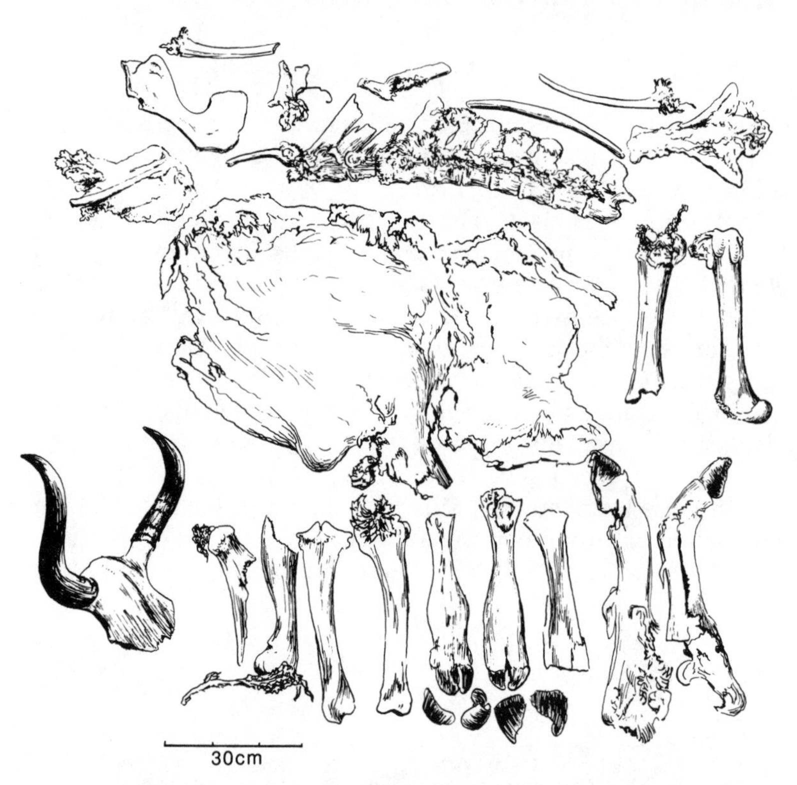

Fig. 1.17. Fairbanks Creek mummy, a female bison (F:AM 2177) found 1952. Judging by the horn annuli, she was 7 years old when she died. The seventh year segment was 13 mm wide; the sixth, 17 mm; fifth, 22 mm; forth, 25 mm; third, 64 mm; first and second combined were 350 mm, for a total of about 520 mm combined length around the longest outer margin. The hoof had three broad segments: 30 mm, 32 mm, and 32 mm (proximal to distal). The hoof and horn annuli appear to be from an autumn or winter animal. The bison showed a great deal of oxidized fat and was moderately well scavenged. She was apparently lying on her left side, with that part on the cold ground, as it is the left side that was best preserved. (head is shown in the next figure)

Fig. 1.18. Head of Fairbanks Creek mummy (F:AM 2177). Skull cap and horns pictured in fig. 1.17 are shown inserted into mummified neck and lower head.

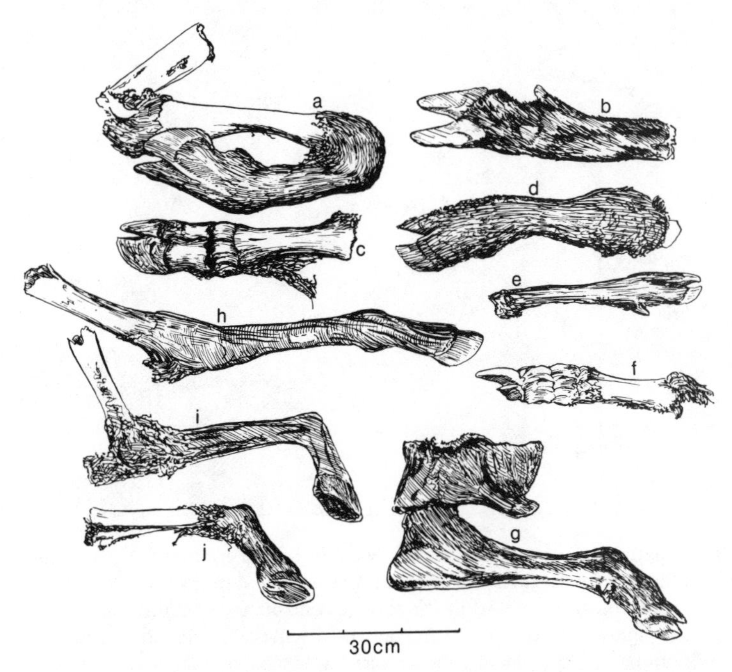

Fig. 1.19. Mummified large mammal legs from interior Alaska. Legs of large mammals are the most commonly mummified parts. These legs were all found in the Fairbanks area: (*a*) *Alces* F:AM A-274-4003 Little Eldorado Creek 1940; (*b*) *Alces* F:AM A-252-8493 Engineer Creek 1939; (*c*) *Bison* F:AM 1641 Cleary Creek 1936; (*d*) *Bison* F:AM 1640 Goldstream Creek 1936; (*e*) *Rangifer* F:AM BX 54 Upper Cleary Creek 1945; (*f*) *Bison* (no number) Upper Cleary Creek 1934; (*g*) *Rangifer* F:AM A-481-4414 Tofty 1948; (*h*) *Equus* F:AM A-119-600A Ester Creek 1937; *Equus* F:AM-A-1638 Goldstream Creek 1936; *Equus F:AM A-1639 Goldstream (no date).*

But we have only the partial skin, one leg, the neck and head, and proximal part of the trunk; we know virtually nothing about it. The two partial mummies of bison found in 1952 are only dried carcasses, and little information was retrieved about their pelage, the circumstances of death, or burial.

This list continues. There are several partial carcasses of the now-extinct stag-moose, probably *Alces* (*Cervalces*) *latifrons*, one (F:AM 274-4002) discovered in 1940 (fig. 1.23) and another F:AM 274-4001) in 1942 (fig. 1.24). Still other moose were discovered in 1980 (fig. 1.25) near Livengood, and radiocarbon dated at 32,040 + 870/−980 yr B.P. (DIC-3090). An almost complete mummy of the

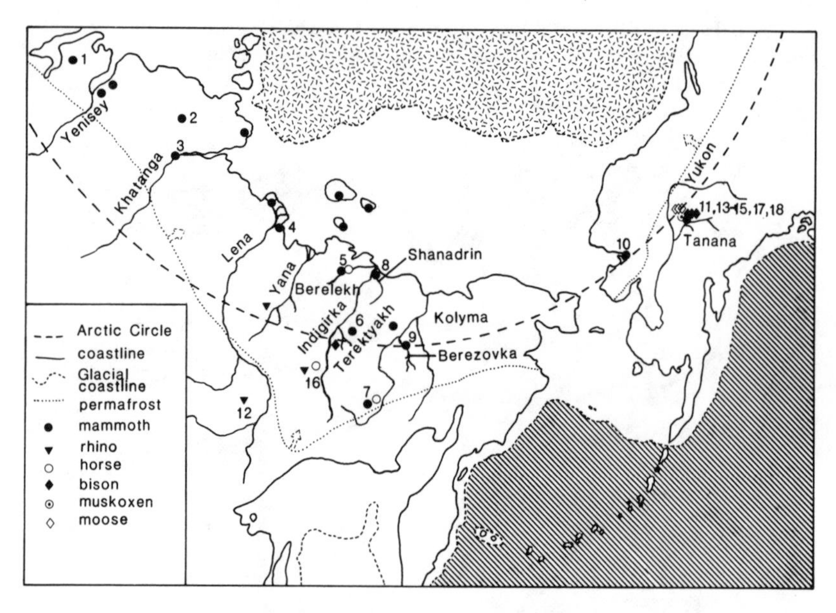

Fig. 1.20. Beringian mummy localities. Most of the Siberian and Alaskan locations of frozen Pleistocene large mammal mummies discussed in this book are shown here. Their distribution clusters within the zone of continuous permafrost and, in the case of Alaska, slightly south of it. In Alaska the discovery of Pleistocene mummies is mainly correlated with placer mining activity in the Fairbanks area. The finds of mammoths are (1) Yuribei, 1971; (2) Taimyr, 1948; (3) Khatanga, 1977; (4) Adams, 1799; (5) Berelekh, 1970; (6) Terekyakh, 1971; (7) Dima, 1977; (8) Shandrin, 1971; (9) Berezovka, 1900; (10) Quackenbush, 1907; (11) Fairbanks Creek, 1948. The bison finds are (12) Mylakhchyn, 1971; (13) Dome Creek, 1952; (14) Fairbanks Creek, 1952; (15) Pearl Creek, 1979. Only one well-preserved horse was found on the (16) Selerikan River. At least three partial moose mummies have been found in the Fairbanks area (17) and one helmeted musk-ox mummy (18) from Dome Creek.

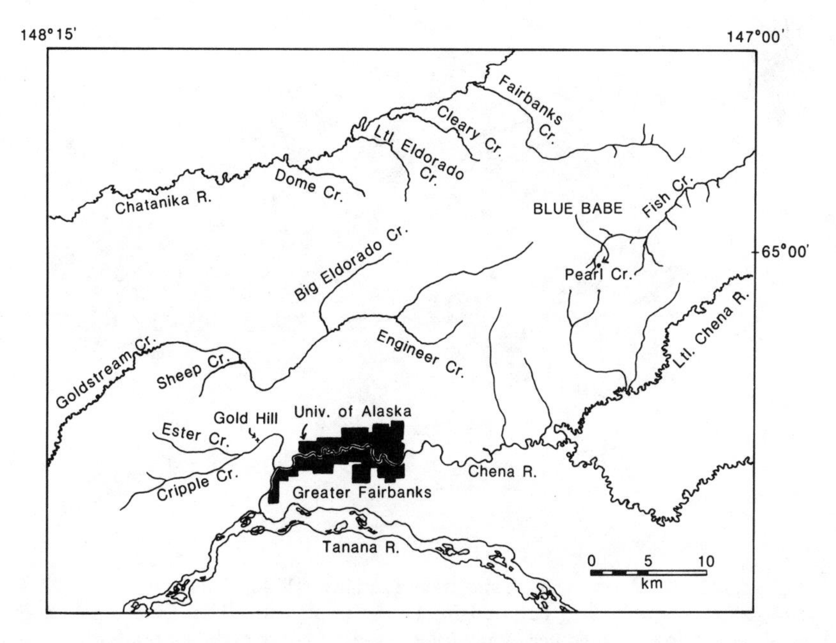

Fig. 1.21. Fossil localities near Fairbanks, Alaska. Most Pleistocene bones and mummies in this area have come from placer mines on or near the creeks shown here. Pearl Creek, where Blue Babe was found, is also shown.

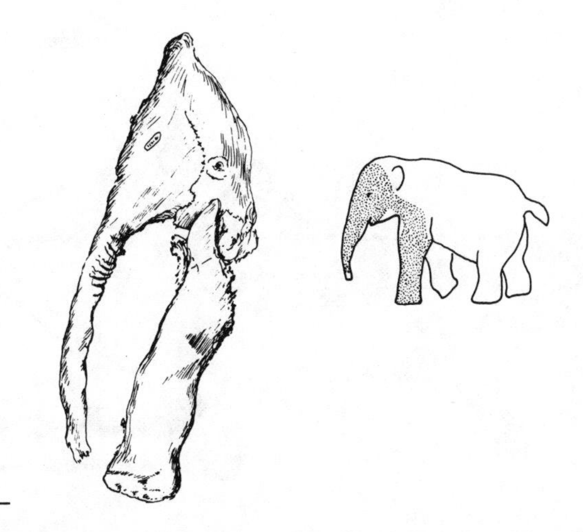

Fig. 1.22. Partial baby mammoth named Effie. the head and attached left leg of a young mammoth (F:AM 99921) was found at Fairbanks Creek in 1948. This animal died in its first year of life, judging from its size (an estimated 1 m at the shoulder—only slightly larger than Dima).

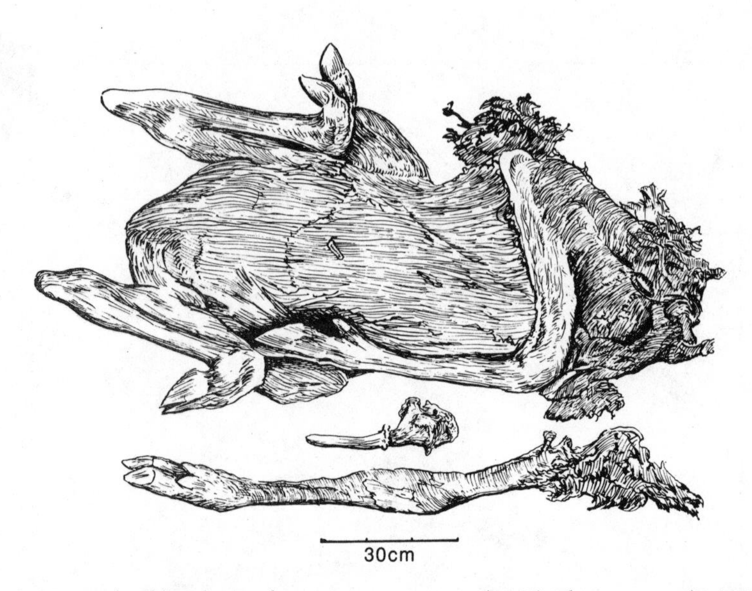

Fig. 1.23. Little Eldorado Creek stag-moose mummy (1940). This mummy (F:AM A-274–4002) was a yearling bull, judging by the small antler size. Large amounts of oxidized fat and the presence of an antler suggest a fall to early winter death. There is no date on this mummy.

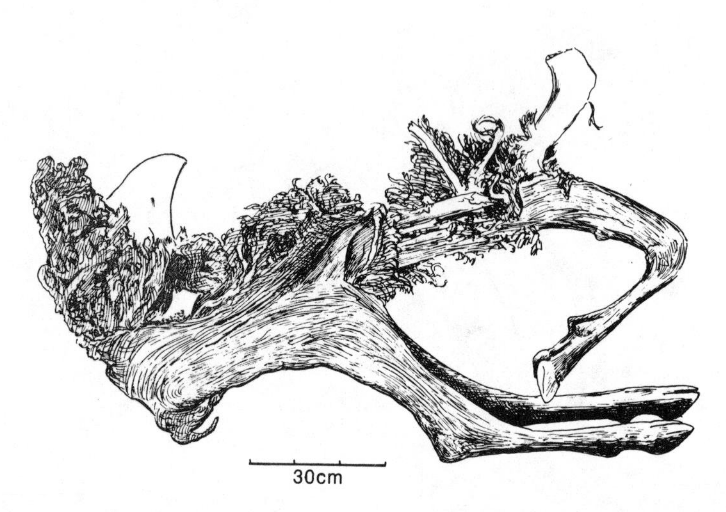

Fig. 1.24. Little Eldorado Creek stag-moose mummy (1942). This moose (F:AM A-274–4001) is a female; four small teats can be seen. The epiphyses had come off the bones of the left leg so we know this was a young animal. The tail is well preserved (sans hair); it is 22.5 cm around the curve. This moose is undated.

extinct helmeted musk-ox (*Bootherium bombifrons*) cow (F:AM A-293-5268) was collected on Fairbanks Creek in 1940 (McDonald 1984). Numerous legs of horses, bison, moose, and caribou were found during the peak gold mining era in Alaska. The mummified carcasses of a number of small mammals such as a ground squirrels (*Spermophillus*), mice (*Microtus*), and a pika (*Ochotona*) have also been found (Guthrie 1973; fig. 1.26), but unfortunately no studies were made of the sediments in which they were located, of their gut contents, or of other elements that might have provided valuable information.

The reason behind most of these lost opportunities is understandable. Miners discovered the majority of these mummies, and

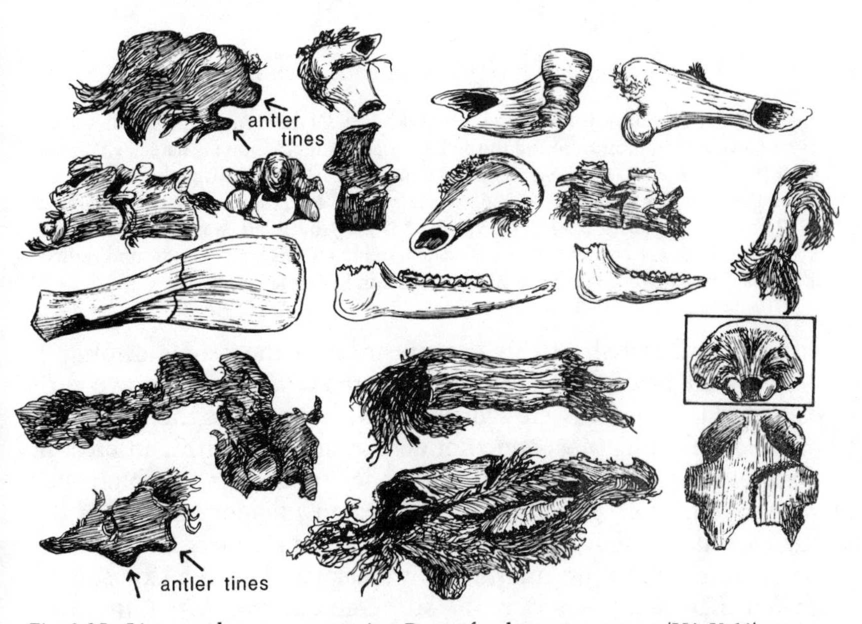

Fig. 1.25. Livengood moose mummies. Parts of at least two moose (UA V-64) mummies were found in the Livengood mining district at 60.5 mi Elliot Highway in 1980. The skull belongs not to *Alces alces*, but to the earlier *Alces* (*Cervalces*) *latifrons* (see box). The skull is broken along the frontal sutures and seems to correspond with the age of the large ramus. Tooth wear on that specimen places it somewhere between 4 to 6 years of age. The arrows point to bumps on antler portions in the "velvet" skin. The antler itself is poorly ossified. Both the incomplete ossification and the bumps which have not turned into tines suggest a May–June death. The M^1 on the small ramus has erupted and is in wear, but the M^2 has not erupted. This indicates a moose of about 5 months in age—an autumn death. All parts have been heavily scavenged; few edible parts remain.

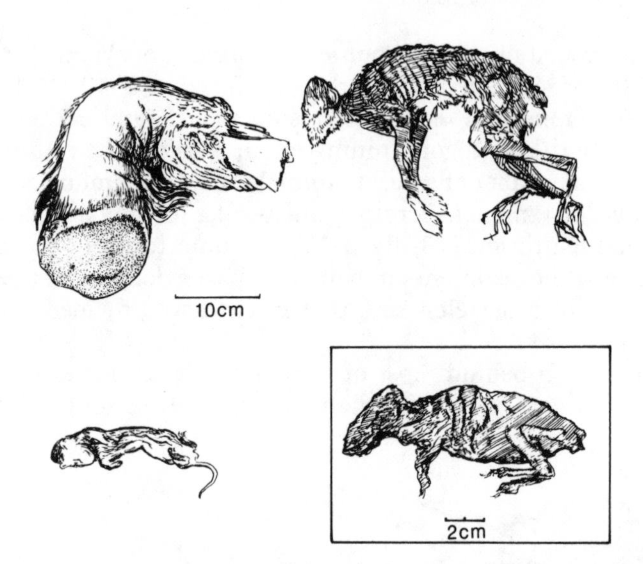

Fig. 1.26. Other mummies from the Fairbanks area. Of three small mammals and the foot of a young mammoth found in the Pleistocene mucks, none is dated. The mammoth foot (F:AM 5001), *upper left,* was found in 1940 in Fairbanks Creek. The snowshoe hare, *Lepus* (F:AM 99926), *upper right,* was found on Fairbanks Creek in 1949. A ground squirrel, *Spermophilus* (F:AM 7177), *lower left,* was collected in 1939 in Eva Creek. In the box, at a different scale, is a pika, *Ochotona.* The ground squirrel is on display at the University of Alaska Museum in Fairbanks.

during the first half of this century most of the people mining in Alaska were new to the north. Mammoths were a little-known curiosity, and few miners were acquainted with the issues that made this material valuable. Most were not even aware of scientific interest in the creatures, much less the importance of the dirt in which they were found. Mining techniques focused on efficient removal of the silt overburden that contained the frozen mummies. Large jets of water from "guns" or "monitors" were sprayed on the exposure to hasten thaw and wash away the silt. One can recognize a mummy coming out of the silt during this process, but to preserve as much information as possible, the jet of water has to be diverted at the very moment the overwhelming inclination is to spray harder and free it from the earth. But a mummy "excavated" this way, washed clean with a monitor, is a little like the dust jacket of the Dead Sea Scrolls without the parchment paper. It is a flashy find, but the essential story is lost.

Fortunately, some Alaskan miners are now aware of the importance of saving secondary information. One of them discovered Blue Babe.

2
UNEARTHING BLUE BABE

The Discovery

In July 1979 Walter Roman discovered the feet of a large mammal protruding from the frozen muck at his placer gold mine north of Fairbanks, Alaska (fig. 2.1). As the local specialist on Pleistocene mammals, I was asked to visit the site, so Dick Reger, a geologist from the Alaska Geological Survey, and I drove out to the mine. Upon seeing the hooves, I recognized that the animal was a frozen bison mummy (fig. 2.2). And not only were the bones preserved, but red muscle and black hair were sticking out of the bank.

I have worked in Alaska for several decades and have excavated more than a thousand fossils. Often I have responded to calls about discoveries that sounded exciting but that turned out to be an isolated tusk protruding from a cliff or part of a bison skull among the gravels. In fact, these fossil finds are usually interesting and informative. But Mr. Roman's find was a first for both of us. He had found a true mummy, and it was a rare event.

From the time of the Alaskan Gold Rush to the late 1950s, thousands of gold miners laboriously washed away acres of silty overburden to gain access to gold-bearing rocks. Several frozen mummies of large mammals had been found earlier in this century, but not many. In recent decades, since the biggest of the gold mining operations had shut down, nothing had been found of much importance, even though there are many small placer mines like the one Walter Roman operates with his wife, Ruth, and son, Ron. Pearl Creek, where the bison was found, is a tributary (fig. 2.3) of the Chatanika River. Latitude and longitude are 65° 59′ N and 47° 19′ W.

Like all good things, the frozen bison was a mixed blessing. It certainly caught me at an awkward time. I was getting ready to leave for a year's sabbatical in Europe, and other constraints made the time

Fig. 2.1. Walter Roman, the gold miner who found Blue Babe, standing next to one of his giant "monitors." The powerful jet of water helps to thaw and wash the silt. Many acres of frozen ground must be stripped away in his manner to reach gold-bearing gravels.

Fig. 2.2. Blue Babe emerging from the muck.

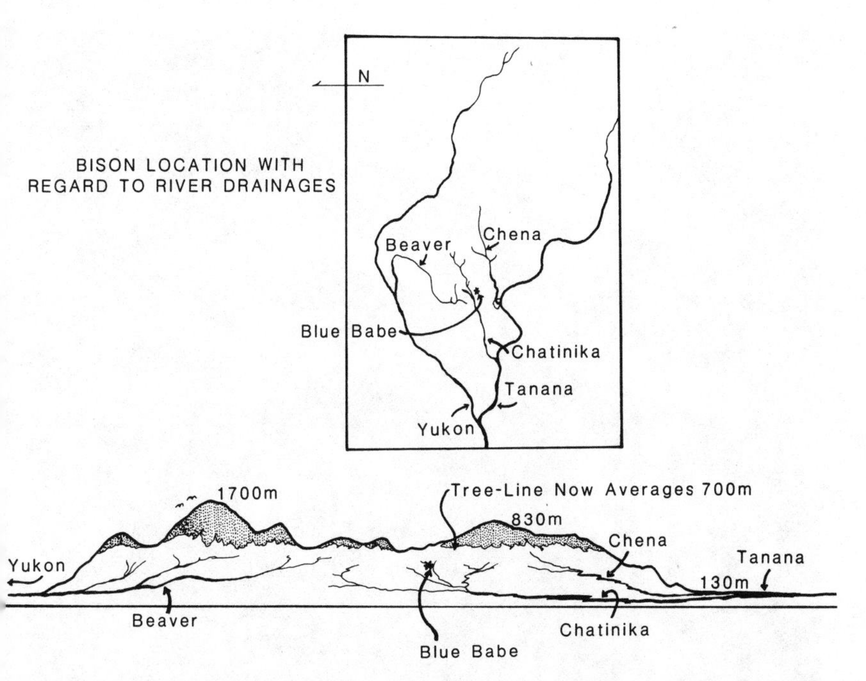

Fig. 2.3. Schematic map of bison site. Blue Babe was found north of Fairbanks, Alaska, in the uplands between two major interior Alaskan rivers, the Yukon and Tanana. The hills around the Pearl Creek locality rise above the present tree line, but Blue Babe, was found well below this tree line.

seem terribly short. Roman was near the end of his summer cleanup, and he needed to finish sluicing before water ran out or froze up. Frozen silt surrounding the bison was in the way, but Roman was able to wash out around the bison and gave us a few weeks' grace to excavate the mummy. The university was not in session, and volunteer help was difficult to find because of the economic boom created by the Alaska pipeline. I finally enlisted my wife and son as assistants. We spent the long midsummer evenings out at the mine, screening silt from the daily thaw zone (fig. 2.4).

The walls of frozen ground cooled the air, and it settled to a cold layer. Approaching the mine was like walking from a warm July evening into autumn. We had to take jackets, boots, and rain gear even on the brightest days. Shedding, thawed ground plipped and plopped from the overhanging walls of muck. A rich, pungent rottenness, like nothing else I have smelled, filled the air. It was a rottenness aged for millennia in the frost—not a stench, but a sweet, in-

Fig. 2.4. Collecting thawed silt. After getting photographs and maps of the exposure, we started to wet screen the silt around Blue Babe which had already thawed and collapsed. Here, Mary Lee, my wife and emergency assistant, collects mud to be screened back at the lab.

Fig. 2.5. Walter Roman's mine. A photo of the Pearl Creek mine and its exposures shows the scene on a gray day. Blue Babe was found to the right of the monitor.

tense tang. We worked in a canyon of ice-cold, blue-black mud, and the water used to wash away the silt was a mud-laden stream at our boots (fig. 2.5).

We had to plan fast. Mud around the bison was melting inexorably, and the miner's big water jet was waiting. First, in a field notebook I tried to record every bit of stratigraphic information. It changed daily. All the silt around the bison had to be washed through a small mesh screen. Soon we realized we could save time by collecting silt now and washing later. We saved every shred of material, but that meant arranging freezer space. And of course we had to plan for enough freezer space to accommodate the mummy once it was excavated. At the same time, I was trying to sort out what to do with the mummy while I was on sabbatical.

Mr. Roman and the owner of the ground from whom he had leased the mine, Alaska Gold Co. (Dan Eagan, president), were willing to donate the bison to the university. When part of the bison began tearing loose from the rest still solidly frozen in the bank, I realized I had to get the exposed portion back to the university and into the freezer. But the museum director was reluctant to sign the donation form because the miner stipulated that the mummy always be displayed. It was an impasse, and I was stuck in the middle of it with the thawing bison. Finally, a solution occurred to me: the biology department could promise to display the bison if the museum did not always want to. The museum director signed. That same evening the legs and torso of the bison came loose from the bank, and we brought it to the freezer at the Institute of Arctic Biology.

A local newspaper reporter had got wind of the find and aggressively sought information. Roman, however, had stipulated that I release nothing to the press because his mining operation was about to start sluicing gold—an understandably delicate time in gold mining—and he did not want a lot of strangers walking by the sluice boxes. Again I was in the middle. I tried explaining the situation to the reporter, but he only became more angry, saying the public had a right to the information. The very night the head of the mummy was freed, the reporter found out via Mr. Roman's wife where the site was located and arrived with his cameras. The photographs of Blue Babe were in the news the same day the bison's head went into the freezer. Still I needed to record more information about other sediments beyond the spot where the bison lay before they too thawed and were washed downstream. We asked for a little more time, which Roman graciously granted.

Although it was frustrating to lock the mummy in a freezer and fly off for a year, the sabbatical provided essential time and access to information. I needed time before the necropsy to organize questions; before I took it apart, I wanted to be thoroughly alert to clues the mummy might contain. And I needed time to obtain funds to hire assistants. Coincidentally, I had already planned to spend time with Soviet researchers who had worked on similar Siberian frozen mummies. I would also be able to study museum collections of European Mammoth Steppe fossils. These European fossils are similar to many found in Alaska because the continents of North America and Asia were once connected.

The bison mummy and numerous bags and buckets of silt were securely stowed away in a freezer at the university, but that big walk-in freezer had broken down in the past. The Institute of Arctic Biology director, John Bligh, suggested a double alarm system and came up with funds for it from the Institute's budget. It would not be the last time Bligh would come through on the Blue Babe project.

Roman's find and the rare opportunity he gave us to carefully excavate Blue Babe meant we were able to collect much information that was later pieced into a complete story about the mummy. The

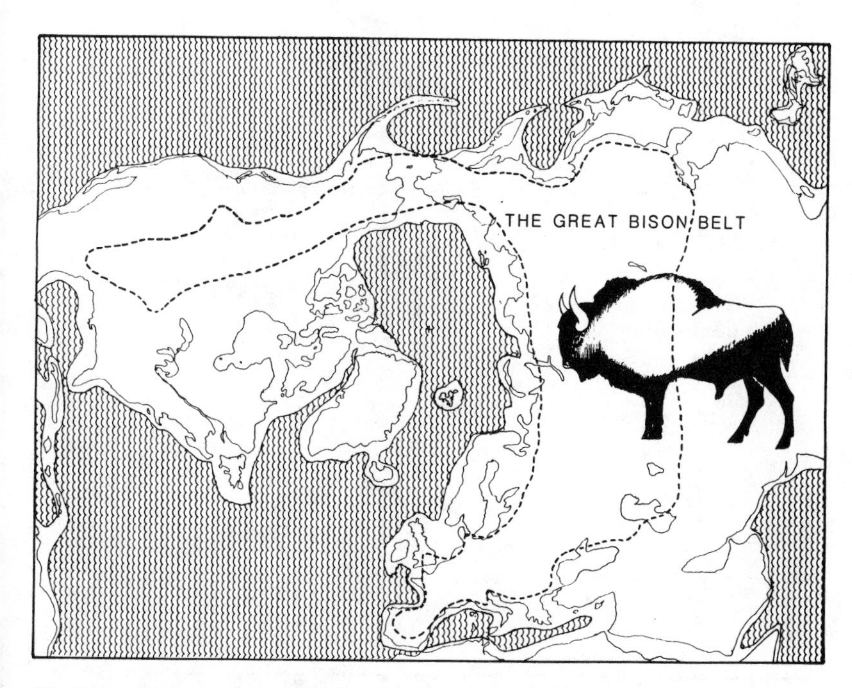

Fig. 2.6. The great Pleistocene bison belt. Fossil remains of Pleistocene bison are concentrated across Eurasia in a wide east-west belt. This is the main axis of the arid Pleistocene Mammoth Steppe. In North America, however, bison remains are concentrated in the rain shadow on a north-south belt. Although bison occurred elsewhere in North America, they were never so abundant as bison in the shortgrass plains east of the Rocky Mountains.

necropsy revealed a fascinating scenario of how and in which season the bison died, how the carcass was scavenged, and finally, how it happened to be preserved. Following the necropsy, the mummy was expertly mounted by Eirik Granqvist, then conservator of the Zoological Museum in Helsinki, Finland. Blue Babe is now displayed at the University of Alaska Museum in Fairbanks.

Although bison are not now associated with the far north, that was their prehistoric homeland. Their distribution across northern Eurasia formed a vast bison belt that ran through Alaska and followed the rain shadow of the Rocky Mountains (fig. 2.6). Indeed, bison may not be the only bovine that consistently used these northern arid regions. A species of cattle, the yak (*Bos poephagus*), ventured into the Asian north on occasion (fig. 2.7). Yak skulls are easily distinguished from bison, but are very similar to cattle. New radiocarbon dates on Alaskan specimens show that these "yak fossils" are

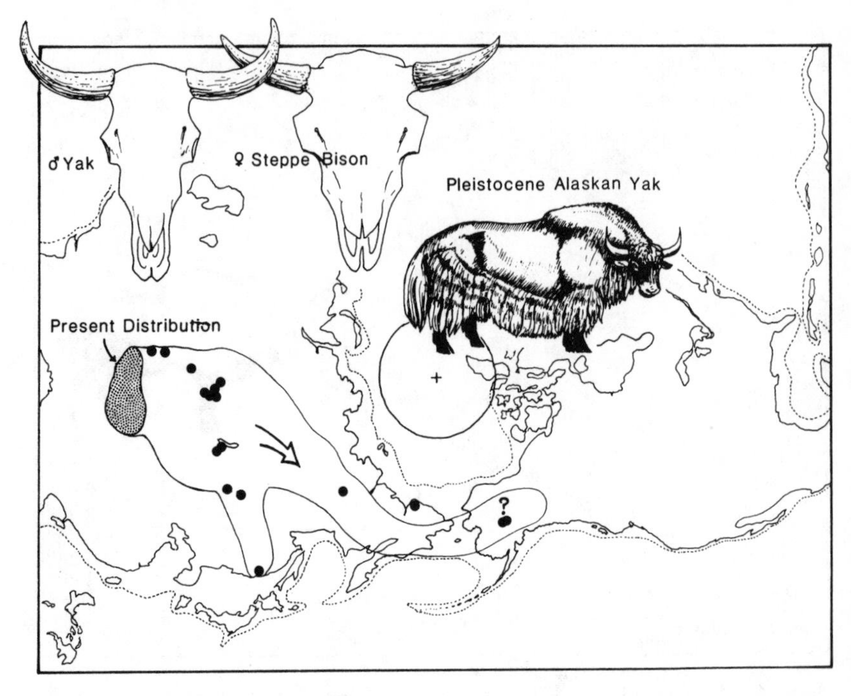

Fig. 2.7. Yak distribution in Beringia. Yak (*Bos poephagus*) bones are found in late Pleistocene sediments in the far north of eastern Eurasia, indicating a range considerably larger than the present. Yak skull identification is complicated because they are so similar to cattle skulls, but skulls of bison and yak are quite diagnostic. Alaskan "yaks" seem to be miner's cattle (see text).

stained skull fragments from cattle that early miners brought to Alaska.

The Geological Context of Blue Babe

During the Tertiary, the last 65 million years, the landmasses of the eastern and western hemispheres have been periodically connected via Alaska and the Soviet Far East when low sea levels exposed the Bering-Chukchi Platform. This vast area comprised a large and complex biotic province known as Beringia. Some of the richest deposits of Pleistocene fossils in the world occur in these unglaciated northern regions. Fortunately this abundance of fossils occurs at the right location to provide clues about animal and plant movements between hemispheres. Because Eurasia and North America have exchanged faunas and floras across this northern avenue, it is sometimes difficult to categorize Alaska as Eurasian or North American in its biogeographic affinities. But these large expanses of Arctic landscape have been more than a turnstyle or corridor for animals and plants. The region has been home as well as highway, sustaining a characteristic fauna and flora (Guthrie and Matthews 1971).

Like most of the Soviet Far East, large expanses of Alaska and the Yukon Territory were not glaciated during the Pleistocene. Because these areas were bounded on several sides by enormous glaciers and glacial outwash streams, today much of Beringia is mantled with a thick deposit of eolian (wind-blown) silt called loess (Péwé 1975a). Loess sediments originated from glacial flow triturating underlying stone into tiny particles. These particles were deposited beyond the glacial terminus by outwash streams. Winds lifted this glacial dust from stream bars and carried it across the landscape, where it slowly accumulated in thick, buff-colored loess deposits (fig. 2.8).

An important aspect of northern loesses for the paleontologist is that when they are washed downslope to valley bottoms, they act

Fig. 2.8. Loess bank at a road cut near Fairbanks. The thick loess which settled in the Fairbanks area during the glacials forms a mantle over bedrock. One interesting feature of these loess deposits is that they stand vertically.

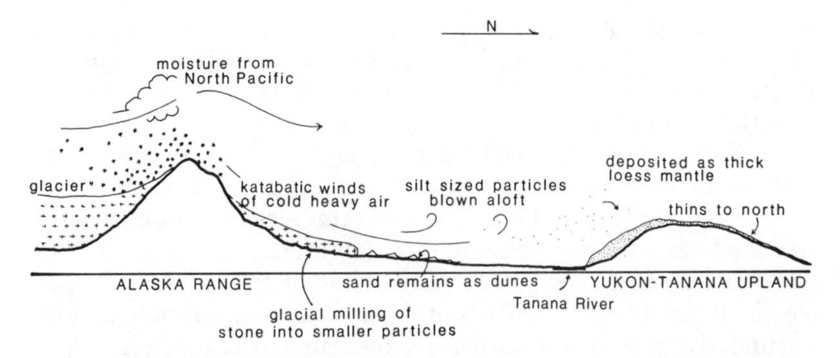

Fig. 2.9. Origins of Yukon-Tanana upland loess. The region where Blue Babe was found has especially thick loess deposits. Although most moisture from the Pacific air mass falls south of the Alaska Range, some is deposited on the north side. In Pleistocene times this moisture nourished great ice fields to the south and fed smaller glaciers that flowed north from the mountains. Glacial trituration grinds rock into sand and even smaller particles of silt. Meltwater carries this glacial flour to river deltas, where wind sorts sand into dunes and blows the silt aloft. Clouds of this dust were carried many miles across the Yukon-Tanana uplands.

as a preserving agent without peer. Although some bones have been preserved in the primary loess, these are rare and lack soft tissues. Frozen mummies and most fossils are found in the reworked silt. Preservation of both fauna and flora is exquisite in the loess that was washed downslope and accumulated in valley bottoms and broad colluvial cones. In such deposits one finds beds of tree roots and stumps, beaver dams and ponds, and countless large mammal bones; many with marrow inside, others with connective tissue, muscle, and skin still attached.

The hills around Fairbanks, which these loesses mantle, are composed of a schist bedrock, penetrated by veins of quartz containing minute particles of gold. Prior to the Pleistocene this schist (1) decomposed to form its own thin soil mantle and (2) eroded along upper stream courses as coarse angular gravel. Heavier gold particles were concentrated at the base of these gravels, on top of bedrock in what we now call gold placer deposits.

Had the geological events of the Pleistocene not occurred, these hilly schist uplands would be a rougher landscape, with only thin soils of decomposed bedrock. But large Pleistocene glaciers in the Alaskan Range, 200 km to the south, indirectly altered the uplands in major ways. The northern Pacific air mass which flows into the state from the Gulf of Alaska carries a heavy load of moisture, but successive mountain ranges—the Chugach, Wrangell, Talkeetna, and Alaska ranges—catch most of this moisture (fig. 2.9). Air that reaches interior Alaska creates a dry, continental climate, and that lack of moisture is the reason the interior remained unglaciated during the Pleistocene. Conversely, moisture falling on the high mountain arcs created large glaciers that coalesced into an ice sheet the length of the Alaskan coast. On the north side of the Alaska Range, valley glaciers extended toward Fairbanks. As these flowed northward out of the mountains they carried with them large quantities of rock.

Rapid Pleistocene uplift of the Alaska Range increased the valley slope, which further accelerated aggradation of bedrock by glaciers and stream action. The sculpting of those north-facing drainage basins by glaciers carried large quantities of stone debris toward the Tanana River, which drains the north face of the Alaska Range. Sediments ground into smaller and smaller particles by glacial movement were deposited beyond the glacial terminus. Meltwater rivers carrying this rock powder fanned into broad deltas toward the north, crowding the Tanana River up against the hills near Fairbanks. Cooled, heavy air rolled over the mountains and funneled down valleys, gaining speed as it dropped into the lowlands. It blew sand into dunes and lofted clouds of silt out across the Tanana Valley onto the Tanana Hills (fig. 2.10). Over a million years, this fine rain of dust formed a thick mantle of loess over the hills and valleys. Loess deposits smoothed the landscape into gentle contours. Judging from exposed sections of mining cuts, deposition occurred mainly during the glacials. Spring runoff and rains carried some of this silt downslope—sometimes rapidly, at times slowly, and during interglacials, hardly at all. Silt movement downslope varied with the amount of wind and vegetation cover, and the quantity and intensity of rainfall or snowmelt.

Cycles of silt deposition and stabilization probably occurred throughout the Pleistocene, but episodes of extreme silt erosion stripped loess deposits from valley bottoms as well as hillsides. We find only patchy remnants of the early Pleistocene silts (Péwé 1975a, 1975b; Guthrie 1968). Most valley bottoms are filled only with late Pleistocene sediments overlying gold-bearing gravels.

Except on lower, well-drained south-facing slopes, these loess and reworked loess sediments exist as perennially frozen ground known as permafrost. Only the upper meter or two thaws in summer. The concept of permafrost is simple: the balance of annual heat gained versus heat lost in the soil becomes negative. Microclimatic dynamics of permafrost are complex. Permafrost is affected by a number of interacting variables, including insulation, moisture, de-

Fig. 2.10. Dust blowing along the Delta River, Alaska. Pleistocene loess deposits were created by similar but larger-scale processes.

Fig. 2.11. Placer miner's deep canyons in permafrost. Miners remove the insulating vegetation; then they use flowing water to cut into the frozen silt. Upstream from where Blue Babe was found, one can see the thin "icing" of thawed ground and vegetation and the much thicker layer of frozen "cake" below.

gree of slope, and aspect. As my wife likes to picture it, we who live in the far north reside on the warm meringue crust of a Baked Alaska dessert. The cold ice cream is never far underfoot. Like a potted plant sitting on top of the home freezer, things can grow in the summer sun despite the frost just beneath (fig. 2.11).

Frozen silt deposits in the Tanana Hills are like an irregular layer-cake, with thin lenses of earlier Pleistocene loess covered by thick layers of the latest Pleistocene. The "cake" part of the sections is the more homogenous glacial windblown silt with few obvious features, while the wetter and warmer interstadial and interglacial periods are usually represented by thinner "icing" layers of fossil soils, often including peat and stumps and roots of woody plants laid down when soils were more stable.

We live on top of this cake. It is only along stream-cut banks and at mining operations, like the one where Blue Babe was found, that a slice can be studied. These vary considerably with topography, but generally form a repeated pattern (Péwé 1975a, 1975b).

The silts are so organically rich that they are black when wet and give off a thick pungent odor of incompletely composted litter. As they thaw, one is aware of decomposition that was arrested by freezing, thousands or tens of thousands of years ago.

Many of the lowland Pleistocene deposits around Fairbanks contain large ice wedges. As I mentioned in the first chapter, intense winter cooling of the ground causes contraction, producing cracks that penetrate a number of meters deep. Ice wedges are formed when surface meltwater flows into these polygonally patterned ground cracks and then freezes. When this pattern is repeated over thousands of years, gigantic subsurface wedge-shaped ice masses form which warp fossil sediments (Lachenbruch 1962). Ice-wedge development still occurs on the Arctic Coastal Plain, but not in the Fairbanks area. Ice wedges in the sediments around Fairbanks are "fossils," that is, they developed in the Pleistocene and are now inactive. Thus Pleistocene sediments around Fairbanks not only contain cross sections of interglacial-glacial sedimentary cycles; they also include complex three-dimensional ice-wedge features, which are present at Pearl Creek where Blue Babe was found.

Another feature complicates this normally simple stratigraphy: valley bottoms occasionally shift within the main valley architecture. Pleistocene sediments moving downslope from one side of the valley can push a stream to the opposite side. Continued downcutting of the new stream channel leaves the gravel of the older channel behind on a gravel bench covered with sediment. Such was the case in the locality where Blue Babe was found (fig. 2.12). The stream

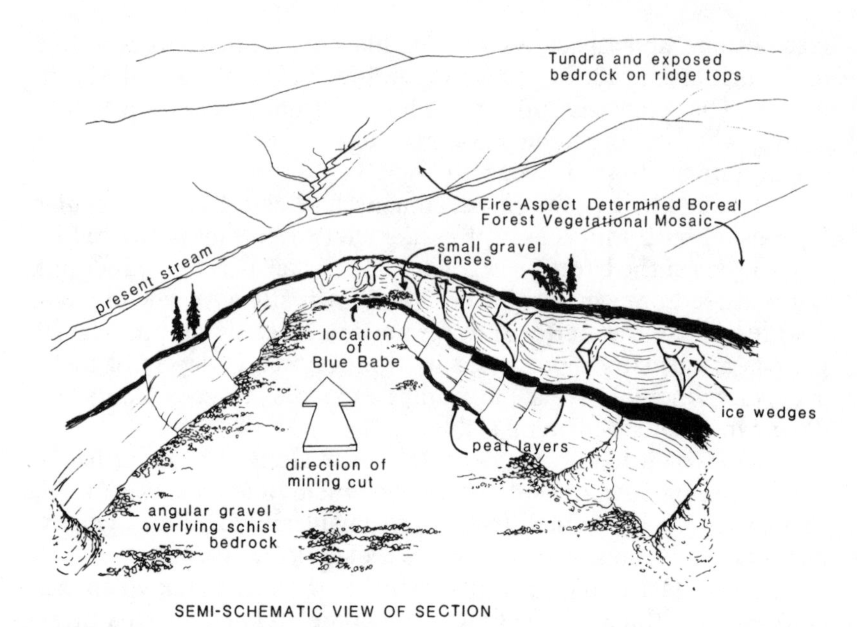

Fig. 2.12. The placer mining exposure at Pearl Creek. The right, or uphill, face of the exposure contains a thick peat layer halfway up the section. This peat tilts toward the present stream, but does not appear in the left section wall. At the head wall one can see where this peat layer thins and disappears near the center of the cut. Blue Babe was found in silts interbedded in this layer of peat. We can presume from Blue Babe's radiocarbon date that the peat is interstadial in age. The Pearl Creek exposure is typical of late Pleistocene sections seen at placer operations in the Fairbanks area.

channel was once near his carcass, but asymmetric movement of silt at right angles to the channel pushed the stream across the valley, leaving Blue Babe on a high bench under a thick deposit of silt.

Stratigraphy in the Blue Babe site is characteristic of later Pleistocene-aged deposits in the Fairbanks area (Péwé 1975a, 1975b; Guthrie 1968). Mining operations created an exposure along both sides of a Pleistocene creek bed, ending in an abrupt cul-de-sac on the upstream end where the miners were working. This U-shaped exposure provided a three-dimensional sample from which we could reconstruct the nature of the deposit (fig. 2.12). Silt deposits overlie Pleistocene gravels in the old streambed, but they angle from west to east, almost at right angles to the earlier streambed and to the present stream farther east. These old gravels, just beneath the fossil bison carcass, are thus a remnant Pleistocene bench of the present stream (fig. 2.13). Sediments that covered the bison are aligned more with the present angle of slope to the west, not with the present valley draining to the south. This accounts for the differences in height of the exposure on the eastern part of the U-shaped section, which bisects the toe of the old slope climbing to the west.

An overall view of the section reveals four striking stratigraphic features. First is the upper peat. From studies on other creeks, we know that this peat represents Holocene deposits after loess deposition and redeposition had ceased or greatly slowed. This peat marks the end of the Pleistocene and recolonization of the interior by hydrophytic vegetation of the present boreal forest from refugia south of the ice sheet. The base of this unit normally yields dates between 8,000 and 10,000 yr. B.P. (Péwé 1975b; Guthrie 1968). The peat consists mainly of compacted moss and tree roots. Its thickness varies, but normally ranges between one and two meters. Péwé (1975b) calls this zone the Ready Boullion Formation (fig. 2.14). Although he includes a portion of his Goldstream Formation in this time unit, the lower part of this peaty zone is roughly equivalent to Hopkins' (1982) Birch Zone interval.

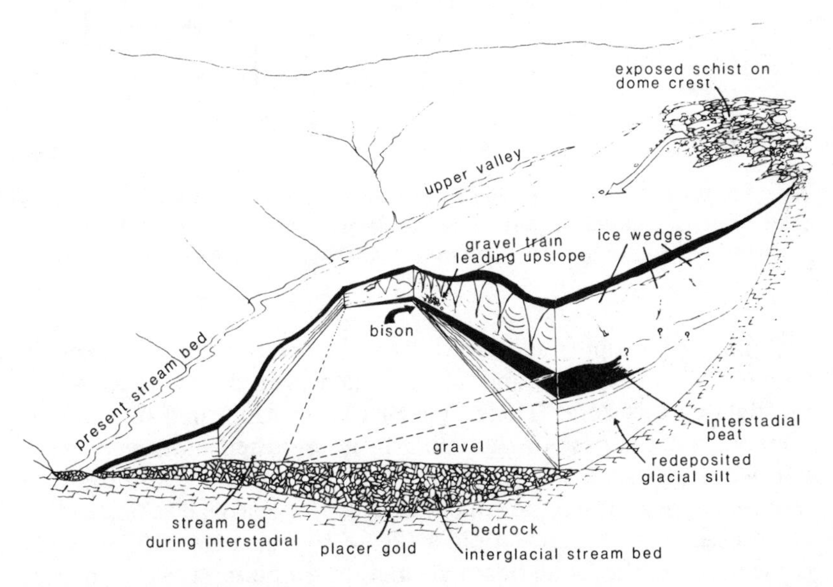

Fig. 2.13. Diagrammatic Pearl Creek section drawing. Gravels exposed at the Pearl Creek mine are portrayed as the former bench of a stream that has migrated and is now stabilized against the opposite valley wall. The fossil side "stream" which brought bedrock material down from the hill crests is also shown. Transport of schist downslope indicates rapid water runoff also capable of carrying enough silt to cover Blue Babe.

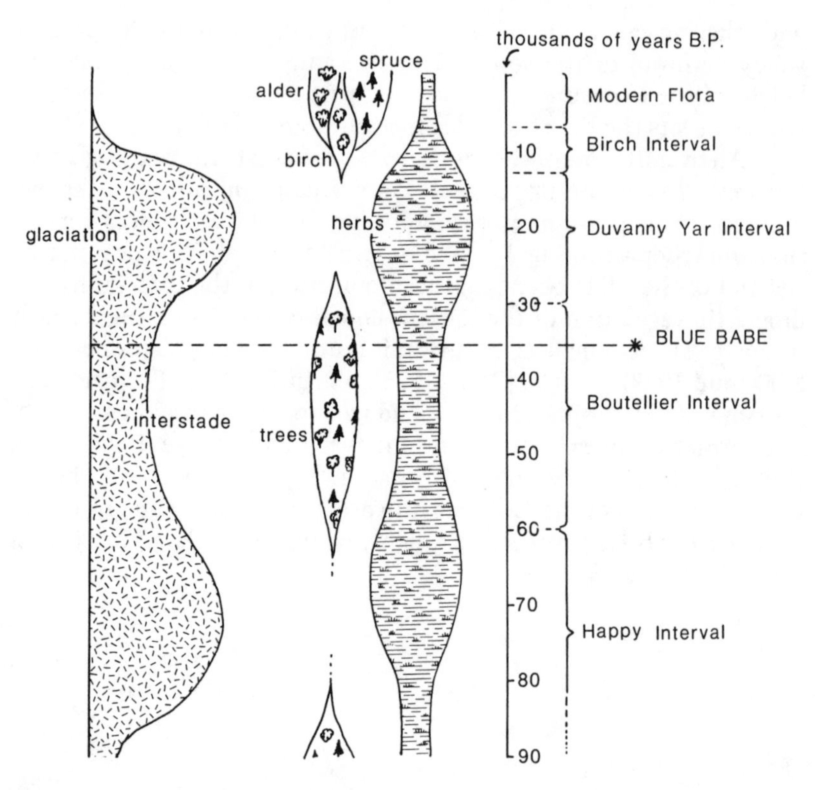

Fig. 2.14. Major vegetation and glaciation patterns over the last 100,000 years. This chart positions Blue Babe in relation to the glacial-interglacial events as well as the vegetational changes that occurred in Alaska over the last 100,000 years.

Below this upper peat is a thick bed of ice-rich silt, gray in color when freshly exposed and black when wet. Ice wedges occur throughout this zone. Their upper parts are truncated one to two meters from the base of the upper peat, and their bottoms feather out near the base. Elsewhere in the Fairbanks area this silt unit represents deposits from the final phase of the last glaciation, and in some cases it may be even older (Péwé 1975b). Péwé (1975b) calls these deposits the Goldstream Formation. Hopkins (1982) identifies them as the Duvanny Yar Interval (which corresponds to isotope stage 2 of the deep-sea cores).

On the right or west wall, another peat unit marks the middle of the section. It is composed of woody material lying in a matrix of compressed nonwoody plant parts that appear to be mostly moss.

This unit is of particular interest because it is near or within this zone that the bison was found. Along the west wall of the section, the peat is continuous, thick, and not penetrated by fingers of silt. As the peat layer dips into the gravels, as seen at the head wall of the section, it changes and is marbled not only with silt but with gravel. The bison was found among this interfingering of different materials. The bison skin collagen was dated at 36,425 +2575/−1974 B.P. (QC-891). A piece of wood taken from above the bison was dated at 30,890 +890/−1000 B.P. (DIC-2417). This peat unit and the bison date from the latter part of isotope stage 3 of deep-sea cores or what Hopkins (1982) calls the Boutellier Interval, roughly corresponding to Péwé's (1975a) Eva Formation in the Fairbanks area.

Beneath this lower peat is another thick bed of reworked loess. Péwé (1975b) calls this the Gold Hill Loess and Hopkins (1982) the Happy Interval; both are more or less equivalent to the isotope stage 4 from deep-sea cores. There were traces of an even lower peat seen only on the western side of the exposure. This peat was on top of the auriferous gravel (fig. 2.12).

These are the overall features of the section, but to understand the preservation of Blue Babe's carcass we have to look more closely at geological features in the immediate area of the bison. The carcass lay within a horizontally bedded silt zone (2a in fig. 2.15) (with a slight tilt toward the east, as with all the beds discussed above) and about 10–20 cm above a layer of compacted peat and silt (1 in fig. 2.15). There was a slight wave-dip irregularity in the peaty material just beneath the bison. Both this peat and the peat above the bison coalesced in a single unit farther north in the section.

Within the silt around the bison (2a) there were thin lenses of relatively clear segregation ice, some gravel, hair, bone fragments, and pieces of skin. These were almost restricted to a single horizon. Farther above the bison, at a little over a meter, 2a grades into what I categorized as another zone, 2b. This zone contains larger distinguishable lenses of gravel (no stone larger than 70 mm in diameter) and then lenses of peat. This latter peat is unlike the lower peat (1) but more like the Boutellier Interval peat in the west wall of the section—very mossy, rooty, and quite uncompacted. There is also "ice-wedge" ice, as much as 1 m thick.

Another more homogeneous silt zone (3 in fig. 2.15), which unlike 2a and 2b was cross-bedded, contained no ice lenses or peat. Above zone 3 was a similar silty zone (4 in fig. 2.15), except that it was horizontally bedded. And then above that was a thick peat bed (80 cm), like the smaller lenses of peat in zone 2b in texture and

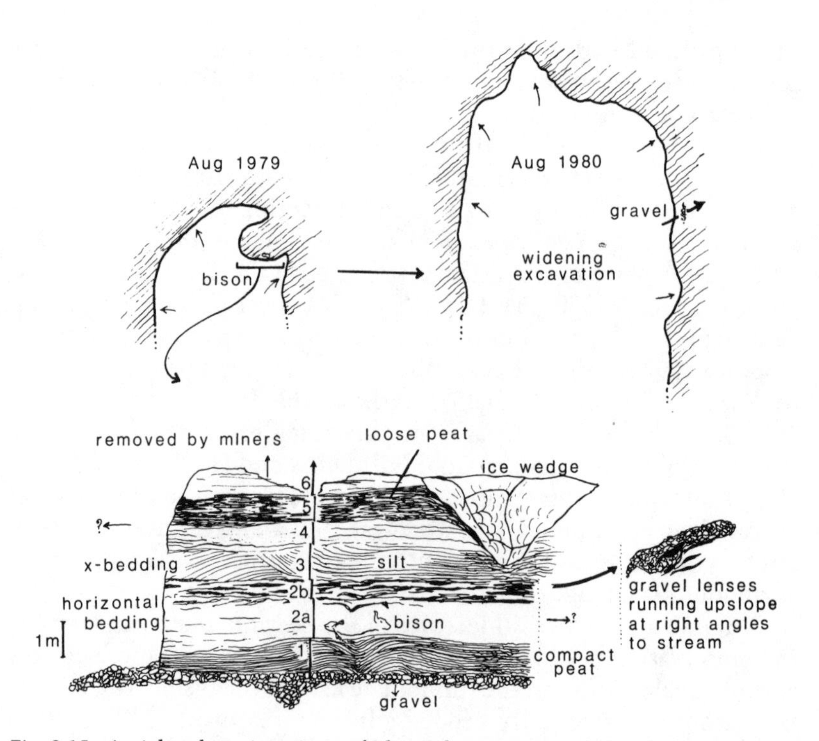

Fig. 2.15. Aerial and section views of Blue Babe excavation. (*Above*) Aerial views of the exposure (*left*) at the end of the 1979 mining season in which Blue Babe was excavated and (*right*) a year later. In 1979 the miners worked around the area in which the bison was located, leaving Blue Babe in a "thumblike" projection. (*Below*) A stratigraphic section is pictured showing the location of Blue Babe. I had the additional benefit of watching this wall of silt as it retreated under the miner's jet of water for the next three years. Numbers represent strata described in detail in the text.

content. Immediately above the bison carcass a layer of silt had been removed by the miners but could be followed horizontally to where the silt (6 in fig. 2.15) was still in place. I found it bedded horizontally, with a slight dip to the east toward the middle of the valley and not down valley.

The bison's location above the gravel and the three-dimensional angling peat bed which passes beneath the body, intercepting gravel farther east, tell us the carcass was deposited slightly upslope from the streambed location of Blue Babe's time.

Within the silt where the bison lay were lenses of vegetation and broken parts of tree or shrub limbs (fig. 2.16). These were identified to species and will be discussed later. The presence of peat beds above and below the bison carcass suggests intermittent silt deposition and revegetation. The mossy peat indicates that this was near the valley floor, where there was woody vegetation and more mesic conditions.

Excavation of the carcass proceeded through July, as each day brought some thaw (fig. 2.17) and revealed more stratigraphic context. Lenses of angular gravel were found in the silt around the bison. These lenses are unusual for silt deposits in the Fairbanks area and indicate rapid flow of water. As mining resumed after the bison was removed, we found that these lenses increased in size and eventually became enlarged uphill to the west, running at a right angle to the old streambed. This gravel pointed toward the nearest ridge crest, indicating that a surface-flushing feeder stream had brought down angular gravel from exposed bedrock on the hill crests above. By surface-flushing stream I mean that, unlike streams today, it had not cut down through silt and peat to bedrock or basal gold-bearing gravels. Instead, the stream was carrying much water in short bursts on top of the thick and probably frozen silt.

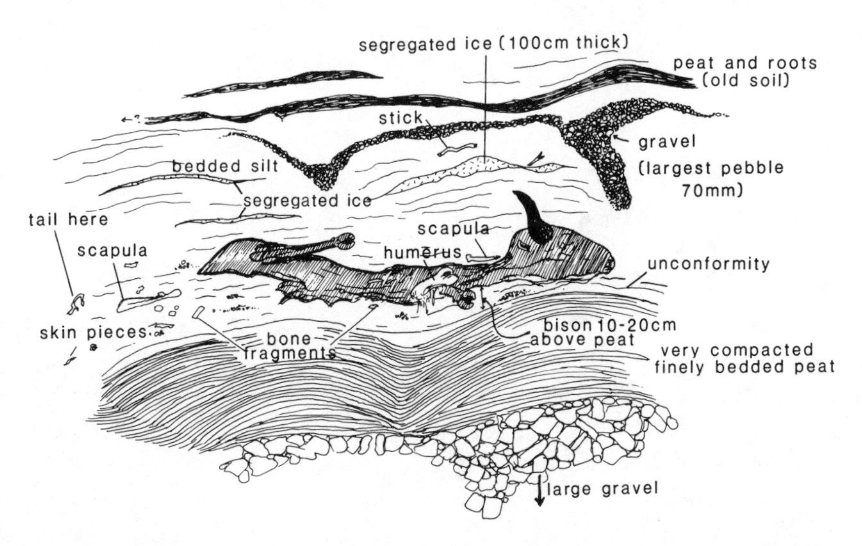

Fig. 2.16. Detail map of Blue Babe exposure. Blue Babe was surrounded by reworked organic silt and thin ice lenses; these, in turn, were surrounded by gravel lenses and peat layers. The gravel lenses continued at right angles up the side slope. They originated from the upper right side of the valley and not from the stream or valley floor. The peats indicate irregular stabilization of the soil in a slightly wooded, near-stream environment with interstadial woody vegetation. This was episodically flushed with water-borne silt from the valley side wall. Blue Babe was buried by these flushes.

Thus the bison was covered with silt from valley slopes and not from the valley bottom stream. In fact, the entire deposit at the site originated upslope to the west, eventually crowding the creek bottom farther to the east after the last interglacial (marine isotope stage 5). During this last interglacial and earlier, the creek bottom was on the old gold-bearing gravels. The area above the stream would probably have been covered with willows and other trees as a finger of riparian vegetation that reached into the uplands.

Because we found parts of the bison, including hair, bone, fragments of skin, and incompletely scavenged bones, frozen in the deposit adjacent to the main carcass, we can say that there was apparently no significant retransport of the bison and its parts. Blue Babe died in this habitat and was covered, not by the main stream many

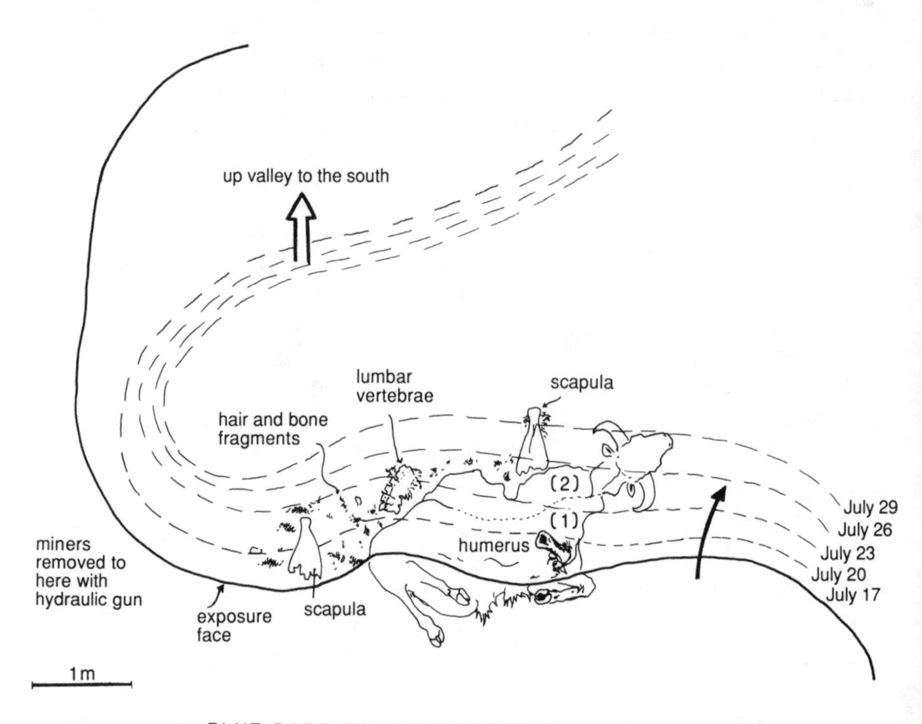

Fig. 2.17. Blue Babe excavation. Based on my photographs, this figure shows the line of thaw as mapped from above. The location of hair and bones around the carcass can be clearly seen. The carcass was removed in two parts because the torso and legs were exposed first and began to tear loose from the bank. The head and neck thawed a few days later.

meters to the east, but by silt moving downslope from the west. The orientation of gravel lenses and the bedding of silt upslope to the west seem to require this interpretation.

It is common in these kinds of deposits to find concentrations of bones at the juncture of such side feeder systems, or "pups," as they are called locally. I have found that extant wild sheep (*Ovis dalli*) bones occur in similar pup streams in areas that today serve as winter ranges for sheep. Wounded or sick animals can go downhill more easily than uphill, so carcasses are often found in these lower drainage funnels. Also, runoff flush may be strong enough to agitate some bones downslope and concentrate them at those apices. The same process covers bones with silt removed from the slopes. But Blue Babe was clearly too large to wash down the gentle gradient at the Pearl Creek site. I think we found him where he died.

The area immediately around the carcass does not show signs of thermokarst deposits in which the bison might have been trapped, nor are there other signs of anomalous preservation contexts. Thin veins of clear ice, 1–3 cm thick, were found in the silt surrounding the bison. These features were apparently segregation ice, the result of moisture freezing out of the soils after deposition.

There is no indication that the side "stream" that covered Blue Babe with silt continued in its bed over a long period because the hill shows neither gullying nor markedly irregular relief. Instead, one sees evidence along the section of sheetwash events of silt removal from upslope and redeposition of thin silt fans several centimeters deep. These horizontally bedded sheets are most apparent during mining operations when a powerful jet of water is sprayed over the frozen silt exposure. Seen in cross section, the thin silt beds fan down the valley sidewalls like book pages. This sheet wash was responsible for preserving many thousands of individual bones of Alaskan mammals during the Pleistocene. A complete carcass, like Blue Babe, could not be covered by such thin sheets, but several centimeters of silt would have been sufficient to cover most bones.

Placer Gold Mining in Interior Alaska

Since Blue Babe as well as other Alaskan and Soviet mummies, and most Pleistocene bones, have been found as a by-product of placer gold mining, it is appropriate to explain the process.

Minute gold particles are so diffusely distributed in quartz intrusions that "hard-rock" mining—going after bedrock gold—has never been very profitable in the interior of Alaska. Instead, most miners have sought placer concentrations of gold beneath old (Pleistocene) stream gravels in headwaters draining the gold-bearing bedrock ridges. The problem in most placer deposits is that the gravels are covered by a thick deposit of reworked silt which froze in place as it accumulated.

Miners in the Alaskan Gold Rush worked throughout the winter, using steam to thaw a shaft down into this frozen silt. So much wood was needed that forests near the mines were often denuded. Once the shaft reached gravel, steam was again employed as miners tunneled in drifts along the top of the underlying bedrock, moving as much gravel as possible to the surface during the winter season. Large piles of gravels were piled outside the shaft until summer, when they could be worked in a sluice box to extract the gold. Such mining was labor intensive, and progress was slow.

Soon much Alaskan mining changed to a method requiring fewer men and more equipment. By completely removing overlying frozen silt, exposed gravels could be scraped away with a dragline or other heavy equipment until a thin gravel zone above bedrock was left. These auriferous gravels were scooped, again with heavy equipment, into large mechanical sluice boxes that sorted the gold from the gravel.

Moving thick deposits of frozen silt was made easier by the use of water. Holocene humus and peat were bladed off, and water was run over the exposed silt. The physical behavior of silt-sized particles is such that they rapidly enter suspension in water, unlike smaller-sized clay particles or larger-sized sand particles. And as long as the water keeps moving, these silt particles remain in suspension. Miners take advantage of this behavior of silt in water by resuspending and flushing it downstream (fig. 2.18).

Alaskans are familiar with this phenomenon in the natural form of milky talc, called glacial flour, which clouds our glacial streams. The Tanana River, running by Fairbanks, is fed by many streams that drain the still-glaciated Alaska Range; its waters are gray and opaque. The Chena River, draining the Tanana Hills north of Fairbanks, is relatively clear because there are no significant glaciers in those mountains.

At a mining site, water running over the frozen silt cuts down toward gravels, leaving exposed sidewalls. To accelerate the thaw, these sidewalls are washed by powerful jets of water. Miners dam uphill streams to get the large quantities of water they need and pipe water downhill to the mine under pressure of gravity. The piped water is sprayed onto thawing silt with large mounted hoses called

Fig. 2.18. Thawed silt flushed downstream. The flowers growing on the rich mud are "mastodon flowers," *Senecio,* which gold miners argue are from fossil seeds. In fact, this species is an aggressive colonizer and is the first plant to take advantage of the newly exposed soil.

guns or monitors (fig. 2.19). Even with waterguns in use the silt thaws slowly, but by removing 4–12 inches (100–300 mm) a day across a wide front of exposure, a large volume of overburden can be moved in a short Alaska summer (fig. 2.20).

As microscopic silt particles thaw and move downstream, they leave behind material too large to enter suspension and too heavy to float. Walking along an active cutbank one can see Pleistocene bones high-graded at the foot or protruding from the freshly exposed bank above. Hundreds of thousands of Pleistocene mammal bones from interior Alaska were found in such a way. And that was how Blue

Babe was found. Its feet were seen protruding from the black, wet, muck face of a freshly hosed sidewall. The rest of the body was still frozen in place, deep in the bank. Walter Roman moved his monitor aside, and Blue Babe's second life had begun.

Taphonomy of Arctic Pleistocene Mummies

Taphonomy is the subdiscipline of paleontology that studies the way organisms are preserved as fossils. When animals die, their soft tissue is usually eaten by carnivores or broken down by microbial decomposers. Many carnivores are adapted to eating skeleton parts when other food is scarce. Uneaten bones remaining on the surface are usually destroyed by physical and organic processes. Insect scavengers eat bone marrow. Collagen fibers in the bone matrix are eventually consumed by decomposers, and the mineral part of the bone is flaked off by drying and cracking. Shallowly buried bones are usually dissolved by mild acids secreted by roots as they extract calca-

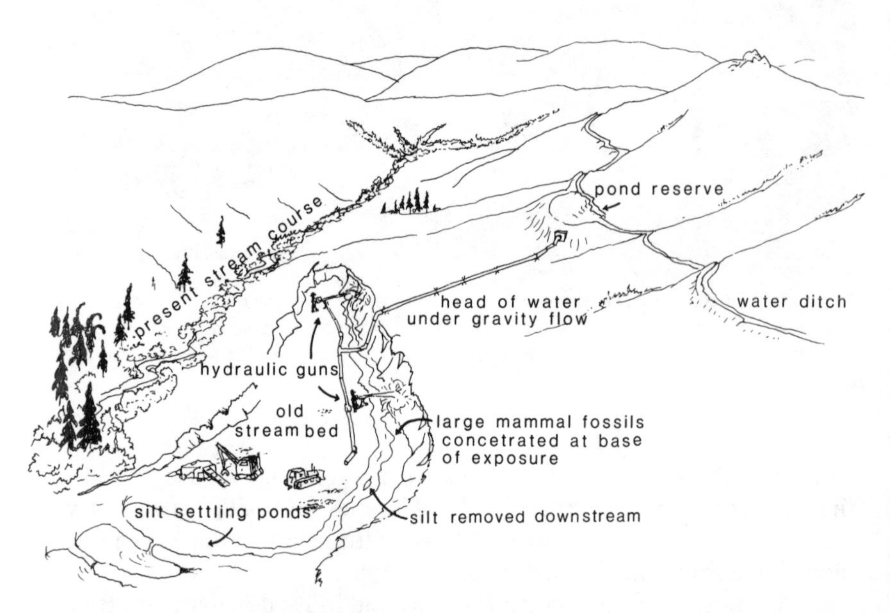

Fig. 2.19. Basic plan of typical placer mine in interior Alaska. After all the silt is stripped away, the upper layers of gravel are pushed aside, gaining access to gravel just above bedrock. This gold-bearing gravel is then moved into a large sluice box at the end of the mining season, not long before freeze-up. Federal and state laws now require settling ponds to remove some of the water-suspended silt and reduce downstream silt pollution.

Fig. 2.20. A brief mining season. When freeze-up comes, most placer mines close for the winter because the water which is their lifeblood stops flowing.

rious minerals from the soil. Very few bones are preserved as fossils (fig. 2.21).

The most common opportunity for preservation is a basic or calcarious environment. For this reason, limestone caves are our most important source for fossil bones. Without limestone caves the Pleistocene record of mammals would be meager. Caves buffer seasonal fluctuations in temperature and moisture, and their floor sediments are seldom acidic. Thus the caves of South Africa, caves and rock shelters of southern France, and limestone sinkhole caves of Florida are important localities for finding Pleistocene mammals. Many caves actively carry water underground. During dry episodes when water pressure decreases, the roofs of these caves sometimes collapse, forming a sinkhole, a natural trap for anything that might step in the hole.

To create most fossils, a bone must be covered so that aerobic decomposers cannot feed on the organic parts. Alkaline ponds and stream oxbows provide such conditions. Bones submerged in the mud of a pond or lake edge are likely to be preserved. Seasonal rains of the Rift Valley have given us a wealth of information about African Plio-Pleistocene mammals in this way.

Fig. 2.21. Few animals become fossils. This elephant died a natural death away from depositional environments out in the open African savanna. Most easily edible parts have been removed by scavengers. Only the thick, dried hide and bones remain. The ears, feet, and tip of trunk are thin-skinned and were the first to be eaten. Eventually, what remains will be scattered by hyenas, but it is unlikely that any portion will become fossilized. This is true for most large mammals.

Other situations, such as oil seeps, also preserve fossils. The loss of volatile short-chain molecules from oil seeps concentrates sticky, long-molecule tars and creates shallow entrapment pools. The La Brea tar pits in California are a famous example. And, as mentioned in chapter 1, a peculiar combination of tar seeps associated with salt domes occurs in Starunia where woolly rhinos were trapped and preserved.

Most conditions that produce fossils preserve, at best, only bones. In very arid regions soft body parts occasionally become so dehydrated that decomposers cannot live and multiply; shriveled mummies are the result. Preservation of Pleistocene fossils in parts of the unglaciated far north is exquisite, but we are only beginning to understand why so many fossils were preserved in Beringia. The cartoons showing woolly mammoths frozen in clear glacier ice misrepresent where mummies are found—they occur in silt deposits across unglaciated portions of the north. Mummification is not just a matter of freezing; carcasses must be buried by some rapid process. The geology of Beringian deposits and the fossils themselves can tell us how this usually happened. The character of vegetative cover is also an important requisite to understanding surficial geology. During former warm and wet episodes (interstadials and interglacials), trees moved north and grew in most of the interior as they do today, and unforested areas were covered by wet tundra. But during drier glacial intervals the Alaskan interior was an arid grassland (fig. 2.22).

The geology of Pleistocene deposits in the Fairbanks area has been thoroughly examined by Péwé (1975a). This information was obtained mostly from placer mine exposures, but also from natural river cut-banks (fig. 2.23). Péwé's interpretation of the reworking of primary loesses into valley bottom deposits of dark, putrid-smelling organic silt, rich in ice and bones, was derived from analogies with what we see today. At present silt moves downslope only above the treeline, where solifluction—a slight annual slippage of thawed soil downslope over the yet frozen portion beneath—is taking place. Péwé (1975a) implied that solifluction created the fossil-rich silt in some areas of the Fairbanks lowlands. Solifluction movements form a characteristic lobate pattern on steep hillsides and can be easily identified by shrubs that grow along the lower edge of the lobe. On the other hand, as Péwé saw, silt from full glacial episodes is evenly and thinly bedded. I argue that these fans of silt, washed from an incompletely vegetated grassy landscape, were the most important depositional agent in the deep deposits of the valley bottom.

Like all ground cover in interior Alaska, except for high-alpine talus and river bars, solifluction lobes are thoroughly vegetated.

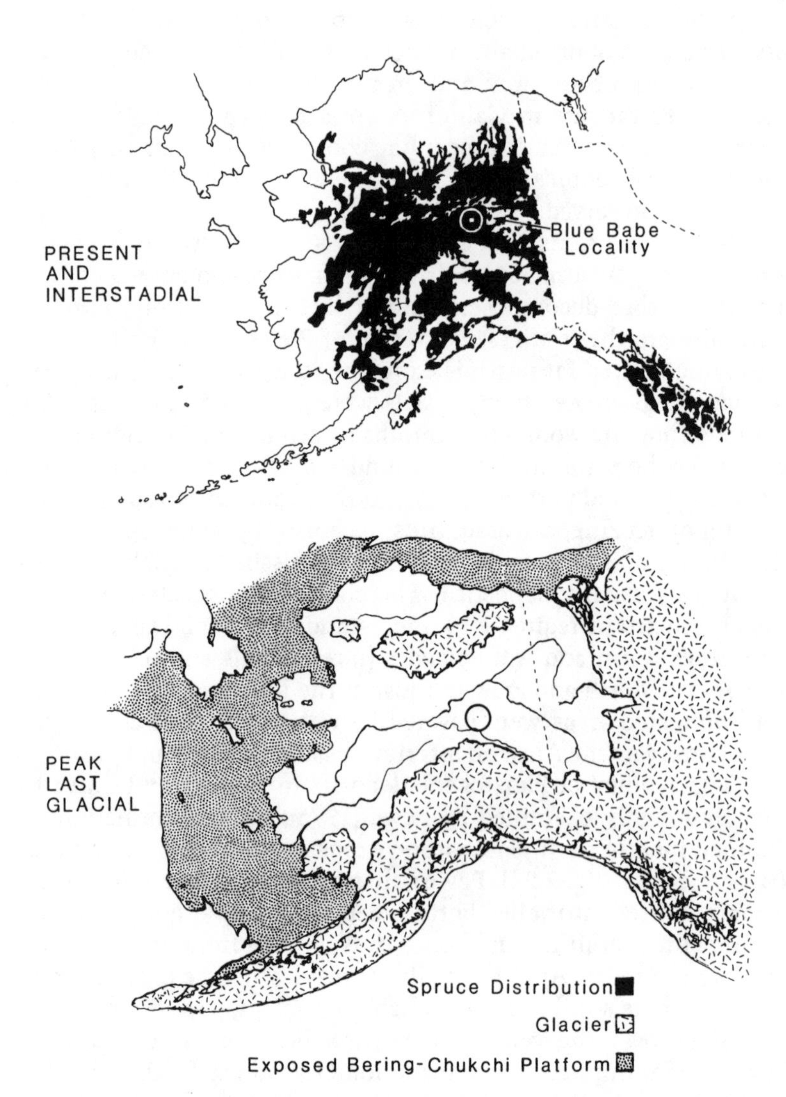

Fig. 2.22. The two faces of Alaska. Alaska has been a critical biotic link between Eurasia and North America. Glacial Alaska was virtually Eurasian; interstadial and interglacial Alaska was the most northern part of North America. During warmer, wetter interglacials and interstadials, arboreal vegetation and fauna (shown here as spruce distribution) recolonized Alaska from the south; at the same time, high sea levels separated Alaska from Siberia. During these wetter times, Alaska was separated by the Laurentide and Cordillerian ice sheets. Lowered sea levels provided a broad connection between Alaska and Siberia. The cool, arid climate pushed tree distribution far to the south.

Bones lying in such a vegetation mat are rapidly incorporated by plant overgrowth (usually by moss), then leached and destroyed by root acids. The rate of movement of the solifluction lobes is not fast enough to incorporate many bones. Additionally, only a small margin of the solifluction lobe would work to cover bones in any one year. It is difficult to imagine a large mammoth bone or skull being

Fig. 2.23. Natural exposure created by Birch Creek. In the dark part of this cliff, large ice wedges, the size of houses, are melting out. The stream removes the thawed slump and keeps the exposure fresh all summer. Many Pleistocene mammal bones are found at sites similar to this one.

incorporated intact. Solifluction movement would take several decades to cross the specimen, leaving one end decomposed, which is not the case with most Beringian fossils. One has to invoke a different, rapid depositional environment for large mammal mummies, and yet there seems to be a continuum between mummies and the smallest bones.

It seems more accurate to say that there are few modern analogues to the downslope silt movement that occurred in the Pleistocene. We should imagine a different soil surface from the present thick layer of moss, lichen, or partially decomposed plants. Today, any silt that begins to move downslope is immediately filtered out and the water runs clean.

Evidence from an array of studies (Guthrie 1982; Hopkins et al. 1982) suggests an incomplete surface cover for Pleistocene soils. There may have been a closed root system underground, but the rootless cryptogamus mosses and lichens, with their above-ground biomass, were much less abundant during cold phases of the Pleistocene. Areas of bare soil seem to have been exposed between plants, like grasses, which retain most of their biomass below ground. A rough analogue is found today on the American High Plains, where soil surface is exposed amid growth of short grasses. Erosion would have been most pronounced during spring or autumn, when naturally poor soil cover was further depleted by grazing or burning.

This exposed-soil grassland is quite unlike most tundras, which have a complete vegetation mat. But thick or continuous ground cover is not a measure of food quality. In fact, the present vegetation in interior Alaska is not subject to heavy grazing by large mammals because it is largely composed of very conservative and highly defended species rich in antiherbivory compounds (Guthrie 1982, 1984b). These same toxins slow decomposition, and old plant tissue accumulates to form an insulating layer that prevents deep thaw and hence restricts drainage, which in turn creates muskeg lowlands. These muskegs seldom burn, but when they do it is a "top burn" that does not greatly affect humus-forming plants. Even after severe wildfires there is little sheet erosion of silt.

During the Pleistocene, however, there seems to have been frequent transport of exposed silt downslope (Wu 1984). We have several measures of the most extreme degrees of this transport but do not know the usual amounts. Most silt in valley bottoms is finely bedded, indicating that the normal amount transported may have been rather modest. But during heavy rain or rapid snowmelt, enormous quantities could be moved over a short time. As previously mentioned, silt has the ability to go into suspension easily and to remain suspended as long as there is water movement. Once movement slows, silt particles come out of suspension and are deposited as fans or bars (fig. 2.24). I can imagine silt washing downslope during the Pleistocene, varying from broad sheets to channeled streams flushing over their banks. Despite the ease with which silt can be eroded, there are few artifacts of steep-walled stream channels in Pleistocene deposits. To me this suggests that most Pleistocene precipitation occurred as infrequent, large-drop downpours—short bursts of rain that exceeded the rate of soil absorption, as it does today in the shortgrass plains. This situation contrasts with the frequent, small-drop rain showers we now have in the interior.

The majority of Alaska Pleistocene fossils were buried by the kind of deposition just described. Nevertheless, judging from the volume of silt required and the season of burial (evidence I discuss later), Blue Babe was probably buried in early spring in a different way. It seems most likely that drifted snow melted rapidly from the uplands, carrying enough silt to cover the bison. Blue Babe was found in the throat of a small basin 6–7 km^2 in size, so the stream

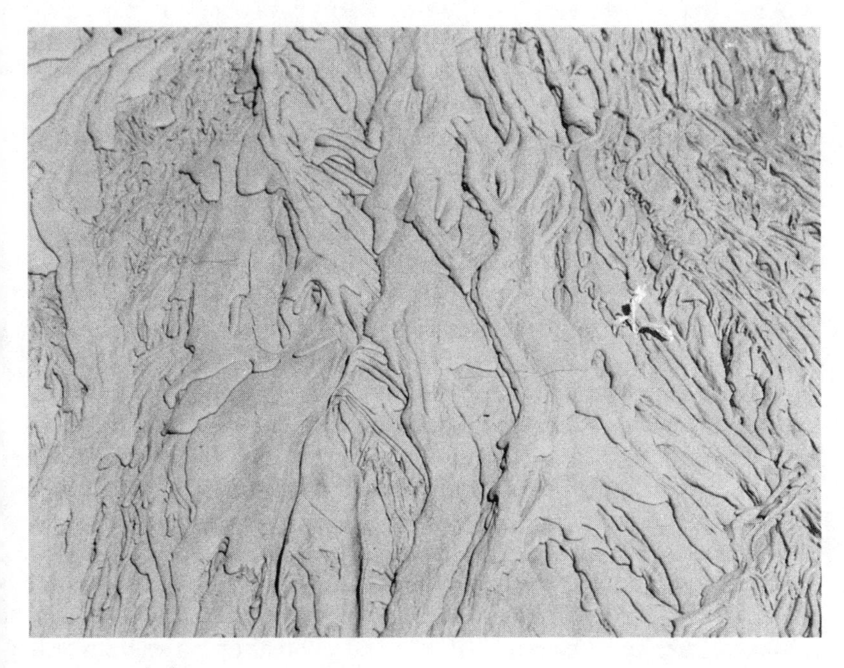

Fig. 2.24. Silt flow.

draining the basin could not have been very large. And if we look immediately above the bison, at the area from which silt could flow, there is only 0.3 km^2 (fig. 2.25).

At Fox, a few kilometers north of Fairbanks, a long tunnel has been cut into the permafrost. Originally made for engineering research, the tunnel provides a unique underground look at the permafrost phenomenon (Hamilton, Craig, and Sellmann 1988). One feature exposed in the tunnel directly relates to how frozen mummies are preserved. The tunnel bisects inactive fossil ice wedges, some of which are capped by ice from a small pond. This pond formed when the top of the ice wedges melted during a warmer climatic episode. The interesting point about this pond ice is that it is so clear; this clear ice indicates that enough silt was deposited to keep the pond below freezing. Any thawing of the pond would have incorporated some overlying silt and resulted in dirty ice. The pond has been preserved as it existed in its winter state. For upslope sediments to thaw sufficiently that they wash downslope and cover a pond, without the pond itself melting, the pond has to be buried with a thick cover of silt insulating it from summer's warmth.

There is a brief time in the spring when all this could happen. I propose that snowmelt from the first warm days of spring, say April, washed down exposed soils on the slopes above, further thawing the silt and carrying it in suspension toward the valley floor. This deposition must be surface wash, because there is a very narrow window of time in which the pond is still frozen yet erosion from melting can occur.

If such a small frozen pond can be buried by an insulative cover so that it would not thaw in subsequent summers, we can begin to imagine how a frozen mummy might be preserved. In Blue Babe's case, however, slipped hair and partial decomposition shows that, unlike the pond, the bison carcass thawed somewhat during the first summer or two.

Snowmelt can be abrupt in the Yukon-Tanana uplands—a result of two factors. The spring sun gains height rapidly at this latitude. Day length increases at a rate of seven minutes a day, yet below-freezing temperatures persist well past the equinox. The product is bright spring days in a frozen world. The inertia of winter frost and springtime insolation finally reaches a tilting point, and the snow melts in a burst. River ice lifted from its anchor by the flush disintegrates as it begins to move downstream, creating a special season known as breakup. Of course, abruptness of breakup varies from year to year; some years it occurs smoothly and gently, while other years it is explosive and violent. Frozen mummies probably originated in

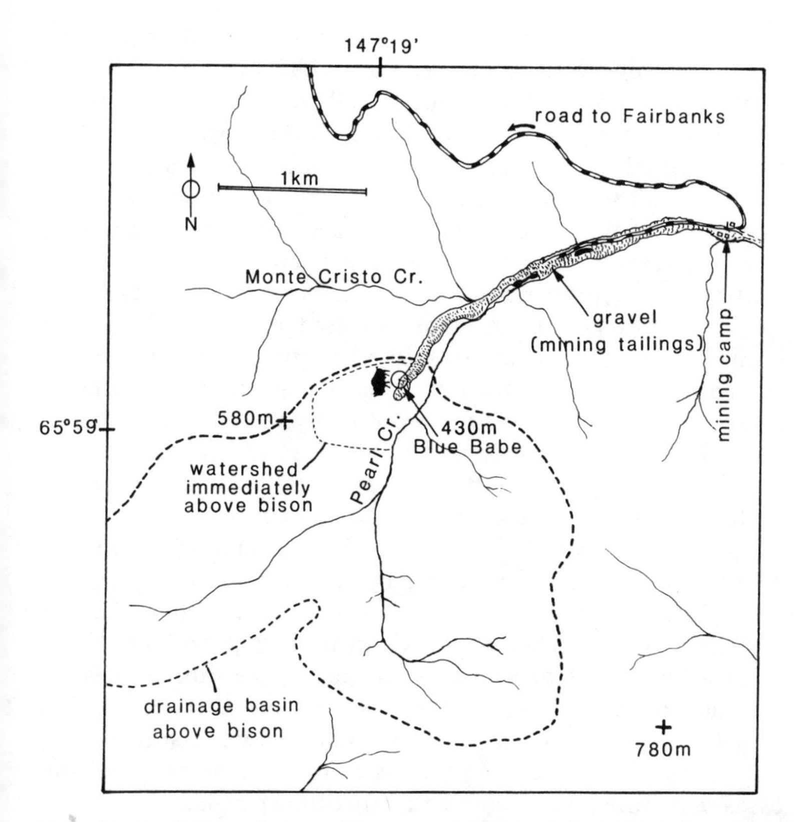

Fig. 2.25. Pearl Creek drainage. This map of the Blue Babe locality shows the drainages that could have carried silt to cover the mummy. The upstream drainage is of moderate size and could have furnished considerable silt if the bison were buried by over-bank alluvium, but this does not seem to have been the case. Apparently, the silt that buried Blue Babe came from a much more limited area, washing down the slope at a right angle to the creek bed. Gold mining activity follows the old Pleistocene streambed; note that the trail of tailings is different from the existing stream, especially up valley.

the latter kind of breakup, as seven month's precipitation flushed across a landscape of exposed soils in a few intense days.

In the drier Pleistocene landscape, a heavy rain and spring snowmelt would leave contrasting runoff patterns. Rain would produce a more even sheet of water flowing down broad flanks of the hills. Modest Pleistocene snows, however, probably accumulated in drifts blown from slopes into swales (Guthrie 1982), as indeed snow does today at slightly higher altitudes in the Tanana Hills. Snowmelt thus flows in concentrated rivulets, or channels, downslope from such drifts. However, Pleistocene snowdrifts would contain dust from the exposed soil surface, and this dusty component would have greatly changed albedo and increased the rapidity of spring thaw.

Hamilton, Craig, and Sellmann (1988) presented another interpretation of interior Alaskan silts, based on their work in the Fox permafrost tunnel. They think silt was deposited by two processes—slow accretion of wind-born dust and massive redeposition of this reworked loess—and propose that eolian accretion occurred during cold, drier, glacial episodes, while redeposition processes were active only during wetter interstades and interglacials. These latter redeposited silts dominate the sedimentary record, in their estimation. I think this interpretation overlooks silt reworking *during* the glacials and overemphasizes interstade—interglacial contribution to the sedimentary record. There are simply too many well-preserved bones and large skulls that radiocarbon date to full glacial (Duvanny Yar). These could not be preserved by a few millimeters of annual eolian loess-fall; their preservation required large quantities of reworked silt. The frozen silt we find enclosing these glacial aged bones is water saturated, not dry, drifted eolian material. For example, the entire skull with tusks of the Colorado Creek mammoth was buried some 15,000 years ago (Thorson and Guthrie, in prep.), prior to Hamilton, Craig, and Sellmann's postglacial episode of sediment reworking. They do portray an interval of mass silt wastage, between 36,000 and 30,000 years ago, that occurred due to increased moisture during the last interstade, which corresponds to my interpretation of the processes that buried Blue Babe during that time period.

I suggest that we are seeing two quite different Pleistocene patterns of silt redeposition in interior Alaska. In one, summer rains produced a broad sheetwash of silt in the millimeter and centimeter scales (on rare occasions much deeper), often over a shorter distance, with perhaps most not reaching the valley floor. In another, rapid spring snowmelt moved in more confined channels, probably taking large amounts of silt all the way to the valley floor. It is near these valley bottoms that one finds Pleistocene mammals buried.

Vereshchagin and Baryshnikov (1982) have proposed that late Pleistocene fossils, particularly the frozen mummies, are most likely to come from two time periods—the first around 11,000 years ago and the second 35,000–40,000 years ago. These episodes may have been unique times in Beringia, when two taphonomic factors coincided: sufficient moisture for silt transport and exposed soil surfaces. During full glacial conditions, there was probably not enough moisture, especially deep snowdrifts, for rapid silt transport. Silt moved downslope, and animal parts, but not many large carcasses, were entombed. During the Boutellier Interval (the last interstadial—isotope stage 3) and during the early Holocene there was an increase in moisture, and as moisture increased, so did erosion and redeposition—as long as bare soil existed. On the other hand, additional moisture favored mesophytic plant species, which quickly created a complete ground cover and prevented silt movement. There must have been a narrow window after moisture increased but before vegetational change slowed erosion when conditions were ideal for preserving large fossils. Erosion may also have increased at the end of the mesophytic episodes. Bison mummies from Siberia and Alaska appear most commonly within these two periods. It is the older, interstadial, window which concerns us in our reconstruction of Blue Babe's death and burial more than 35,000 years ago.

Blue Bones

Although Pleistocene bones from silt deposits in Beringia are themselves stained dark from organic chemicals and minerals (they range in shades from ivory to jet black), their surface is often covered with a dusting of brilliant blue. As our bison mummy dried, its gray surface also turned blue. This blue was the same color as the dust used to chalk billiard cues or to powder a mason's line—an incongruous hue on the remains of so old an animal.

The blue that forms on Beringian fossils is actually a mineral known as vivianite, an iron phosphate; whitish gray in its unoxidized condition, vivianite soon turns blue when exposed to air. This mineral occurs where organic remains of animals, low in iron but high in phosphates, are buried in damp silt that is relatively rich with iron but phosphate poor. Palynologists see vivianite layers in pond sediments (H. E. Wright, pers. comm.). Presumably, this situation arises from a similar combination of iron oxides (which precipitate out in standing water) and phosphate-rich organic chemicals from the remains of pond organisms. Such bands of vivianite are white when the pond is first cored, but like Pleistocene bones, they turn blue when exposed to air. Chinese potters used iron phosphates in their glazes to obtain the blue-green celadon which imitated jade.

Vivianite appears as an irregular dusting on many fossil bones I have collected. Sometimes this coating is so thin as to be barely visible; on other specimens it is a thick crust that totally masks the underlying bone. In the latter case all one sees is a blue form in the shape of a bone. Blue Babe was not only covered with a dusting of vivianite, but additional wartlike growths (fig. 2.26) of vivianite crystals occurred in clusters on the skin (the size of 0.5–1.0 cm). These

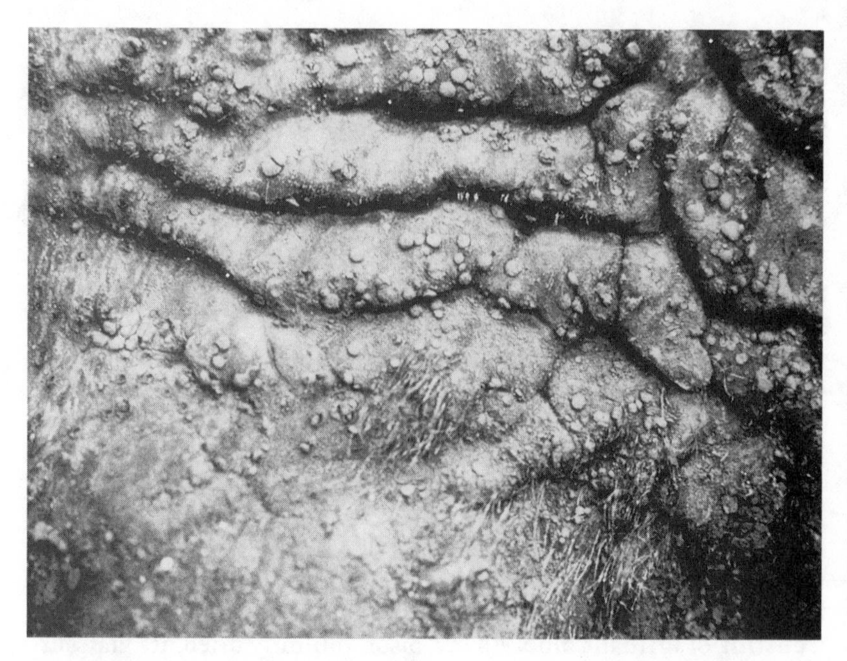

Fig. 2.26. Close-up of vivianite on the furrowed scalp between the horns.

blue warts were especially apparent on the head. When they were removed in the preservation process, the underlying skin showed pock-shaped erosion craters, undoubtedly the result of chemical breakdown of phosphorus-containing skin proteins, particularly collagen, in the underlying dermis.

After Blue Babe was mounted we restored his color by taking vivianite collected from other bones and dusting it over the skin. A thick mixture of vivianite and shellac was also dabbed into the craters remaining on the head skin to re-create the blue warts.

3
RECONSTRUCTING BLUE BABE'S DEATH

Season of Death

Several features of Blue Babe's carcass provide clues about the season in which the bison died. The first clue was evident when we excavated the carcass. Although hair had "slipped" from the skin, frozen muck held much of the pelt in place, at least on anterior body parts. This allowed us to sample the hair, mapping its length and color (fig. 3.1). It was obvious from these early samples that both the guard hairs and the fine underfur of the winter pelt were present and mature. Clearly, the bison had not died between spring, when pelage is shed, and early fall, when new underfur grows in, so June, July, and August were excluded as time of death.

The second clue came later, during the necropsy. Although much of the carcass had been eaten by a carnivore or scavenger, thick fat deposits remained, particularly around the sternum. The accumulation of fat in autumn or early winter is a universal characteristic of northern ungulates; during the long winter these fat reserves are depleted. From observing Alaskan animals over several decades, I can say that no ungulates enter the spring with significant fat reserves. In fact, of the dozens of sheep (*Ovis dalli*), moose (*Alces canadensis*), caribou (*Rangifer tarandus*), and bison (*Bison bison*) I have seen killed in mid- and later winter, few have much fat anywhere in their body. Roe (1951) quotes Richardson as stating that in early winter he had killed many plains buffalo that were fat, but beyond midwinter, buffalo were invariably lean. Large deposits of fat in the mummy strongly suggest the bison died during autumn or the first part of winter.

After Blue Babe was excavated, I sent an incisor from his lower jaw to Matson's, a private lab that specializes in sectioning mammalian teeth to determine age and season of death. Gary Matson's exam found summer growth complete but the winter annulus not yet fully developed, so he concluded that the bison died in autumn or early winter.

Horn annuli can also indicate the season of death. Northern bovids begin laying new horn from the base each spring. Growth continues until autumn, then stops until the next spring (Fuller 1959). During this dormancy a constriction in the horn forms a winter annulus similar to that in tooth root cementum. Blue Babe's last horn segment was complete, and a winter annulus was beginning to form. The final segment was 22 mm wide on the dorsal surface; the segment of the penultimate summer was 26 mm, while the one before that was 35 mm and the one before that 46 mm. The final horn segment was the size predicted for a complete summer of horn

Fig. 3.1. The necropsy. Blue Babe was taken out of the freezer, and we began to carefully clean the carcass, looking for clues to the bison's death and mode of preservation. (Photo by Don Borchardt)

growth. In contrast, a final segment 5 mm long, for example, would have indicated death early in summer.

Other characteristics of the carcass affirmed these early assessments that death occurred soon after freeze-up. Blue Babe was incompletely eaten, and the carcass was not fly-blown. Braack (1986) showed that, in South Africa, flies reduced a carcass to bones in five days during summer and fourteen days during winter. Muscle scraps found in the adjacent muck were actually still red, as were the muscles that were still attached. The upper or dorsal part of the body was eaten. Most of the vertebrae were missing, and I suspect they were eaten. Neural spines on the few remaining vertebrae were gone, and only scraps of ribs remained. Ventral body parts as well as the head were relatively untouched. This pattern of scavenging and preservation led me to think the bison was probably frozen after a predator had eaten part of the carcass. When a large carcass freezes, the skin becomes almost like sheet steel; a heavy, frozen hide is difficult for a predator or scavenger to penetrate.

A Modern Blue Babe Experiment

While I was working on Blue Babe's carcass in the laboratory, I heard that fifteen bison from a herd introduced at Big Delta, a hundred miles south of Fairbanks, had died after drinking melted snow produced by a spill of liquid urea fertilizer. With permission from the Alaska Department of Fish and Game, I obtained the largest animal, a bull about 3 years old. The weather had been cold −30° F (−34° C), and the bison was frozen. A man with a wrecking truck helped us hoist the body into a pickup, and we returned home to Fairbanks. There we tied the bison to a tree and drove the pickup out from under the stiff carcass; it landed in packed snow. I had no detailed plans; I just hoped to learn something from watching the Big Delta bison decompose in the spring.

During that first night there was a light snowfall. I noticed that snow on the bison was melting, which seemed an odd thing to happen on a frozen animal. After three days, and another light snowfall, it was obvious that this bison was not totally frozen but was continuing to ferment inside, like a compost pit within a blanket of thick hair. By this time the hair was starting to slip. I decided to take out the viscera to allow complete freezing. At −20° F (−12° C), I removed the steaming rumen and emptied the rest of the abdominal cavity. Within two days the carcass was stone hard.

Ravens and neighborhood dogs began to work on the frozen bison without much effect. They slowly cleaned exposed meat and ribs in the body cavity, but the bison's legs and head remained untouched, encased in the impenetrable skin. Most scavenging occurred in late April during the warmth of breakup, and soon as the snow was gone, flies began to assemble. Very soon thereafter, before much green was showing on nearby plants, the carcass was a pulsating mass of maggots. A little more than a week later it had settled into a bed of black blowfly pupae cases about 50 mm deep. Nothing remained but bones which had taken a light brown stain that I judged to be a by-product of the maggots. Local dogs no longer bothered the bones.

This smelly exercise taught me several lessons about neighborly relations and decomposition. An unopened animal continues to decompose after a fresh kill, even at very cold temperatures, because the thermal inertia of its body is sufficient to sustain microbial and enzyme activity as long as the carcass is completely covered with an insulating pelt and the torso remains intact. Therefore, it is unlikely Blue Babe met a natural senescence or disease-caused death and was later scavenged, because the preservation of remaining muscle, the solidly preserved epidermis and dermis, and the firmness of connective tissue and fat all indicate the bison carcass was opened immediately after death.

The especially good preservation of the mummy's head is also best explained by a winter kill. No part of the head was more than a few centimeters from the cold, and the bison's large horns would have acted as radiators, cooling the head rapidly after death. Rapid cooling would slow decomposition and render the head inaccessible to predators and scavengers, freezing it solidly soon after death. Finally, the head and lower limbs are the last parts of a large carcass that carnivores consume, so these would have had more time to freeze (fig. 3.2).

Apparently, Blue Babe died with his legs tucked under his body. Perhaps the legs did not fare as well as the head because snow is a good insulator. Nevertheless, they did preserve relatively well, with small patches of hair clinging at least to the forelegs. As the hind legs were almost totally exposed when I first saw the carcass, I cannot say if leg hair had been present in the adjacent silt, as it was over most of the head.

My observations of the decomposing Big Delta bison were thus consistent with ideas I was beginning to have about the nature of Blue Babe's death. To forestall all decomposition and preserve so well, the carcass had to have been opened quickly. Also, the body had to be buried quickly, quite early in the spring, before blowflies were active.

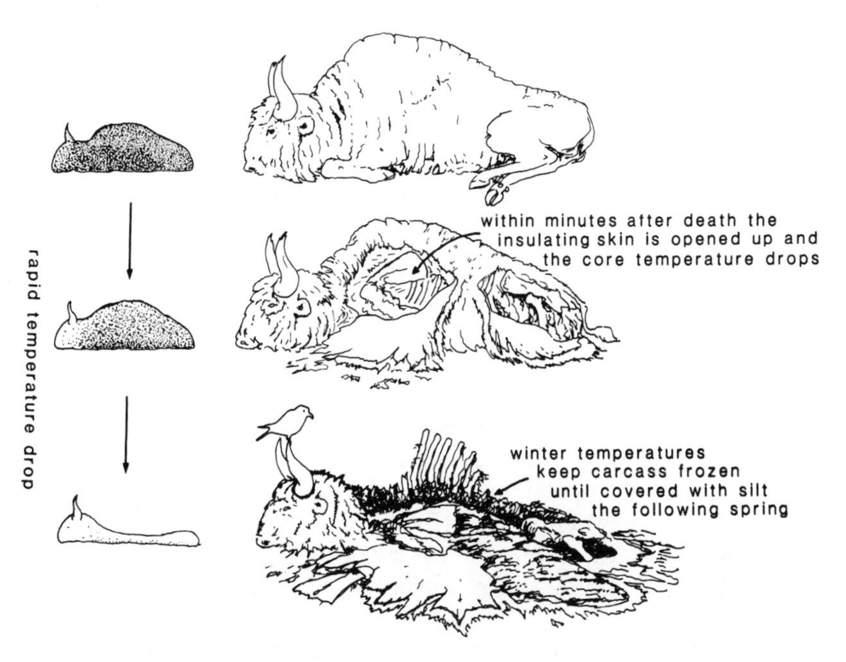

Fig. 3.2. Heat loss and decomposition. At winter temperatures the head and legs of a large mammal cool rapidly after death and soon freeze. Tough, frozen skin makes these parts difficult to scavenge. Anaerobic activity in the rumen, however, continues to generate heat and keeps the torso warm even at low temperatures. This is aided by the insulation of dense body hair. If the body is opened by predators soon after death, it cools rapidly and muscles and viscera freeze completely within a day or so.

Blue Babe's Season of Burial

In addition and closely related to the clues delimiting Blue Babe's season of death, the carcass yielded information about why Blue Babe died and when he was buried. We know that preservation of a relatively fresh-looking mummy such as Blue Babe required burial and temperatures near or below freezing. For instance, the carcass could not stay above ground all summer, exposed to microorganismal decomposers or insect scavengers. These would have quickly destroyed soft tissue.

If our conclusion about an early winter death is correct, the carcass was exposed all winter to avian and mammalian scavengers, because the bison could not have been buried when the ground was frozen. Today the ground is hard in Fairbanks between October freeze-up and spring thaw in late April or May, and there is no reason to suspect this pattern would have been much different when Blue Babe lived.

Once spring arrived, if the bison carcass was exposed for even a week it would have attracted insects, particularly blowflies. Pupae cases are common in Alaskan Pleistocene fossil bones—they are found wedged into foramina and in brain cases—so we know that blowflies were present in Pleistocene Alaska and that they attacked large mammal carcasses. Despite considerable search, no pupae cases were found with Blue Babe, either in the carcass or in the silt immediately surrounding it. Likewise, identification of insect parts from silt around the bison did not show an unusual assortment of scavenger beetles, which are abundant around decaying carcasses. (See identification of insects in Appendix A, by John V. Matthews, Jr., Canadian Geological Survey, Ottawa.)

During the spring, bears and other northern mammals regularly search for winter-killed animals, locating them with remarkable acuity when the carcasses begin to smell from enzyme action or microorganismal decomposition. Today a carcass as large as the bison does not long escape the notice of mammalian or avian scavengers and probably would not have lasted long in the Pleistocene either.

The high degree of tissue preservation, incomplete scavenging, lack of concentrated remains of insect scavengers, and geological information all indicate that Blue Babe was buried in the spring by the runoff of snowmelt. Today the ground is still frozen when snow melts, and runoff water moving through the vegetation mat carries virtually no inorganic particles. However, under conditions in the Pleistocene, with far more exposed soil, spring runoff was likely very different. At first I had assumed that snowmelt could not wash frozen silt so early in the spring; I assumed Blue Babe must have been buried by later spring rains. But during the spring of 1985, I noticed wholesale erosion through snowmelt runoff of naked soil at several construction projects left unvegetated during the previous building season. Because of these observations and the absence of blowflies, I now rule out burial from silt transport by spring rains.

Many fossil skulls and bones in the paleontological collections at the University of Alaska Museum do have fly pupae associated with them. In these cases, it is possible that soil removal downslope from summer rains rather than from spring snowmelt must have figured prominently in their burial.

Additionally, I believe Blue Babe was completely buried in a short time because there is little difference in the way various body parts are preserved. There is no evidence of horizontal gradations of preservation that would be caused by successive layers of silt accumulation. All parts, including bones scattered around the carcass, are equally fresh looking, as one would find them in winter and not after a summer of decomposition and weathering. That so many hair tufts, strings of connective tissue, small and large fragments of bone, entire bones, pieces of skin, and small bits of muscle were still scattered around the carcass strengthens the above point and tells us there was negligible retransport of the carcass by physical processes. These scraps were located as they would have appeared after several months of winter scavenging, with some red muscle buried in the snow and head and legs protected by frozen skin.

Haynes (1982), in his studies of bison killed by wolves in Wood Buffalo Park, Canada, has shown that wolves eat the bison in a fairly consistent pattern. A large wolf pack is capable of dismembering and consuming the soft parts of an entire bison. Although Blue Babe's remains do not conform to this pattern, they are similar to one of Haynes's mature bull bison that froze (fig. 3.3) before the wolf pack could complete its consumption of soft parts. In this particular case, portions of the visceral cavity and upper rump muscles were eaten before they froze. One month later, most of the ribs had been eaten, and parts of pelvic bones and many projecting parts of vertebral

Fig. 3.3. Bison in Wood Buffalo Park, Canada, partially eaten by wolves before it froze. (Photo by Gary Haynes)

bones had been chewed off. The head and lower limbs still retained their skin; they were essentially untouched. This sounds similar to Blue Babe's appearance.

Cause of Death

At first, during the excavation process, I guessed that the bison's death was not caused by predation because the carcass seemed relatively intact when it was still embedded in the silt. Much edible flesh remained, but signs of scavenging were obvious too. Perhaps, like many present-day carcasses I had seen, Blue Babe died in the winter, froze, and was chewed on and pecked at by an array of scavengers. Late winter is a common time for ungulates to succumb to accumulated winter debilitation. I pictured a late winter death for the bison, and the winter underfur I collected was consistent with that interpretation.

It was not until the necropsy was underway that evidence inconsistent with my first idea emerged. I found thick deposits of fat that were quite unexpected. I had never seen a fat animal, especially one with thick subcutaneous fat, die from internal causes in springtime. Deaths early in winter from causes other than predation are not common. Maybe Blue Babe had been killed by a predator, but what predator could kill a strong bull bison?

The next surprise during the necropsy was finding long scratch marks on the rear of the hide (see figs. 3.4 and 3.5). These scratches scored deeply through the epidermis into the dermis but failed to break through the thick hide. Occurring in clusters of three and four parallel lines, the scratches looked just like those made by claws of a large predator. Today many large mammals in Alaska (my basis of comparison) are killed by wolves (*Canis lupus*), but wolf kills do not show such large scratch marks. Canids do not use their claws to kill or to dismember a carcass. A large bear was the next culprit I considered, but grizzly bear (*Ursus arctos*) claws are mainly used to excavate fossorial rodents, not for clinging to large prey. Bear claws are powerful, but they are not retractile like a felid's and hence are not very sharp. Bears are probably incapable of making deep scratch marks like those we found in the bison hide.

Another large ursid in the Fairbanks area during the late Pleistocene, the giant short-faced bear (*Arctodus simus*), does not seem to have been a rodent excavator like the grizzly. This Pleistocene short-faced bear may have been a more active predator (Kurtén and Anderson 1980), and perhaps it was capable of killing a bison. Ursids

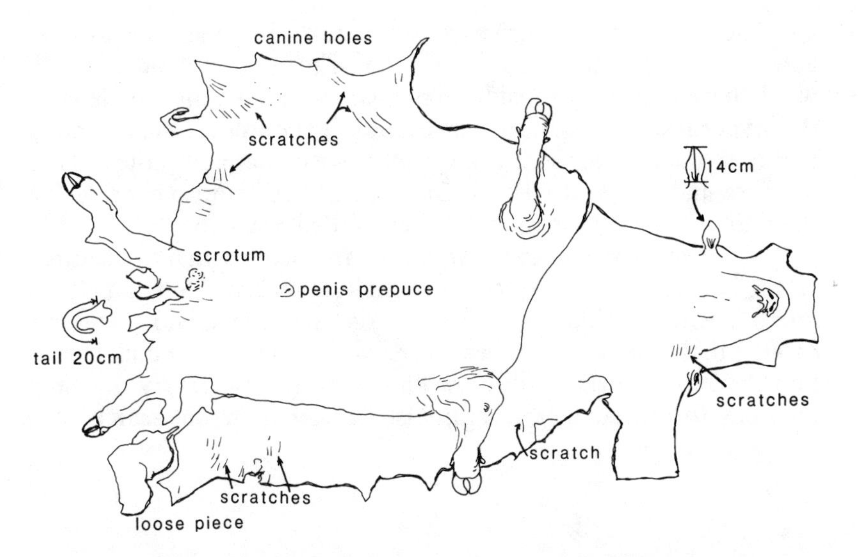

Fig. 3.4. Blue Babe's flayed skin. Predators opened the carcass from the top, so genitals and sternum are in the middle of the hide. A thin line indicates the cut made during excavation when the exposed torso and legs were cut away. Scratches are also shown. The tail was found adjacent to the carcass, presumably bitten or torn off from the bison when fresh.

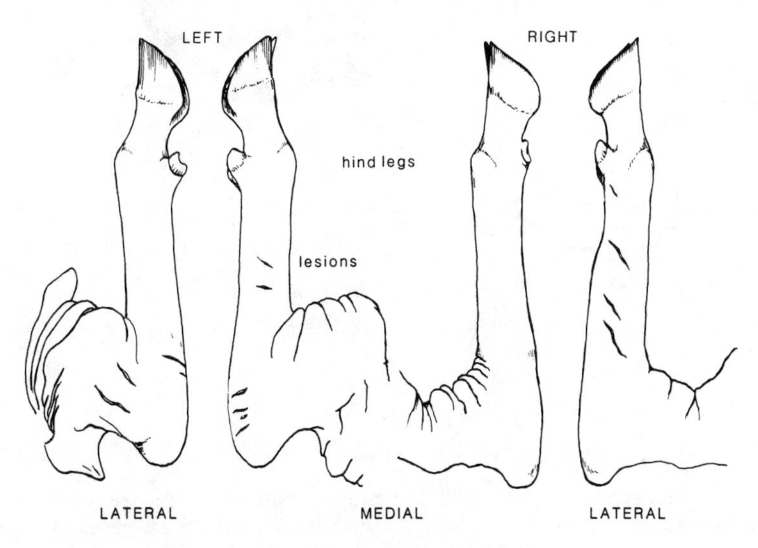

Fig. 3.5. Additional scratches found on the lower hind limbs.

are generally solitary hunters, however, and a bison as large as Blue Babe, in healthy condition, with sharp horns to defend himself, would have been a formidable challenge. I was not comfortable with the hypothesis that a bear caused Blue Babe's death. The scratch marks strongly suggested a large felid with sharp, sheathed claws (Gonyea and Ashworth 1975). Gradually, I backed into considering the unlikely: Blue Babe may have been killed by a lion.

Continued mapping showed that the scratch marks occurred on the bison in the same places as Schaller (1972) and Sinclair (1977) found lion (*Panthera leo*) scratches on African buffalo (*Syncerus cafer*). Sinclair found scratches on the back and flanks of buffalo that lions had tried to grab and throw off their feet, a necessary first step for lions to kill such large animals. Further literature search con-

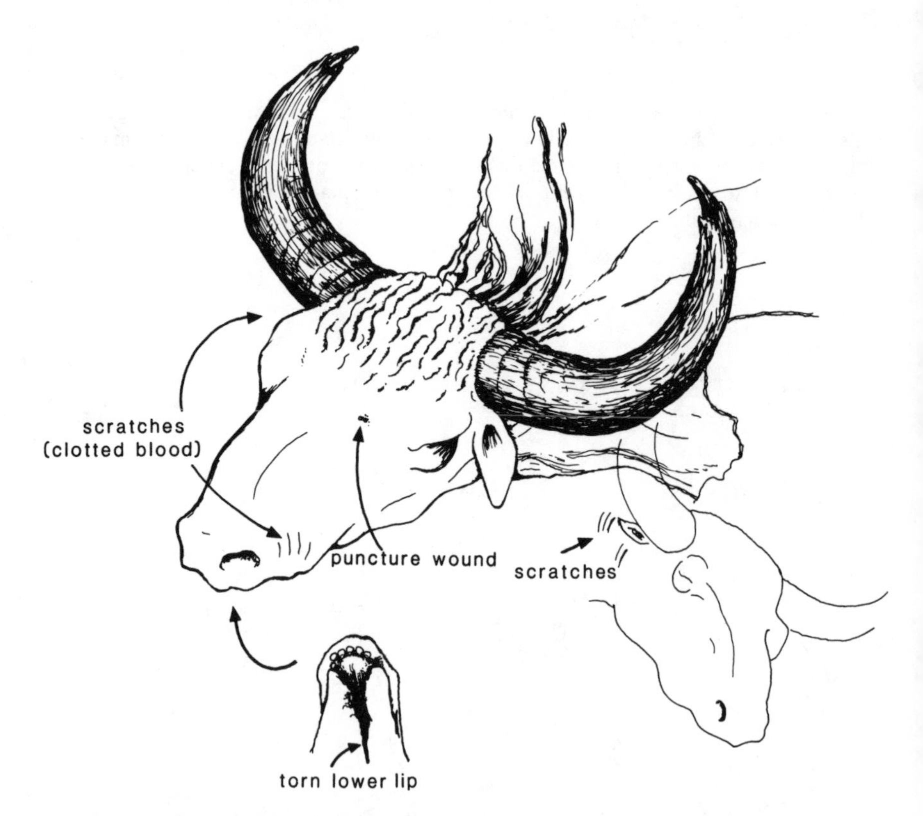

Fig. 3.6. Blue Babe's head wounds. Scratches and punctures could be seen on the head before we removed the skin. The lower lip was torn and partially scavenged. Whether this was done at the time of death or later is uncertain.

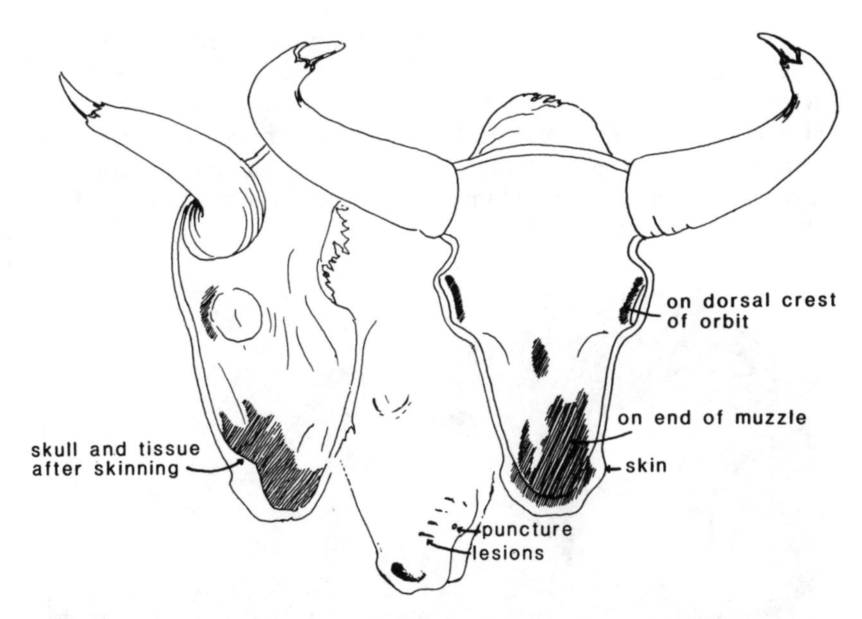

Fig. 3.7. Clotted blood on inside of skin. Clotted blood in the middle of the nose and face skin shows trauma occurred soon before death. This clot and breaks in the facial skin support the idea that Blue Babe was strangled by a hold on the nose. Other traces of bruising were evident over the forward part of the orbit, possibly from violent handling by the lion.

vinced me that these claw marks are a characteristic indicator of large felid kills or attacks. Sinclair (1977) found that many buffalo carried such scratches from their past encounters with lions.

I was stunned—an Alaskan lion—dare I even think it? Later in the necropsy we indeed found a puncture mark on the snout (fig 3.6) and blood clot stains on the interior of the nose skin (fig. 3.7). This hemorrhaging indicated the injury had occurred while Blue Babe was still alive. Because large bovids have exceptionally thick skin and large muscles on their necks, lions and tigers cannot use their regular neck bite to kill these animals; instead lions actually kill bovines by strangulation. Using claws for a secure hold, a lion will throw a buffalo down and clamp the buffalo's entire nose and mouth in a firm bite (fig. 3.8) or clamp the trachea closed (fig. 3.9). The lion must hold the buffalo this way until it suffocates (Schaller 1972). Evidence was mounting. Incredible as the idea first seemed, Blue Babe apparently was killed and then partially eaten by one or several lions.

The hypothesis changed: Blue Babe was attacked by a large cat or cats, from which he successfully escaped; however, the bison eventually died from wounds sustained in the encounter. I thought Blue Babe must have escaped because a large pride of lions would certainly have stayed with the carcass until it was totally eaten. The successful escape explained incomplete use of the carcass and signs of small animal scavenging.

Fig. 3.8. Muzzle bite. In this drawing from a photograph, a lioness strangles a blue wildebeest by seizing the wildebeest's mouth and nares. Blue Babe's nose had a canine puncture on the side, suggesting a kill of this nature.

Fig 3.9. Throat bite. Lions also strangle large prey by biting the animal's trachea. The male in this picture has a firm clamp on the trachea, even though the zebra is still on its feet. The female claws at the rear end, attempting to throw the animal off its feet. Note the female has her claws in the same place as the scratch marks on Blue Babe's legs. (Drawn from photo)

While thinking over my new hypothesis involving Alaskan lions (*Panthera leo*), I found indications that northern lions may have been more solitary in their social behavior than are most lions today. In winter, even two hungry lions could not have eaten an entire bison before it froze; they would have left a carcass just like Blue Babe's.

There is some evidence that lion prides in the far north were small. First, European cave paintings of the same Pleistocene lion species show males without a large, contrastingly colored mane (fig. 3.10). Clutton-Brock, Guinness, and Albon (1982) have observed that a male lion's social paraphernalia is most extreme in situations where males compete for large numbers of females. In African lion populations where prides are quite small (Tsavo Flats), males normally are maneless (Schaller 1972). Thus the lack of a contrastingly colored mane in Pleistocene lions suggests that prides were smaller. Second, in the several instances where multiple lions are pictured in Paleolithic art, the group is never large.

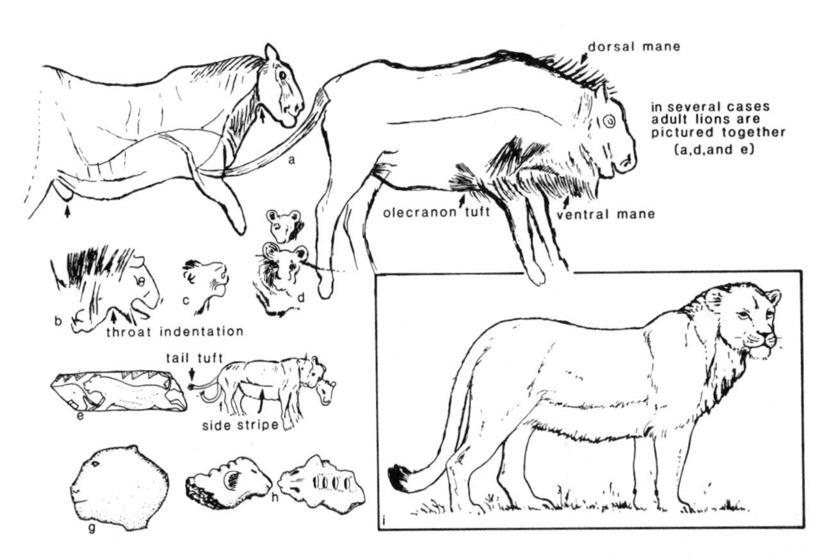

Fig. 3.10. Reconstructing northern lions from Paleolithic Art. Northern Pleistocene steppe lions, or "cave lions" as they are called in Europe, are almost identical to living lions in skeletal structure, but Paleolithic art indicates cave lions had a slightly different appearance. Male European lions probably did not have an enlarged, constrastingly colored mane; instead, they wore discrete dorsal and ventral manes, including a throat indentation at the hinge point of the jaw. Tails had the familiar dark-tufted tip, and some sort of side stripe or break in color pattern can be seen in the drawings. No large prides are pictured, but several drawings (*a, d,* and *e*) show more than one animal. Apparently these northern lions were social, like living lions.

My working hypothesis that the bison was attacked by lions, had escaped, but later died and froze, and was subsequently scavenged began to seem contorted. Perhaps the bison was simply killed by felids who, for some reason, did not eat all of the meat. Two more pieces in the puzzle supported this later conclusion. When the mummy's skin was split and stretched for mounting by the conservator, Eirik Granqvist, we found canine tooth punctures in the flank skin. These probably were not from bites intended to kill the bison but were punctures made in pulling the hide from the carcass to gain access to the meat, a characteristic method lions use (Schaller 1972). The puncture marks occurred in pairs, averaging 8.5 cm apart (fig. 3.11), almost exactly the width of the upper canine teeth measured on a Pleistocene lion in collections at the University of Alaska Museum. The width of the upper canines of the two short-faced bear (*Arctodus*) skulls available for measurement was under 7 cm. Both of these skulls were from very large bears. Wolves, wolverines, coyotes, and foxes all have much narrower distances between their canines and could thus be excluded.

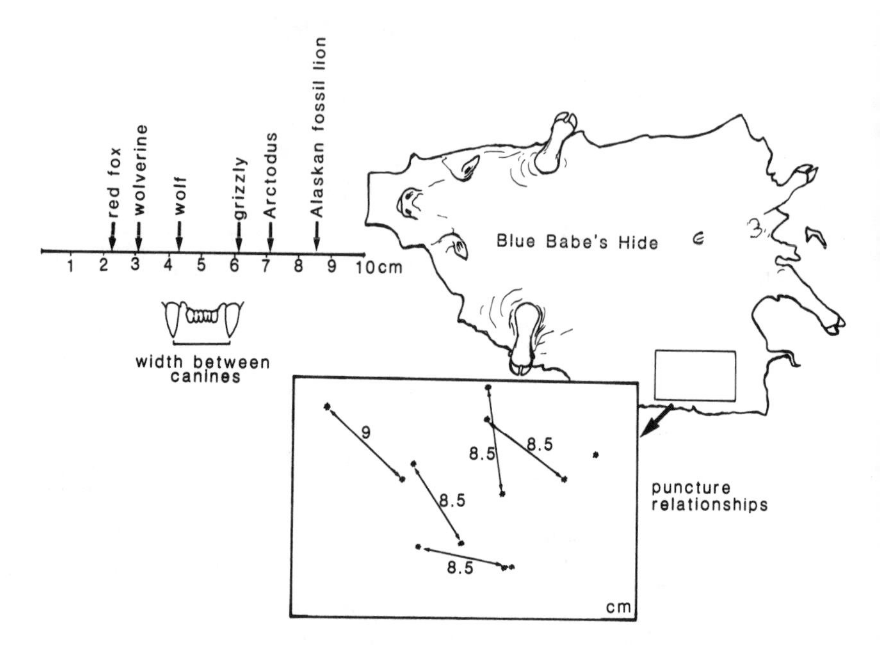

Fig. 3.11. Clues to the killer's identity. Distance between punctures is shown in the enlarged square. Width between canines of Alaskan predator species contemporaneous with Blue Babe are shown in the upper left corner.

Evidence pointed to a large cat as Blue Babe's most likely predator, but was it really a lion? Lion fossils in Beringia are comparatively common (Harington 1969; Vereshchagin 1971; and Kurtén 1985). What about another large species of cat? There is no evidence that American jaguars colonized the Old World via Alaska during the Pleistocene. There is circumstantial evidence that the cheetah reached very far north at this time, and its bones are found in northern Europe and Asia. But cheetah fossils have not been found in Alaska. Additionally, cheetah canines are not widely spaced and could not have made paired holes over 8 cm apart. Cheetahs do not prey on large bovines, nor do they scavenge much. Thus, I think it most likely that Blue Babe was killed by a lion.

The saber-toothed cat (*Homotherium*), or dirk-toothed cat as it is sometimes called, was in Alaska during the late Pleistocene (no precise dates). But those aptly named teeth would most likely have left cuts instead of punctures on Blue Babe's hide. Saber-toothed canines are perfectly shaped for slicing; they are knife like in cross section and serrated on both anterior-posterior edges (fig. 3.12). The more conical canines of lions often do not break the skin, but rather secure the strangulation hold described earlier. Saber-toothed cats seem to have taken a different evolutionary route, relying on a penetrating neck bite, as do smaller felids today. I suspect the "saber" shape of the saber-tooth's canine enabled the animal to penetrate the thick neck skin of a large ungulate.

To appreciate the different ways cat canines function, we must look at the physics of skin penetration. The epidermis, or outer cornified skin layer, can be penetrated without great force. Blood vessels lie immediately under the epidermis, so even shallow wounds bleed. By contrast, the leathery collagen fibers of the dermis can only be penetrated by sharp, pointed objects (such as a hypodermic needle). A cone-shaped tooth has limited penetration because these collagen fibers interlace. As a conical tooth continues to penetrate, dermis fibers around the puncture are stretched, increasing friction on the tooth. The thicker the skin, the more difficult it is for the fibers to expand around the puncture; as a result, the depth of penetration is limited. Early hunters were aware of skin's resistance to puncture and used a sharp projectile point larger than the shaft of the spear itself. In some cases they inserted stone microblades in a bone shaft to function in the same manner. Stone points are designed to cut an opening rather than punch a hole, allowing the shaft to penetrate much more deeply. Again, I think the long canines of saber-toothed cats worked this same way, slicing deeply into the neck of their prey.

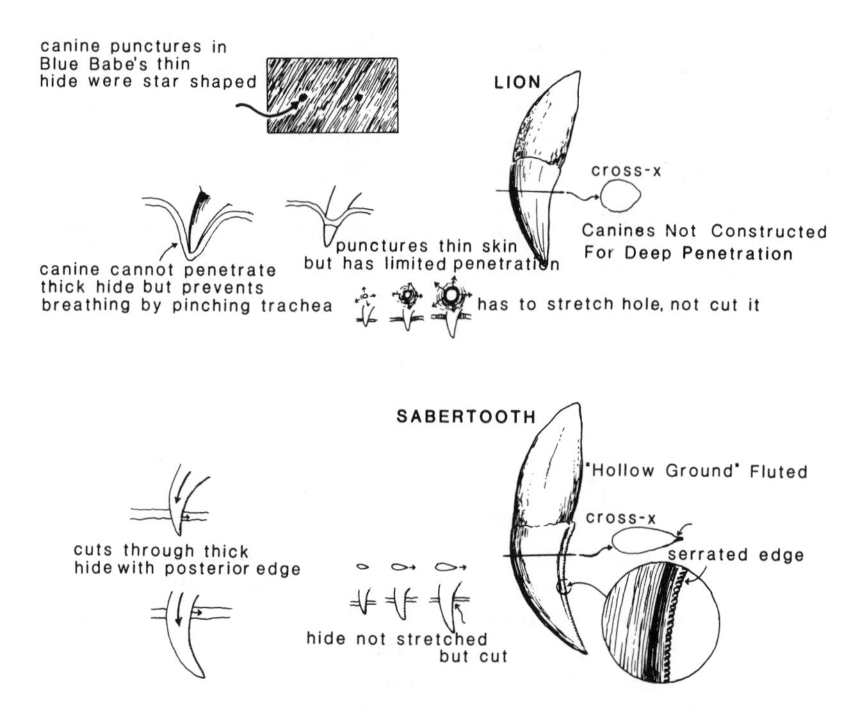

Fig. 3.12. Lion versus saber-toothed tiger canines. The serrated edge on the sabertooth's canine sliced through thick skin and allowed deep penetration. The more conical lion canine can only puncture. The uncut collagen fibers restrict penetration.

Fig. 3.13. It is unlikely a single lion killed Blue Babe. Judging from African analogues, it takes at least two lions to bring down a large bovine. Large scratch marks on the rump and hind legs suggest that after several tries, lions threw Blue Babe off his feet from the rear.

The saber-toothed cat's canines (1) puncture the skin with a sharp tip, and (2) having a sharp posterior edge, they cut a wider opening instead of stretching a hole. This edge is not only sharp but (3) serrated. Serration concentrates the force alternately on a small surface, repeatedly severing collagen fibers as the canine penetrates. In addition, (4) the canine is flattened in cross section, to reduce friction and to penetrate deeply yet keep its strength. Together, these features allow deep penetration of a thick hide, and (5) once through the skin they sever critical tissue. A saber-toothed tiger's teeth would thus leave characteristic cut marks on the carcass. This was not the case with Blue Babe; instead, we found puncture marks.

Working with evidence from the necropsy and sorting through these various considerations, I felt more certain—although still amazed—that a lion or lions (fig. 3.13) had killed Blue Babe. Lions were once found across the far north of Eurasia and North America (fig. 3.14), and lion fossils have been found in Alaska that date from the same time as Blue Babe.

Blue Babe's carcass later provided another bit of information to confirm that lions were involved in his death. When his skin was split in preparation for mounting, the taxidermist found a large tooth fragment embedded in the neck. Microscopic examination showed that this was a 1.01 mm thick fragment of enamel. I suspected this fragment was part of a tooth broken by a scavenger, but the enamel was too thick to belong to a wolverine or to any canid or ursid. The only Beringian late Pleistocene carnivore tooth that was of similar enamel thickness (0.92 mm) was a lion's cheek tooth (canine teeth have thin enamel). An African lion skull in our collections had a carnassial enamel thickness of 1.15 mm. Size and surface texture of the fragment corresponded to that of a lion's carnassial (fig 3.15). Lions use their carnassial teeth to shear through skin of ungulates and gain access to the meat (Ewer 1973), so it seems unlikely that a carnassial would have broken so dramatically on a soft hide. More likely, the lion broke its tooth while working on the bison after it had frozen. If this interpretation is correct, either soon after the kill or in a later scavenging episode, a lion broke its carnassial trying to get more meat from the carcass. Judging from the tooth fragment and the amount of meat remaining on the carcass (I estimate about 20 kg, fresh weight), the lion was not completely successful.

This information explains why the bison carcass was apparently abandoned while still incompletely eaten. Schaller (1972) found that African lions finished every carcass (unless chased away by hyenas), usually eating all but the long bones. Let us assume it

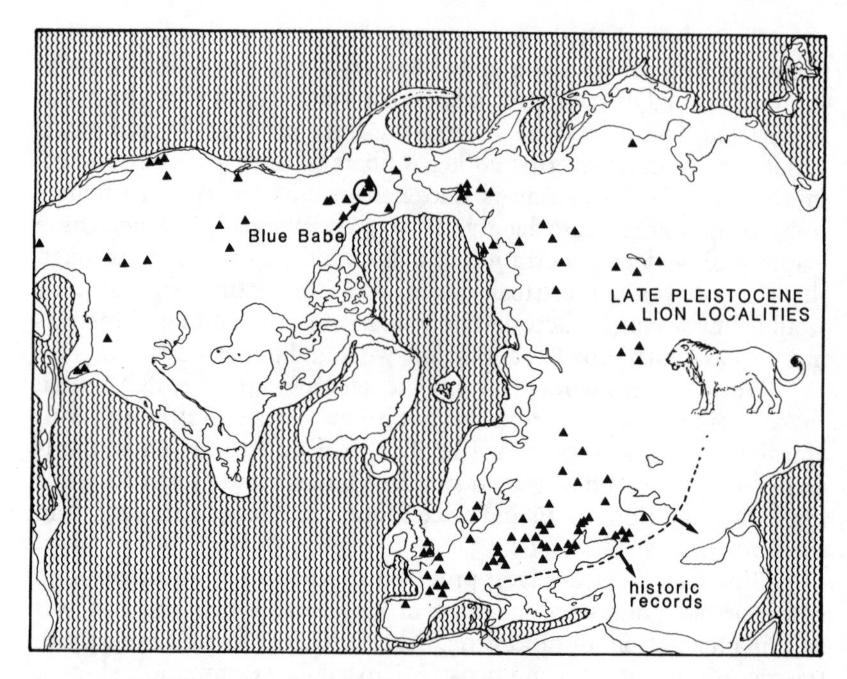

Fig. 3.14. Late Pleistocene lion localities. Lions were one of the most widespread Pleistocene large mammal species. The relative abundance of lion bones found in Alaska suggests they were a dominant large predator. Lions seem to have lived throughout the steppes of northern Asia during most of the late Pleistocene. Though seemingly exotic, the association of lions with Blue Babe's death is not unlikely when Pleistocene distributions of lion and bison are overlaid.

would take, at most, two to five days for a skinned and partially eaten bison carcass to freeze solid. (I have watched carcasses freeze in a wide range of winter temperatures, and this seems a reasonable rate.) Now we must take into account how much lions can eat. A fat bison bull provides considerable calories. Normally an animal will yield about 50% of its live weight in meat (Spiess 1979). Although there is no way of knowing exactly how much Blue Babe weighed, comparing his size with living bison I would guess his weight at about 700–800 kg. Schaller (1972) found that two male lions and one female lion ate a total of 150 kg of meat in a little less than two days. From other observations he calculated that a male lion required 35 kg and a female 22 kg of meat to gorge. A thin lion would gorge itself regularly over the first day or two, but satiety would slow consumption after that. Schaller estimated that over the long term, male lions

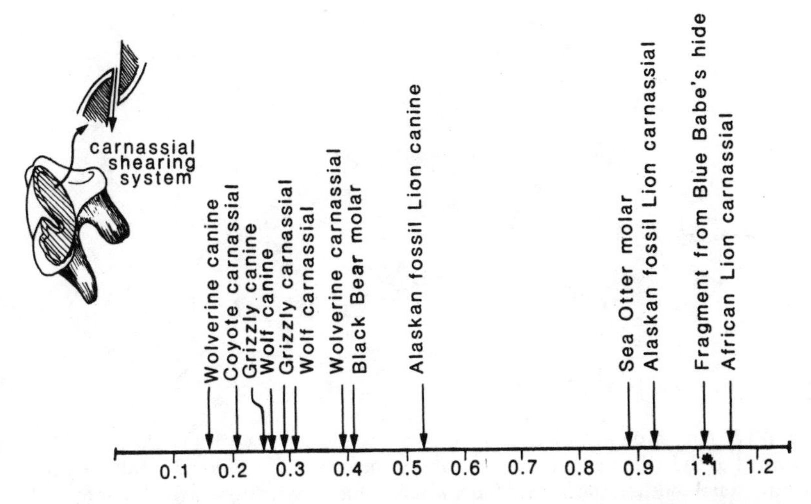

Fig. 3.15. Comparisons of enamel thickness in northern carnivores. The thickness of the tooth enamel in the chip embedded in Blue Babe's skin corresponds most closely to the enamel thickness of a lion's carnassial. The enamel of lions' large carnassial cheek teeth is quite thick because the wear pattern of its scissorlike action produces a self-sharpening, fragile edge.

Fig. 3.16. Blue Babe's carcass was incompletely eaten before it froze. Quite a bit of fat and red muscle remained frozen beneath the hide. Three or more lions could have finished the carcass before it froze, so it is likely that three or, more probably, less than three lions fed on Blue Babe. He was opened from the top while laying propped on his legs in an upright position. Bison often go down in this way when they are killed by wolves (Haynes, pers. comm.).

Fig. 3.17. A lion or lions fed on Blue Babe after the carcass froze. Perhaps a lion returned to the kill site and tried to scavenge more meat. A lion did break the tip of its carnassial tooth in the bison's hide, which would not have occurred in a soft, fresh hide, but could have easily happened in frozen skin.

require 7 kg and females 5 kg of meat daily in the Serengeti. Using these figures we can see that it would take a minimum of four or five adult Alaskan lions to consume Blue Babe before his carcass became completely frozen.

These calculations cannot, of course, determine how many lions were actually there when Blue Babe died, but they suggest that it was around two or three (fig. 3.16). Lions normally occur in adult sex ratios of one male to two females (Bertram 1975; Smuts 1976). Lions could and, as we see from the tooth fragment, probably did continue to feed after the bison was frozen (fig. 3.17). And judging from the scatter of fragmentary remains, other scavengers also used the carcass.

Paleobehavior of the Alaskan Lions That Killed Blue Babe

An obvious question that arises at this point is why Blue Babe—a mature male, 8 to 9 years old, in good condition—was susceptible to predation. Today, predators in the north usually take more vulnerable ungulates: individuals that are old, very young, diseased, injured, or in poor condition (e.g., Mech 1966). Studies of prey selection by African lions can help us resolve this puzzle. Sinclair (1977) showed that interaction between lions and large African bovines results in a nonrandom selection of prey related to sex and age, but it is a pattern different from the one we see between wolves and moose.

Social defense is important among African buffalo. Thus the buffalo females and young which make up most of the herds are less likely to be preyed upon than are solitary old bulls. Sinclair found that bulls 8 years old or older were twice as likely as females to be taken by lions. Lion predation in the Serengeti accounts for 23% to 28% of total buffalo mortality, which means that almost half the males are ultimately killed by lions. Much of this vulnerability is due to the behavior of bull buffalo. Old bulls are normally solitary because younger bulls oust them from year-around herds (Sinclair 1977). Additionally, perhaps because of their social aggressiveness, old bulls are not tolerated by females or often by younger males. Such aggressiveness, however, has reproductive rewards in acquiring dominance for mating privileges, so it appears to be selected for, despite costs that later increase exposure to predation.

When not confined to restricted enclosures, mature plains bison bulls are also likely to lead a solitary existence outside the rut season (Petersburg 1973; Shult 1972). Geist and Karsten (1977) proposed that, like African buffalo, female bison cluster together as an antipredator strategy because their smaller size makes them more vulnerable to predation than the bulls. Geist and Karsten further contended that bulls remain solitary or in small groups to avoid competition for forage from a large group. This allows them to stay longer in a single locality and still enjoy relatively high-quality forage.

Although African buffalo bulls are formidable opponents, with sharp horns and massive build, they can only defend their anterior end. This defense is almost impenetrable by a lone predator (fig. 3.18), but its effectiveness declines against several lions (Schaller 1972; Sinclair 1977). Judging from both Schaller's and Sinclair's observations of ineffective attempts by lion prides to kill male buffalo and the numbers of living buffalo bulls carrying healed scars from lion attacks, killing success is far from certain even if several lions are involved in an attack.

Again from Schaller's observations, it seems unlikely that female lions regularly take healthy buffalo bulls. Schaller (1972) proposed that one advantage of including male lions in a pride was the group's ability to take larger-bodied prey, due to the male lions' greater size and strength. Also, the larger and more pugnacious males play an important role in protecting carcasses against spotted hyena packs.

The implications of these African analogies for Blue Babe's death are many. First, Blue Babe was of similar body and horn size as

Fig. 3.18. Lioness turns to run from a buffalo bull. Although lions are formidable predators, a single lion, especially a female, does not commonly tackle an African buffalo. Several lions are required, and normally a mature male lion assists the females in hunting large bull buffalo. (Drawn from photo)

African buffalo bulls, and Alaskan lions were in the same size range as African lions. Thus the Pleistocene Alaskan predator-prey relationship may have been similar to the situation we can now observe in Africa. It seems unlikely, therefore, that single Alaskan lions regularly killed adult bison bulls. I have argued elsewhere (Guthrie 1980) that the sharp tips and long horns of Pleistocene Beringian bison were probably antipredator devices. Thus equipped, bison, with the weaponry and power to defend themselves, could whirl and face a single lion.

On the other hand, a pair of lions, one fore and one aft, poses a much greater defense problem. This is how African lions manage to kill buffalo. If the African-Alaskan analogy is appropriate, we can say that Blue Babe most likely was killed by several lions. As stated earlier, morphological evidence from cave drawings suggests that these northern lions did not form large prides. Perhaps two females hunted together, or maybe a lioness hunted with her almost-grown offspring. A male and a female could hunt in mated pairs, possibly in concert with several additional females. Also, two unmated males could form a team and attack larger prey not regularly taken by single lions. In fact, lions portrayed in Paleolithic art are usually shown in pairs, and in one instance the pair is clearly a short-maned male and a maneless female.

Taking the African analogy even further, I would argue that one of Blue Babe's attackers was probably a male, since African male lions are normally involved in killing large bovines. I think not less than two lions, including at least one male, killed Blue Babe on Pearl Creek.

If this chain of logic is correct, it allows us to say something about the social structure of Pleistocene lions (fig. 3.19). Information

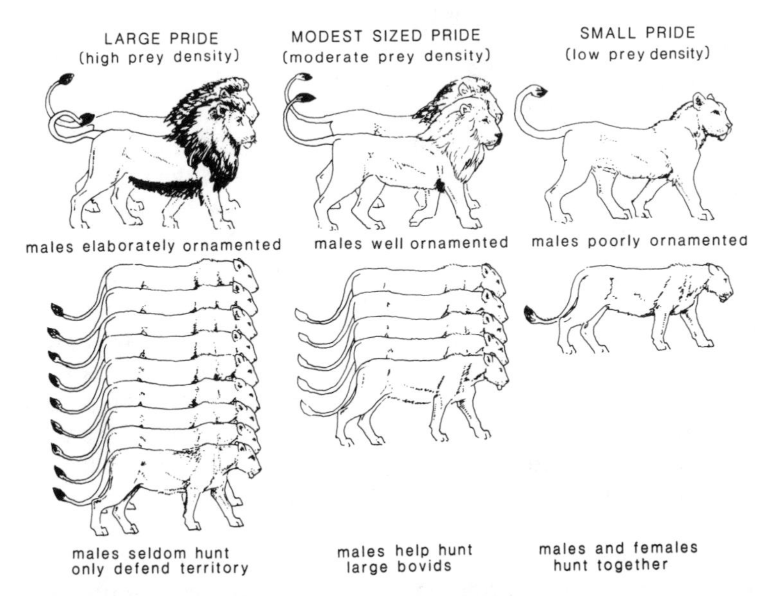

Fig. 3.19. The relationship of prey density, pride size, and male ornamentation. Anatomical ornamentation of males is exaggerated with increasing polygyny. A few males fertilize most of the females in a given population. This principle seems to explain some geographic differences in lion coat color. Male lions, often brothers, will share territory and females, thus creating a formidable defense against lone male contenders. In areas with large prides (eastern and southern Africa), pride males have large manes and there is a high frequency of dark-maned males. Where prides are small or dispersed during the dry season (Kalahari and Tsavo Park), manes are smaller and few males have large, dark manes. Pride size seems to relate to prey abundance and density. The less dramatic ornamentation and smaller prides of northern Pleistocene lions were probably related to prey density. Males in small prides must assist females in the hunt, and gaudy ornamentation can be a considerable debit.

from Blue Babe and European cave paintings can help us place Beringian lions on a continuum of social cooperation. Tigers are at one end of that continuum: adult tigers (*Panthera tigris*) seldom hunt together. Female and male tigers have separate territories (Schaller 1967), and their social paraphernalia is almost identical. Tigers hunt by stealth in deep woodland or woodland clearings, where a pair of animals would often be more of a detriment than a help. At the other end of this social spectrum is a large lion pride, composed of two or even more males and a number of females. African lions are dimorphic in their social paraphernalia. Male lions are readily identified by large contrastingly colored manes. Large lion prides hunt socially and in more open areas than do tigers. Lionesses do most of the hunting; males help in hunting larger species.

Paleolithic drawings of European Pleistocene lions show males had a slight ventral and dorsal ruff on the neck, different from females but not nearly so exaggerated as the mane worn by males today in the larger prides of African lions. We know from reconstructions of the northern Eurasian habitats in which Pleistocene lions lived that they hunted in open landscapes not unlike that of African savannas (Guthrie 1982). These northern lions probably had very large territories. We know that at its best, the Mammoth Steppe was not nearly as productive as most African or Asian savannas, and since the territories would have been large to support even a small pride, probably little energy was exerted on territory defense. Van Orsdol, Hanby, and Bygott (1985) have shown that African lion range size is negatively correlated with prey abundance during the lean season. They also showed that pride size and lion density are correlated with the abundance of prey during the lean season (figs. 3.20 and 3.21). Bertram (1973) had earlier suggested that pride size and density were regulated by food availability. The large ruff of an African male lion is an encumbrance in hunting (Schaller 1972); selective pressures against a ruff probably derive from greater pressure on male lions to participate in hunting. These later pressures would have been even greater among lions living in a small pride or as a mated pair. The small ruff of Pleistocene males we see in Paleolithic art would have been a compromise between the social advantages of a large ruff and its encumbrance in hunting. The slim ventral and dorsal mane on Beringian males served as an artificial enlargement of areas wielding critical weaponry (neck) and as an indicator of maturity, while still being somewhat inconspicuous so that the lion did not stand out like a haystack when attempting to get within attack

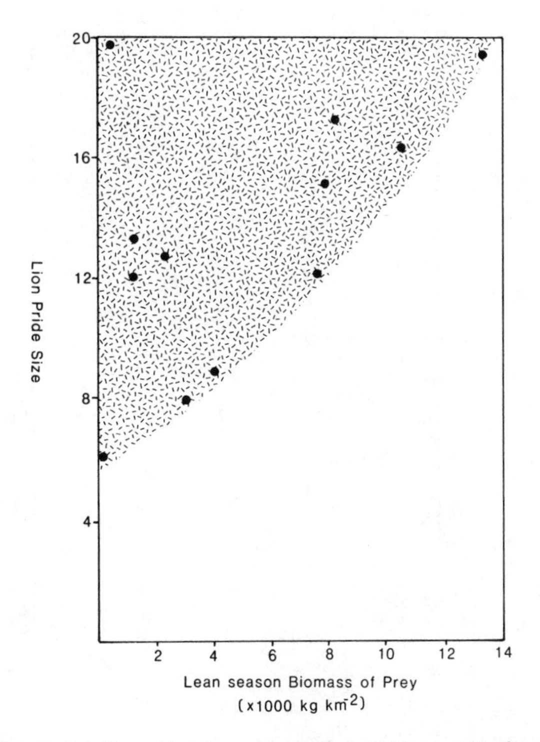

Fig. 3.20. Pride size is affected by the amount of available prey. Both small and large lion prides can occur with large numbers of prey, but only small prides occur when prey numbers are low. The data shown here are plotted from Van Orsdol et al. (1985) for African lion. They found that lean season biomass was most highly correlated with pride size. I use this well-known phenomenon to argue that northern lions probably did not live in large prides.

distance of potential prey. This modest mane still allowed a male to display his sex and age to his peers.

Obviously, the African lion is only a rough analogue to the Alaskan one, as many life history features would have been different. African lions are very aseasonal in their reproduction (Bertram 1975), the males have short tenure (Bertram 1975; Smuts 1976), and they occur at the high density of 10 to 15 per 100 km² (Smuts 1976). But for other features African lions may serve as an informative analogue.

In areas of Africa where prides are quite small, male lions today have small, noncontrasting manes. Both in the heavy bush of Tsavo Park (Schaller 1972) where there is low game density and in the Ka-

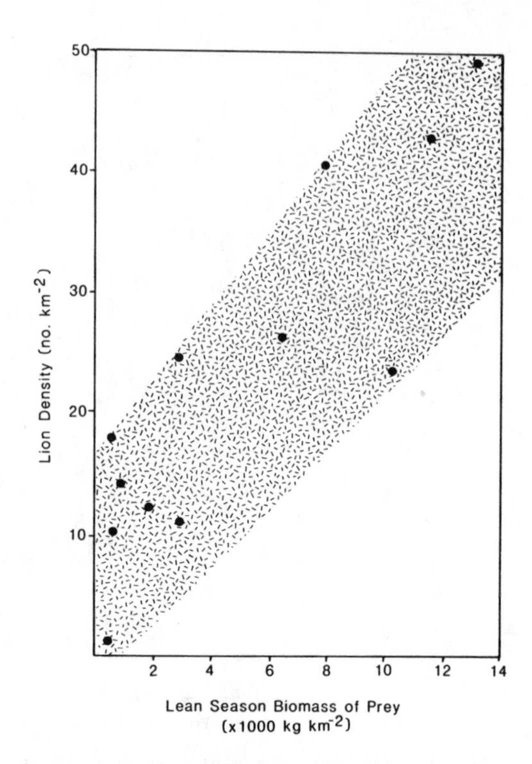

Fig. 3.21. Lion density is highly correlated with prey density. There is, of course, a lower threshold of prey abundance on which lions cannot exist. It appears in this plot, taken from data in Van Orsdol et al. (1985), that lion density reaches zero at the same time as prey density. The single point on the plot which suggests this is based on prey and lion density in Kalahari-Gemsbok Park, which is an unusual situation. Although the entire park was used in density calculations, virtually all the park's large mammals occur in the valleys of two dry rivers where boreholes have been drilled. Thus the functional density of lions and prey is actually greater than overall averages indicate. The point at which prey density will no longer support lions is difficult to derive, but it exists.

lahari (pers. obs.) where prides disperse during the dry season for a more solitary life, lions have small tan manes.

Sinclair (1977) found that the majority of African bull buffalo were killed along rivers where lions had some cover, allowing the lions to approach within attack range. Since Blue Babe was found in the lower part of a valley and there were small sticks and twigs in among the silt matrix, it is possible that the Alaskan lions also used valley bottom shrubs for cover before attacking.

For that same reason—requiring cover for their attack—lions

are mainly nocturnal hunters (Schaller 1972). Lions are not long-distance coursers, but rather rely on their ability to accelerate faster than their prey in a short chase. A great amount of muscle distally in the limbs gives lions their acceleration, but this heavy musculature is metabolically costly and severely limits the lion's distance once full speed is attained. The opposite is true for ungulates; their anatomy emphasizes efficiency once maximum speed has been attained, with some sacrifice of initial acceleration rate.

An ungulate tries to avoid capture by feline predators by maintaining escape distance—distance sufficient that the lion tires before overtaking the ungulate. For an attack to succeed, the lion must slip fairly close, to within the ungulate's escape distance, before giving chase.

Ungulates are at their greatest disadvantage when lions can gain proximity under cover of darkness. In Chobe, Botswana, with Petrie Viljoen I followed (in a vehicle) a pride of lions for several nights. Our sample of those hunts was consistent with his larger experience that lions have a much higher success rate on cloudy, windy nights. On one such night a pride made multiple kills of tsessebe from a single stalk. Tsessebe are one of the fastest antelopes and normally are difficult for lions to approach and catch.

Taking this African-Alaskan analogy to its logical conclusion, it is likely Blue Babe was ambushed and killed in shrubs along the valley bottom of Pearl Creek by a small group of lions during the lengthening dark of early winter.

Postmortem Scavenging Activity at the Carcass

After lions abandoned the frozen bison it appears that scavengers arrived and continued to visit the carcass until spring (fig. 3.22). Today any large mammal carcass that lies out all winter is visited by a variety of large and small predators and scavengers. Many of these species, such as wolves (*Canis lupus*), foxes (both white, *Alopex lagopus*, and red, *Vulpes vulpes*), and wolverines (*Gulo gulo*) occur as fossils from late Pleistocene deposits in the Fairbanks area, and there is no reason to presume they were not part of the local fauna 36,000 years ago when the bison died. Although few avian bones have been found in Alaskan Pleistocene deposits, some scavenger birds, such as ravens (*Corvus corax*), which are today ubiquitous in the Arctic and subarctic, as well as magpies (*Pica pica*), which now occur discontinuously across the Holarctic, were probably present during much of the Pleistocene.

Fig. 3.22. Blue Babe's carcass in late winter. The manner in which bones and tendons were chewed indicates Blue Babe had been used by avian and mammalian scavengers. Scavenging would have continued all winter. The frozen head and lower-leg skin kept scavengers from using these portions. Other parts were heavily used; most of the vertebrae, femora, and pelvic girdle were either consumed at or carried from the site.

The mummy's general appearance shows scavenging took place, but the carcass was not completely utilized. This is not surprising as such a large frozen carcass presents quite a challenge to small scavengers. When we screened the silt around Blue Babe we found many bone fragments, probably from mammal feces deposited by scavengers. (Today scavengers commonly mark a carcass with feces accompanied by secretions from anal glands.)

There were other signs of scavenging. Several vertebrae were missing; neural processes of other vertebrae had been chewed. Easily eaten bones such as the pelvic girdle were completely missing. The hard, outer bony cortex is thinnest on innominate bones of the pelvic girdle and vertebrae, so these bones are usually among the first that scavengers eat. The scapulae were found near the carcass, both with chewed dorsal margins. These margins are made of soft cartilage, and my past experience with recently scavenged carcasses affirms that these soft edges are a choice target.

On large mammal carcasses I have watched in past winters, the initial flurry of scavenger activity seems actually to decrease accessibility for subsequent scavenging. Snow on and around the carcass is trampled into a dense icy floor that covers many otherwise edible parts. Then, because the carcass is not insulated by snow and is exposed to the full effect of winter air temperatures, it freezes even harder.

On Blue Babe we found "feather brushes" of tendonous fibers attached to a number of bones, such as the vertebrae, which indicates avian scavenging. Mammalian carnivores use their carnassials to scissor these muscle parts completely off. Birds leave strands of this tough connective tissue behind as they peck away at the meat. Few bones were cleaned completely in the way a large felid will do, using the rough surface of its tongue like a rasp. The absence of these characteristically cleaned bones further supports our interpretation that the lion(s) quit the carcass as it froze. One wonders about the consequences of sticking a wet tongue on a −40° F(−40° C) bison tibia.

Some chewing marks were found on the bison's lower lip, but in general the head remained unscavenged. Red muscle was attached to the occipital part of the skull. The face and lower jaw muscles were protected by thick hide.

Once we had cleaned and split Blue Babe's skin (in preparation for the taxidermy work), we spread it out for mapping. At this point it became apparent that the bison had been opened from the dorsal surface, just to the left of the midline. To gain entrance to moose, most wolves and bears open the ano-genital area or the thinner posterior-ventral abdominal skin. I learned, however, from Gary Haynes (pers. comm.) that bison in Wood Buffalo Park normally go down on their abdomen when killed by wolves and are customarily eaten from the dorsal side downward. This behavior may explain why the legs of the mummy were so well preserved. When skinned and eaten from the dorsal surface, a bison's legs are covered by a blanket of skin draping down from both sides. In fact, Blue Babe's front legs were found in a collected position, so it is likely they were contracted under his body during all the scavenging.

The lower parts of the limbs (distally from the calcanium and the olecronon process) were essentially unscavenged (fig. 3.23). The sternum and its articulated ribs likewise remained in place. The uneaten ventral ends of the ribs (averaging around 12 cm in length) were still attached to the sternum and ventral skin, in their bed of oxidized fat. Skin was folded over this ventral part, making it less accessible, especially if one imagines the hide frozen before the carcass was thoroughly eaten. Once the skin had been peeled down the sides, the carcass would have frozen in that condition, forming a gigantic saucer-shaped block. Further trampling would have cemented this large disc to the ground, making the entire carcass difficult for

Fig. 3.23. A relatively unscavenged lower front leg. Blue Babe's preservation suggests that the front legs were wedged in under the frozen skin of the torso and unavailable to scavengers.

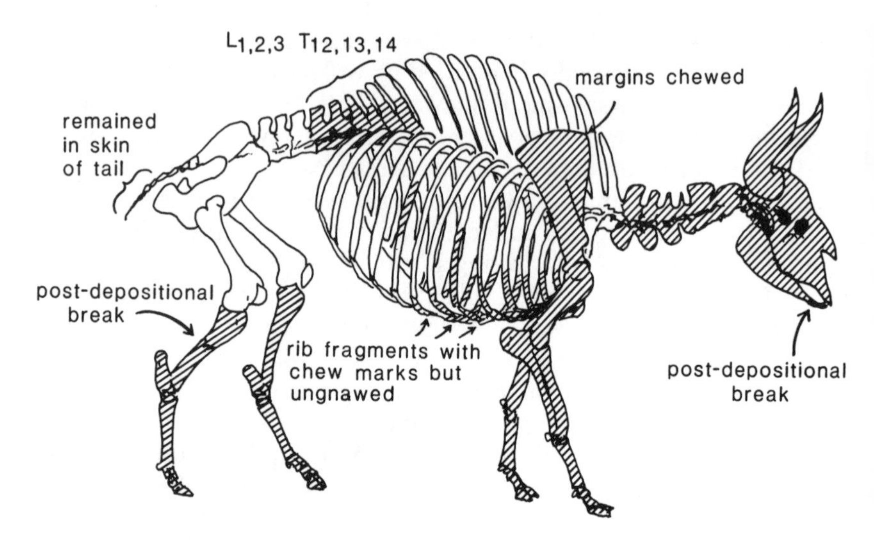

Fig. 3.24. Bones found with Blue Babe's carcass. Some of the missing bones were probably dragged away from the carcass. Others, like the pelvic girdle, may have been eaten.

carnivores to detach, overturn, and eat from the other side. The following bones of Blue Babe remained with the carcass (see also fig. 3.24): skull, with horn sheaths, and mandible; anterior six cervical vertebrae; last three thoracic vertebrae; first three lumbar vertebrae; last three caudal vertebrae; left and right scapulae; all the bones of the pectoral limb; all the bones of the pelvic limb, minus femora; most of the sternum; and numerous rib fragments, particularly those parts attached to the sternum.

Postdepositional Changes

The bison mummy, while well preserved for its age, underwent certain changes during its 36,000 years underground. While it is not easy to separate condition at burial from postdepositional diagenetic factors, in Blue Babe's case some clearly identifiable changes occurred well after burial.

The most obvious change is in the color of the bones, which had acquired the staining of surrounding sediments—a mixture of oxides and organic browns. The hair, however, seems to have retained its original color despite the fact that the hair of other Alaskan mummies has either been bleached by the soil or stained (or both). Mummified legs of caribou, moose and mammoths from Pleistocene Alaskan deposits have hair of a peculiar reddish pink color, perhaps a result of a bleach-staining processes, or more likely a differential bleaching of the melanins and eumelanins. Other specimens, such as Blue Babe, show no such modification: blacks are glossy, and reddish tans still look like fresh hair.

The posture of the mummy at the time of excavation may not have been identical to that before its burial. Blue Babe could have been contorted during burial or by ground movement after burial. Ice has peculiar characteristics that allow it to change shape (as in a glacier), and these are sometimes difficult to separate from forces affecting frozen mammals before they were buried.

Most of Blue Babe's broken bones, ribs, and zygopophyses, for example, were clearly broken by carnivores or scavengers, but two large bones do not appear to have been broken during the killing or scavenging process. The right tibia was broken with both pieces in place, the distal part still attached to the carcass. The bone does not appear to have been broken when fresh because (1) it would most likely have been dragged away from the carcass by a scavenger if it had been broken then, (2) the raw edges of the break show unscavenged white marrow, and (3) the break looks more shattered than is typical of carnivore breaks that occur in a clean sharp-edged spiral fracture.

The same is true of the left mandible. It has a rough, jagged fracture near the symphyses, and evidence indicates that this too is a diagenetic change. Both mandibles were held in place by dehydrated muscle and skin and were permanently skewed to the right side. This degree of distortion cannot be achieved in a freshly killed animal by the weight of the head itself, nor even by the weight of a dozen centimeters of silt. It is because of this contortion that the left mandible was broken. I think both the tibia and mandible were broken by diagenetic postdepositional processes rather than by predators or scavengers.

Since most of the hair was found in sediments immediately surrounding the bison and not attached to the skin (it is unlikely the hair would have been plucked out by scavengers; at least no known scavenger exhibits this behavior), I suspect the hair was still affixed to the hide when Blue Babe was buried but later slipped as part of postdepositional processes. This suggests that the skin of the buried bison may have thawed during the first or even subsequent summers, allowing residual enzymes to break down protein in the hair follicles.

Fig. 3.25. Blue Babe's burial. The lack of fly pupae cases and other evidence suggests Blue Babe was buried in early spring by silt moving down an adjacent side slope, probably from snowmelt water running over exposed soil. As this silt in suspension reached the more gentle slope of the valley floor, it was redeposited at sufficient depth to cover the bison which lay on still-frozen soil. The newly deposited silt insulated the frozen soil, preventing significant decomposition. Subsequent years added to this cover and eventually the bison lay in permanently frozen ground.

In summary, we can conclude that Blue Babe died in early to mid winter, before his large reserves of fat were significantly reduced. He was killed and partially eaten by one or, more likely, two or three lions, which fed for several days until the carcass was frozen. Freezing slowed consumption by these lions, who then left the kill, allowing other scavengers to pick at the carcass throughout the winter. Indeed, the bison probably was scavenged by an array of mammalian and avian species, judging from bone fragments, feces, and a characteristic pattern of tendon connective tissue left by avian scavengers.

At the close of winter, the bison carcass was buried by silt carried in rapidly moving snowmelt water (fig. 3.25). As the mummy lacked blowfly pupae cases and concentrations of scavenger beetles, burial must have occurred prior to the emergence of these insects in spring. The carcass probably thawed the first summer, but remained bedded on frozen ground and covered by cold silt. He was refrozen in subsequent winters and, as silt accumulated year after year, was gradually interred beneath the lower reaches of annual thaw within permafrost.

4

RECONSTRUCTING BLUE BABE'S APPEARANCE

Vertebrate paleontologists work mostly with bones. Reconstructions of extinct animals rely on our anatomical understanding of modern animals to make analogous decisions concerning the form of an animal for which we may have only a handful of bones. Often times, however, a strong dose of fancy is also added. The fossil record rarely preserves details of external appearance. The skin and hair preserved in a mummy such as Blue Babe provide a unique view into the past.

Some groups of mammals have relatively conservative coloration and body form, for example field mice (*Microtus*). A frozen Pleistocene-aged *Microtus miurus* carcass has been found near Fairbanks; the little mummy simply looks like a gray mouse. Because the appearance of *Microtus* has changed so little over time, mummies of this genus are unlikely to yield many surprises. Bison, however, are quite different. The two living species, *Bison bonasus* or wisent (fig. 4.1) in Eurasia and *Bison bison* (fig. 4.2) in North America, differ from each other in general body form and color. Additionally, Pleistocene steppe bison (*Bison priscus*) drawn on European cave walls (fig. 4.3) and portrayed in mobilary art are unlike either extant bison species. I use the term *species* advisedly, as late Pleistocene and living bison seem to form a single species complex.

Studies of fossil bones had already told us much about steppe bison before Blue Babe was found. By reconstructing skeletal shape, we could see that bison in Beringia were different from living bison (Guthrie 1980). Beringian bison were a little larger overall than extant bison (Skinner and Kaisen 1947), with disproportionately larger horns. But these subtle skeletal differences may be a poor clue to external appearances. For example, living bison species show much more variation in their coat patterns and hair color than in their skeletons. Within a related group, such as a genus, skeletons are comparatively similar. A bison mummy with hair provides much

Fig. 4.1. Wisent or European bison (*Bison bonasus*). This species is found in Poland and the Soviet Union.

Fig. 4.2. American bison (*Bison bison*). This species is quite variable. The individual pictured here is a typical plains bison (*Bison bison bison*) with large hair bonnet and pantaloons.

Fig. 4.3. Late Pleistocene steppe bison (*Bison priscus*). Many bison portrayed in Paleolithic art are exquisitely rendered. From this artwork we get a good idea of the appearance of the steppe bison. This picture is from Santimamine in Spain.

more information about overall appearance than one can ever get from a skeleton alone.

Woolly Bison, Woolly Mammoths, Woolly Rhinos, and Unwoolly People

We characterize the northern, Pleistocene rhino and mammoth as "woolly" because their nearest analogue, living in more temperate regions have short hair. If bison were extinct and all we had for comparison were extant forms like cattle and Asian and African buffalo, we certainly would be talking about woolly bison too. Bison are part of modern natural history, so the adjective *woolly* is unnecessary, but they are one of the Pleistocene woollies. If the bison and musk-ox had become extinct 11,000 years ago, along with the mammoth and northern rhino, the title of this book would have been different and I would have been writing about a mummy of the extinct woolly bison.

In fact, northern horses were also woollies. Some horses portrayed in Paleolithic art have long "beards," thick belly hair, and "feathered feet," woolier than domestic horses raised at northern latitudes. Likewise, lions would have been very thickly coated, as are other felines adapted to northern climates. That northern subgroup could legitimately be referred to as Pleistocene woolly lions.

This train of thought about the woolly bison and its compatriots on the Mammoth Steppe is not all facetious. Pleistocene northern large mammals were adapted to extraordinarily low temperatures: wind chill factors must have regularly brought effective winter temperatures down to −60° F and lower. The animals' tails and ears were short and well furred. Mammoths, rhinos, and horses had long hair even on their lower limbs.

The balance of the amount of energy taken in over the amount lost is central to the equation of whether a northern animal lives or dies each winter. Energy input can be reduced if the animal is more capable of conserving energy. Pelage is not simply a coat needed to keep warm. Heat is produced as a by-product of body metabolism, and in the case of a large herbivore, heat is a product of food composting by microorganisms. Most food we consume goes into maintaining our body temperature rather than meeting other basic requirements. Reducing heat loss, for a northern-adapted animal, directly affects use of fat reserves. Effective pelage can extend a little further the meager calories in winter food, which invariably are below maintenance. Woolliness can mean the difference between life and death.

The woolly pelt and runty tails and ears of these Pleistocene creatures are a signpost to us as we try to imagine what kind of ecological limit could have prevailed for about 20,000 years prior to the Holocene and probably why unwoolly humans did not inhabit Alaska until the "warmer" time during the end of the last glacial, as discussed in chapter 10. But pelage does more than retain warmth. It is an amalgam of a variety of functions that include the trappings of stature and social position.

Hair Length

When Walter Roman first discovered Blue Babe, only the posterior portion of the carcass was showing. Hair was not attached to the exposed skin, but tufts of black hair were found in the surrounding silt. By contrast, when the bison's anterior end was excavated, we found the hair had slipped but was still held in place by frozen muck on the head and in patches over the forequarters and forelegs. I had often wondered how Pleistocene steppe bison must have looked, so finding hair preserved with the frozen carcass was really exciting for me; we carefully sampled and mapped this hair before the carcass was moved. Concurring with Geist (1971b), I had often observed

(Guthrie 1980) that hair patterns of European Pleistocene bison (as portrayed on cave walls) were unlike either species of living bison.

Living bison from different geographic areas exhibit considerable differences in hair patterns. For example, "wood buffalo" (an unfortunate name because it suggests deep woodlands instead of the parkland clearings characteristic of their Canadian habitats), *B. bison athabascae,* have a high hump, small hair bonnet, and almost no chaps on the forelegs (Geist and Karsten 1977). Bison from the Caucasus, *B. bonasus caucasicus,* were different from Polish and Russian subspecies (Flerov 1977). Since Beringian bison occurred across such a vast area during the Pleistocene, it is likely they had geographic variations in hair color and pattern comparable to those of extant bison. Thus we could expect that Alaskan Blue Babe would resemble bison portrayed thousands of miles away by French and Spanish Paleolithic artists, but differ in certain details. Today ungulates occupying that geographic span exhibit subtle differences east and west. For example red deer, *Cervus elaphus,* moose, *Alces alces,* and caribou, *Rangifer tarandus,* vary between western Europe and eastern Asia, but are still justifiably considered members of their respective species.

Hair pattern among the three bison species *B. bonasus, B. bison,* and *B. priscus* differs most on certain parts of the body: the head, beard or ventral manes, foreleg pantaloons or chaps, and the hump. Fortunately these are located on the anterior part of the body—the very portion preserved on Blue Babe. But since our bison carcass was eaten from the dorsal side, skin and hair on the hump region were torn away. Likewise, thoracic vertebrae and their long neural arches were removed from the carcass and were not preserved. These parts would have told us about the hump. Eurasian *B. bonasus, B. priscus,* and North American *B. bison* all differ in neural arch contours and in hump hair patterns (Poplin 1984). We can use other steppe bison fossils to reconstruct the underlying bony part of the hump, and indeed this is the subject of chapter 5, but for now I confine my interest to hair and compare the hair contours of Blue Babe's forelegs, head, beard, and tail with that of other bison.

Blue Babe retained sufficient hair to show that he lacked well-developed chaps or leggings. In this regard he is similar to the wood bison, *B. bison athabascae,* and the Eurasian *B. priscus* (fig. 4.4). Geist (1971b) pointed out that bison in cave paintings do not have long hair on the forelegs. The living European bison, *B. bonasus,* does not have pendulous leggings but does have long hair on the forelegs, which exaggerates the apparent thickness of the upper leg

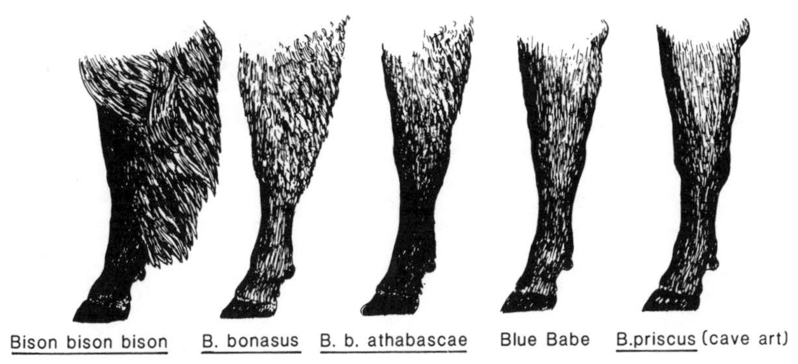

Fig. 4.4. Variations of "leggings" or "pantaloons" among bull bison. A comparison of pantaloon development shows Blue Babe is closest to the European Pleistocene steppe bison (*B. priscus*), which had negligible pantaloons. American plains bison (*B. bison bison*) has the largest. Both the European bison (*B. bonasus*) and the wood bison (*B. bison athabascae*) have traces of pantaloons.

(fig. 4.5). Hair on the posterior side of the leg down from the olecronon process is 30 mm on the Alaskan mummy, compared to the much longer hair (up to 300 mm) of mature plains bison bulls.

The length of Blue Babe's bonnet and forelock, hair over the frontal and parietal bones, is comparatively short for bison (fig. 4.6). It is about the same length as hair found on *B. bonasus* and, judging from cave paintings (fig. 4.7), must have been similar to the Eurasian *B. priscus.* North American bison have long hair above the eyes, producing a large bonnet. However, as first pointed out by Geist and Karsten (1977), *B. bison athabascae* has a smaller bonnet than plains bison, *B. bison bison,* and it is a rather different shape. The Alaskan bison mummy is quite unlike either extant American subspecies in this regard.

Only two parts of Blue Babe's beard were preserved—just behind the lower lip and, farther posterior, just forward of the jaw angle. Both are short in comparison to *B. bison bison* and resemble other bison with short beards, especially *B. priscus* in cave art (fig. 4.7). There was simply not enough hair on Blue Babe to determine the overall shape of the ventral mane.

Blue Babe emerged without a tail. His tail (bones and skin) was later found attached to a patch of rump skin. Tail hair also differs among living bison (fig. 4.8), but unfortunately Blue Babe's tail was hairless, so no comparisons of hair length are possible. We can evaluate color, however, because the broken bases of coarse black hairs can be seen embedded in the skin.

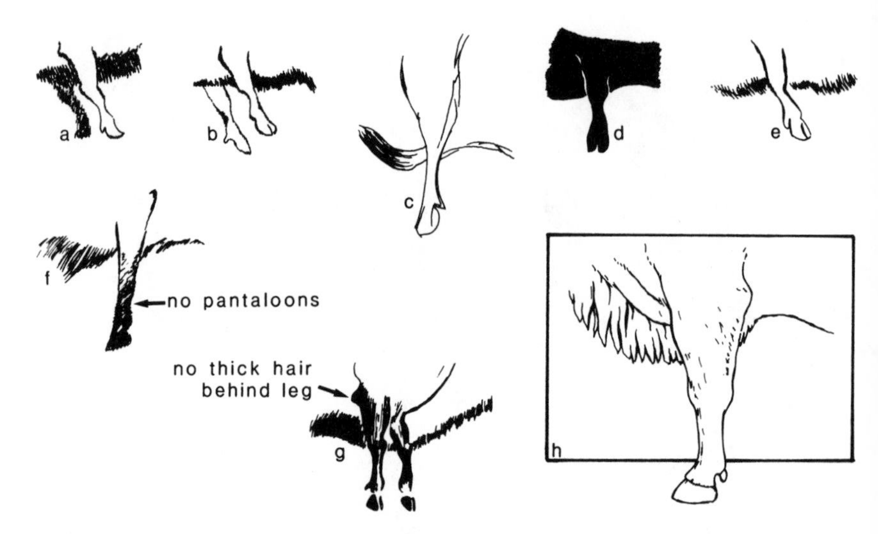

Fig. 4.5. Foreleg hair patterns of steppe bison from Paleolithic art. All living bison have some degree of hairy pantaloons on the foreleg, especially on the posterior part of the radius-ulna region (below the elbow). Pictures of steppe bison painted by Paleolithic artists show no indications of foreleg pantaloons. These examples of Paleolithic art are from (*a* and *b*) Niaux, (*c*) Candamo, (*d*) Le Portel, (*e*) Les Trois Frères, (*f*) Font-de-Gaume, and (*g*) Altamira. The hair on Blue Babe's forelegs was similar to the pattern shown in these Paleolithic pictures of steppe bison.

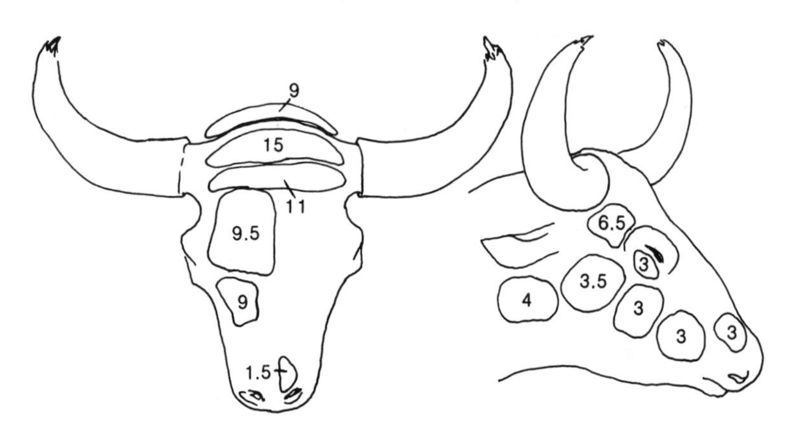

Fig. 4.6. Head and facial hair. The hair removed from Blue Babe's face and head was mapped and measured. The hair had "slipped" its attachment from the underlying skin but was still held in place by the surrounding wet silt. Hair length is shown in centimeters.

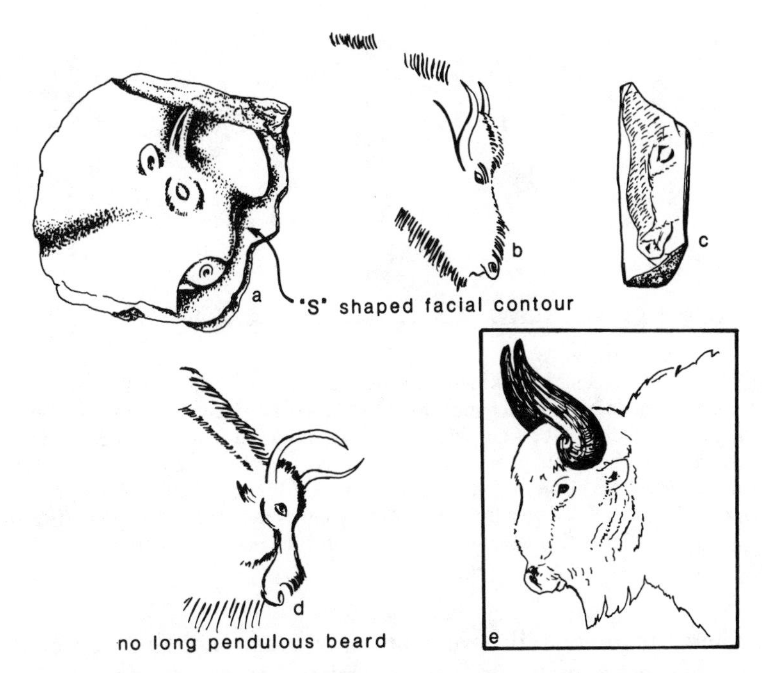

Fig. 4.7. Reconstructed face of steppe bison from Paleolithic art. Unlike American plains bison, steppe bison in Paleolithic art are pictured with short facial hair, an S-shaped profile, and a shorter beard. These Paleolithic bison portraits are from (*a*) Angles-sur-l'Anglin, (*b*) Le Puy de Lacan, (*c*) Isturitz, and (*d*) Les Trois Frères. Remnants of facial hair on Blue Babe (*e*) show a similar pattern.

Blue Babe's tail without hair is around 20 cm long, much smaller than *B. bison's* tail. Van Zyll de Jong (1986) shows that the mean length of tail, without tassel, for *B. bison bison* was 42 cm and that that of *B. bison athabascae* was 50 cm (both male samples). Tails of living European bison are much longer (Flerov 1979 illustrates the bison from the Caucasus as having a small tail). Flerov (1977) estimates the tail of a female Siberian bison mummy at 30 cm, longer than Blue Babe's but shorter than living European bison.

Geist (1971a) has shown that tail length is important socially and seems negatively correlated with the degree of cephalization of social display paraphernalia. But tail length is also related to severity of winter. Ungulates living in the far north generally have shorter tails than their warmer climate counterparts. Long tails (speaking of the structural bone-blood-skin part of the tail and not hair) are a source of heat loss. Additionally the insect season in the far north is so short that the advantage of a fly switch for hind quarters is small.

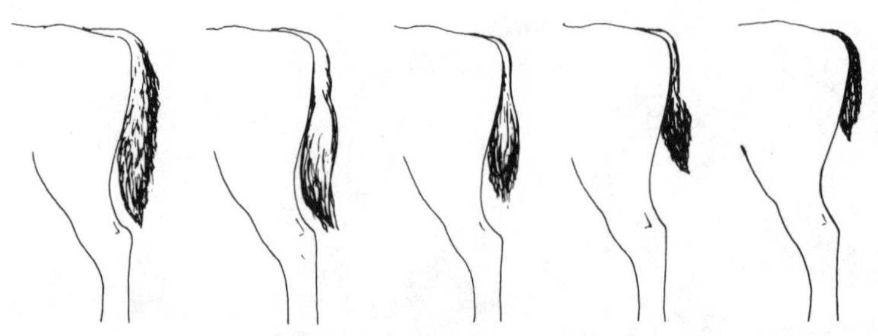

Fig. 4.8. Bison tail lengths. Tail length varies among bison groups. European *B. bonasus* have the largest tails and plains bison (*B. bison bison*) have the shortest of living bison. However, Blue Babe's tail was even shorter. No full-length hairs remained on the tail, so the tail pictured here is a tentative reconstruction based on bone and skin remains. The steppe bison in cave art have long, well-furred tails like European bison. Presumably tail length varied from west to east across the Mammoth Steppe because the continental climatic effects in the east produced (and still produce) colder winters.

Northern ungulates all have short tails—just enough to cover the ano-vulvar area as an insect-proof seal and an insulating cap. Consistent with this, the woolly mammoth and woolly rhino pictured in Paleolithic art have smaller tails than their living analogues in more southern latitudes.

Although Blue Babe was a steppe bison (*B. priscus*), his tail was shorter than tails of European steppe bison (*B. priscus*) shown in Paleolithic art (fig. 4.8), probably due to the milder winters European steppe bison experienced, especially during the warm and wet Pleistocene interstadials. Winters were never warm for Alaskan and Siberian bison, even during the interstadials, so those bison have short tails. American bison, *B. bison*, which are derived from native Beringian bison, seem to have retained short tails as they moved south. Winters on the High Plains are not as severe as those in Alaska, but they are not mild.

In addition to these hair patterns, other areas on the bison mummy were not preserved. For example, lack of hair on the shoulders did not allow us to assess whether a shoulder mantle was present, as in *B. bison bison*, or almost absent, as in *B. bonasus*, *B. bison athabascae*, or Eurasian *B. priscus* (in Paleolithic art). Also missing from the specimen was the penile tuft, which is most developed in the plains bison, *B. bison bison*, and least developed in the European bison, *B. bonasus*.

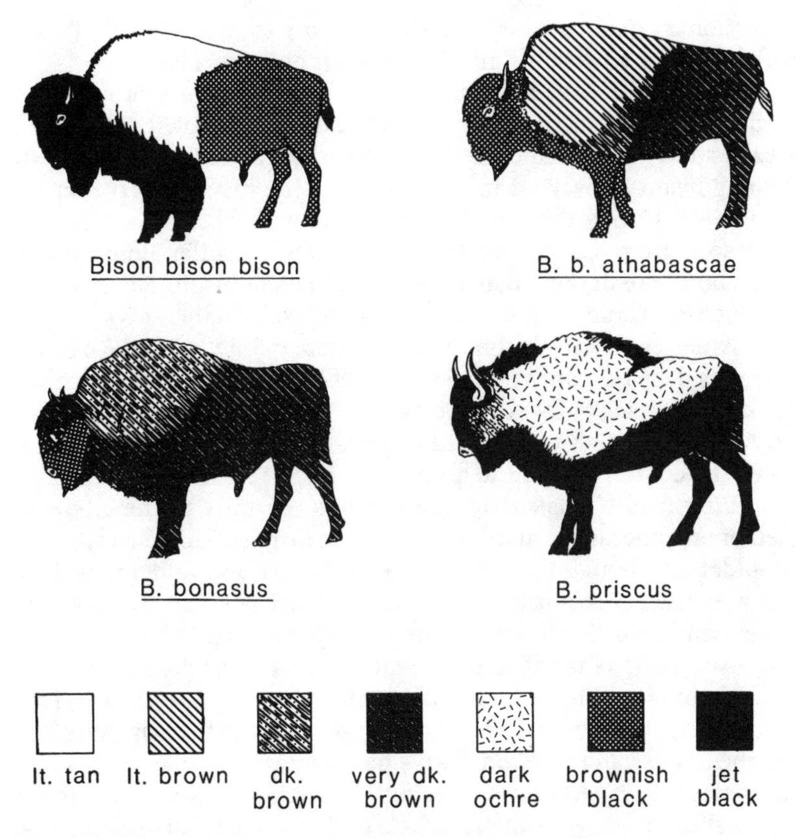

Fig. 4.9. Bison color patterns. A bull typical of each "species" is shown here along with a reconstruction of steppe bison from Paleolithic art.

Blue Babe and Bison of Other Colors

The color patterns of Blue Babe and subspecific variants of living bison species are compared in figure 4.9. There is some variability in pelage within bison populations as well as some geographic variability, so these comparisons are of the more typical forms. Coat color is a fairly labile character among mammals, unlike teeth or feet. The evolution of teeth and feet are tied to conservative and exacting environmental constraints, but external appearance is linked more closely with less conservative traits such as social behavior; as a consequence, external appearance is one of the more variable characters between groups. Just as external appearance is the character that differs most among living bison, so it would be expected to differ on an evolutionary gradient. Because the mummy is an adult bull, I compare him only with adult bulls from other bison species.

European bison, *B. bonasus*, are the most homogenously colored bison. Their coat is a uniformly dark rusty brown. The hump area is subtly lighter and the face forward of the ear a darker brown. Among bison I observed in the Prioksko-Terransny Reserve on the Oka River, USSR, there were some bulls with black faces. Skins I have seen from Caucasian subspecies had a similar homogenous color, but were darker than Russian and Polish bison. Flerov (1977) also pictures Caucasian *B. bonasus* as uniformly dark.

American plains bison are more dramatically colored; they are, overall, a deep rust brown. Most bulls have jet black hair on the head, neck, and foreleg (a few individuals are more "cow" colored, with a very dark brownish tone in these parts). A stripe of dark hair runs down the central part of the hump, in contrast with the light sand of the hump itself. This sandy hair extends ventrally to the elbow, or olecronon process, as part of a thick mantle of hair covering the shoulders. Although European bison bulls have a shoulder mantle, it is not exaggerated, nor does it have a discreet posterior border as in American bison. Posterior to this shoulder mantle, hair on American bison bulls is much shorter and quite dark, ranging from black to dark brown. These contrasting tones produce a striking late-autumn appearance before the pelage is bleached by sun and weather. Geist and Karsten (1977) have shown that the wood bison subspecies (*B. b. athabascae*) is not so dramatically colored as the plains subspecies: wood bison have a darker, more reddish hump that is less dramatically colored than that of the plains bison.

In contrast to these variants of living bison, Blue Babe looked like a dark bay horse; he was mainly a rich dark brown with black "points" on his face, legs, and tail. His forelegs were dark brown, but hair on the lower leg, from metacarpals to the hoof, was so dark it was almost black. The face, or anterior part of the head, was black, reaching posterior only to the eyes and dorsally toward the frontals on a line between the anterior edges of the horn bases (fig. 4.10). The rest of the head, bonnet, cheeks, and ears was a rusty brown. The beard or ventral mane was black, at least on its anterior parts (where hair remained on the mummy). Only a few hairs remained on the shoulders. These were a light reddish brown. No light sandy-colored hair, characteristic of living American plains bison, was found with Blue Babe. The hair that came out of wet-screened silt from the site was either very black or reddish brown.

Geist and Karsten (1977) showed that the ventral name of *B. b. athabascae* was poorly developed, while that of the plains subspe-

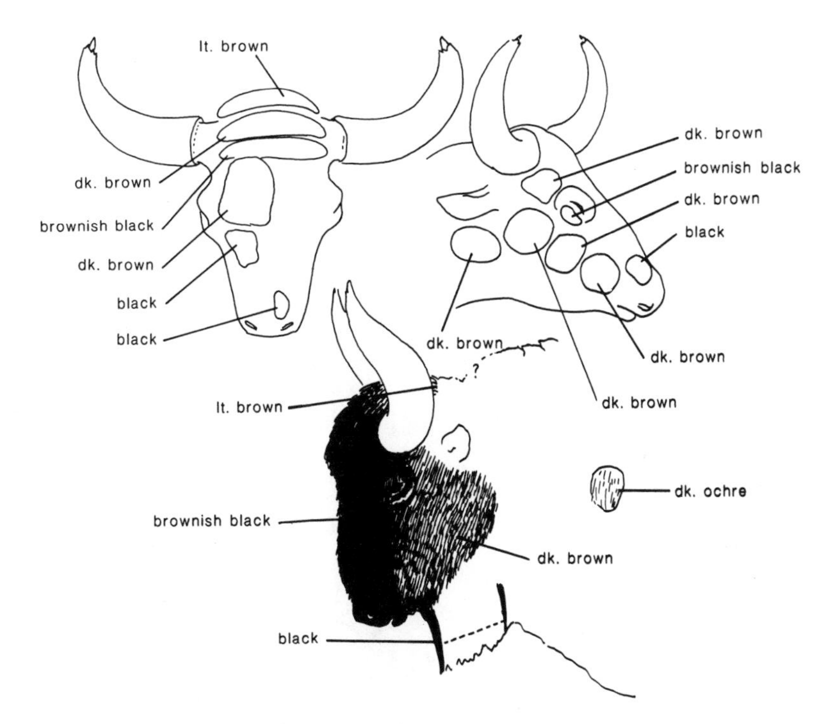

Fig. 4.10. Blue Babes's head color. Hair samples were collected as the bison emerged and during necropsy. Most of the hair had already slipped, but some facial hair was in place. On the front of the face the hair was almost jet black. On the sides of the head it was dark brown, and one sample from the shoulder was a reddish brown.

cies, *B. b. bison,* and the European wisent, *B. bonasus,* was quite long in both males and females. Unfortunately Blue Babe's ventral mane posterior to the beard was not preserved. As already noted, the mummy's tail hairs were not preserved, but broken bases of hair still in the tail skin were very black. This tail hair resembles that of American bison tails (which ranges from black to very dark brown) and is quite different from the long reddish brown tail hair of European bison.

We can compare this picture of Blue Babe with the images of his European contemporary, the *B. priscus,* drawn by Paleolithic hunters. Of course one must be cautious when using art works to determine actual pelage pattern (Guthrie 1984c), just as a single portrait offers only suggestions, not proof, about details of appearance. But cave pictures of bison are like a set of portraits done by different artists of the same model. Given certain caveats, one can argue that if a general pattern exists, it very likely relates to what was actually seen.

Although bison are among the most common animal portrayed in Paleolithic art (Leroi-Gourhan 1982), only a few of these drawings and paintings are fully colored (figs. 4.11 and 4.12); most bison are drawn in outline and lack body tones or color. Also, there is a wide variability in the pictures which surely did not correspond to an actual variability in nature. The bison are always positioned laterally in the paintings, and the interior of the body usually has the same color. Among the bison that are fully colored, the body interior is dark reddish, while the periphery, particularly the cervical and posterior thoracic hump, the tail, legs, and face are regularly darkened or painted black. This coloration differs from living bison, but is very similar to Blue Babe's, at least to the portions for which we have hair.

Flerov (1977) discussed hair color and hair length of a two-and-a-half-year-old female Pleistocene bison mummy found on the Indi-

Fig. 4.11. Color pattern of steppe bison in Paleolithic art. A dark wedge runs from rump to elbow, and a smaller dark wedge appears at the posterior base of the hump. A dark mane, legs, and head leave a light side panel in the shape of a tobacco pipe. These samples are from (*a*) Le Portel, (*b*) Altimira, (*c*) Niaux, (*d*) Lascaux, and (*e* and *f*) Santamamine.

Fig. 4.12. Steppe bison in Paleolithic art. (See also those of fig. 4.11.) Further examples from (*a*) Fontanet, (*b* and *c*) Les Trois Frères, (*d*) Le Portel, (*e* and *f*) Niaux, (*g*) Font-de-Gaume, (*h*) Lascaux, and (*i*) Altimera.

girka, USSR. Unfortunately, females have different color patterns than males among living American bison, and this dimorphism probably held true for Beringian bison as well. The color patterns Flerov reports are roughly similar to those I have described for Blue Babe except his are lighter in every individual area. The most notable difference is a very light leg coloration ("light yellow ochre") of Flerov's female mummy; no living bison has lightly colored legs. This could be the color in life, or the leg hair could have been bleached in a manner I described earlier.

Flerov (1977) also describes a slightly different pattern of hair length from that of Blue Babe. Unfortunately he does not directly describe the hair length for the young female, but uses her to reconstruct how bulls must have looked. He proposes short pantaloons on the rear of the forelegs; there is no trace of these in Blue Babe. Flerov does describe the "rusty-ochre" and "rusty-cherry" color of the neck and shoulders that I saw on Blue Babe.

Horns and Hooves

Unlike most fossil bison skulls found in Alaska, which have poorly preserved sheaths on their horn cores, Blue Babe's horns were in almost original condition. Horn keratin had undergone some chemical deterioration and was brittle, but the surface remained, appearing as it must have in life. The horns are a deeply pigmented black and show none of the porosity that appears as a white frosting among some living black-horned bovids, including some individual bison. The tip end, or distal half, of the horns is highly polished, and one cannot see the more distal annual growth rings. There is a broken terminal portion on each horn, which is unusual among bison fossils from Alaska; most terminate in unbroken slender sharp points (Guthrie 1980). Despite Blue Babe's broken tips, the horns still have sharp points, because the lost portion was like a cone pulled from the distal end, exposing the tip of an underlying sharp cone. (Large sharp horns are an important clue to the behavior of steppe bison, which I discuss in chapter 7.) Blue Babe's horns are much larger than those of any living bison, but fall in the lower size range of steppe bison.

Hooves

Blue Babe's hooves came off during the excavation. They were bagged and frozen, but later, when thawed and dried, the hooves curled, cracked, and came apart. Nonetheless, I have photographs taken of them during the excavation as well as the terminal phalanges that fitted inside the hooves. These differ in no noticeable dimension or shape from living American bison.

Flerov (1977) proposed that the hooves of Beringian Pleistocene bison were considerably larger than those of living species, although Vereshchagin and Baryshnikov (1984) disagreed. Flerov shows the forehoof of an old European bison bull, *Bison bonasus*, and compares it with a two-and-a-half-year-old Pleistocene cow. The cow has hooves the exact appropriate size and shape for an American bison of that age (Murie 1954), but Flerov's illustration of the foot of the European bison bull does not fit my photographs of *B. bonasus* feet. Either Flerov mistakenly used a hind foot of a European bison (which has the same size and shape as the one shown by Flerov), or the European bison he used had dramatically different-shaped hooves than ones I have examined. From a search of many slides of European bison the latter does not seem to be the case.

I think that in fact the slightly overlapping, inward-curved hooves Flerov describes (1977) as being adaptive for swampy environments, are simply late winter hooves showing lack of wear, as occurs in winter among living bison. Bison on snow-covered ground do not wear down their hooves, nor do they regularly use their hooves as horses do to dig through snow for food. I have seen musk-oxen (*Ovibos moschatus*) with similar, and even more extreme, overlapping hooves in the spring. Hoof shape and size are very important in understanding an animal's adaptation to different substrates; they are discussed in chapters 8 and 9.

Skin Thickness

Although skin thickness is not directly related to Blue Babe's appearance, it tells part of the story of how behavior affects external anatomy. Geist (1971a) noted that patterns of skin thickness vary considerably from one ungulate species to another and, further, that skin thickness is correlated with mode of combat. Geist compared sheep (*Ovis canadensis*) with Rocky Mountain goats (*Oreamnos americanus*) and found that while sheep have thick skin around the head

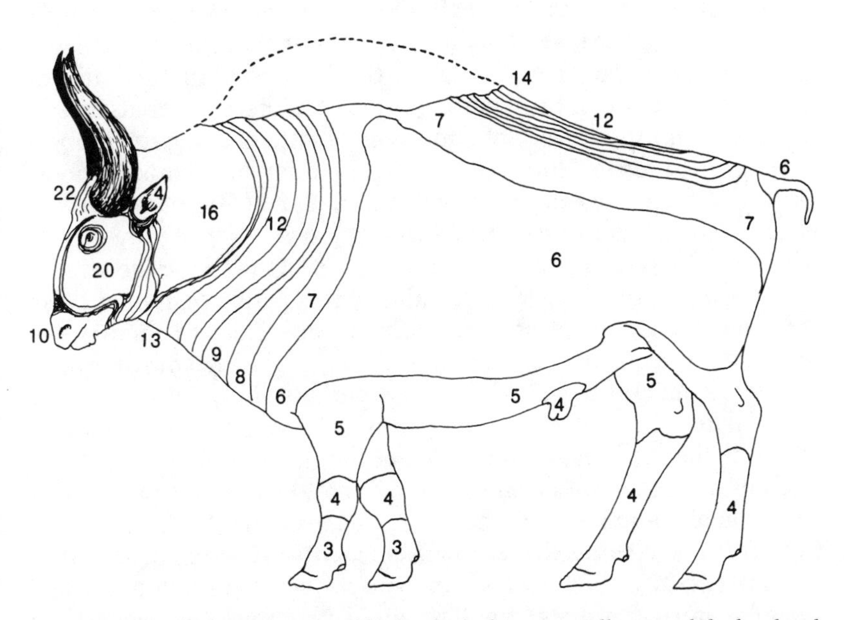

Fig. 4.13. Skin thickness. Presumably the thick skin, especially around the head and neck, was a defensive armor against an opponent's sharp horns. The thinner torso and leg skin allowed supple movement. (Thickness shown in millimeters.)

and thin skin on the posterior of the body, this pattern is reversed in Rocky Mountain goats. Their face and shoulder skin is thin, while the skin on their flanks and posterior is thickest. Sheep are head clashers, and occasionally their horns slip onto the forward part of the opponent's body. Rocky Mountain goats fight standing side by side, facing opposite directions, and dig their sharp horns into one another's flanks. For both sheep and goats, thick skin thus acts as defensive armor, decreasing the damage of an opponent's blow.

Even when the mummy's neck and head were first exposed during the excavation, it was obvious that skin in this area was unusually thick. I have plotted skin thickness in figure 4.13. Skin around the distal part of the legs is 3–4 mm thick, and over the lateral part of the body skin averages 6 mm. Along the dorsal surface, over the sacrum and lumbar area, it is thicker, ranging between 8 and 16 mm. Skin on Blue Babe's head is by far the thickest, 22 mm—measuring nearly one inch.

Age

Much of Blue Babe's appearance would have been determined by his sex and age. We know from horn size and the presence of male genitals that Blue Babe was a male. I used several methods to determine Blue Babe's age because age at death is so critical to other elements in the analysis. For instance age is the key to estimating rates of tooth wear, which, in turn, has important ecological repercussions.

Determining the age of a mummy is rather easy, like finding a freshly killed bison. However, aging techniques are approximations. Northern mammals have an obligate period of dormancy during winter. All northern ungulates will not grow—cannot grow—even when ample food is present; they "turn off" growth for the winter. The horns and teeth leave an annulus of this period of dormancy, so the growth segments represent summer only. Thus on the basis of tooth and horn aging, it would probably have more meaning to speak in terms of an animal being so many summers old.

There are five well-delimited growth zones (including the last, which makes up the proximal edge of the horn base) near the base, but annuli in the middle and distal parts of the horn are obscured, a common phenomenon even among living bison (Fuller 1959). Listed from the base distally, these growth segments measure, on their dorsal surfaces, 22, 26, 35, and 46 mm, and a long 430 mm from the last segment until the distal tip. The tip is "broomed" somewhat, so there was an additional missing length beyond this. Judging from those measurements, and comparisons with horn sheaths of other

Pleistocene bison, I estimated Blue Babe's age to be 8 or 9 years. The large horns with their annuli are probably the best evidence that the bison is in the 8-year range. They assure us that Blue Babe is not a young six-year-old; nor is he a ten-year-old.

Teeth cementum annuli can also be informative, but they too are approximate. When the bison was excavated the incisors fell out and were preserved. The first incisor (I1) was sent to a commercial tooth-sectioning laboratory in Montana. While not necessarily the best indicator tooth in bison, the first incisor was available before the bison had been rethawed and necropsied. This tooth erupts at about 2 years of age (Fuller 1959); bison do not normally lay down root annuli on the incisors until about 4 years of age (Navokowski 1965). The lab read the first incisor's annuli as four; thus the annuli reading would make the bison about 8 years old.

Among populations with the same rate of tooth wear, stage of wear can be a rough indicator of age. Fuller's sample of 1,800 bison from Wood Buffalo Park showed that at age 8 the lower canine has more than 2 mm of wear. That tooth in Blue Babe appears to have at least 2 mm of wear, but not much more, which would place it at about 8 years of age according to Fuller's sample.

Skinner and Kaisen (1947) constructed wear stages of 1,322 mandibles from interior Alaskan Pleistocene sediments, including some from Pearl Creek. Blue Babe corresponds to S-2 in their diagram. At that stage the enamel of the first lower molar (M1) is worn about halfway down its vertical height. Their measurements and mine on Alaskan material indicate that a bison from age 14 to 16 shows complete wear of the M1 enamel. Also their increase in mortality due to senility begins with age class S-3, which is to say that death from aging debilitation begins in bison at around 10 years of age. So age estimation (just prior to S-3) also points to the 7- to 9-year range.

Epiphyses, which usually fuse in an age-related sequence, have been mapped for European bison (*Bison bonasus*) by Duffield (1973). He found that vertebral epiphyses fuse during the seventh year; during the eighth year those of the pelvis, ribs, and scapula fuse. All of the epiphyses of this mummy seem to be well fused, again suggesting it is at least 8 or 9 years old.

At 8 to 9 years old Blue Babe would have been at his prime, having reached his maximum body size, social station, and fullest pelage development. In fact, his age is probably one notch past that point, just past the break in slope of the mortality curve, when older males begin to be lost from the population. In short, Blue Babe is a fine specimen of his ilk.

5
TRACKING DOWN BLUE BABE'S MISSING HUMP

Hump shape may sound like an esoteric issue, but dorsal contours of different species relate in important ways to their behavior and ecology. Living bison and fossil Eurasian steppe bison all have subtly different hump shapes; I propose that this is because humps are a structural adaptation to the environments in which different bison groups live or lived. For this reason, I think that the hump shape of the Alaskan Pleistocene bison is significant and that it can be compared with humps of other bison for clues to behavior, locomotion, diet, and a whole suite of biological features. What sort of hump Blue Babe had is important.

I was unable to reconstruct hump contour directly from the mummy because critical thoracic vertebrae were missing. There are, nevertheless, hundreds of fossil bison vertebrae from Alaskan deposits, and, although few of these have been dated, we know most bones are from the latest Pleistocene, well within radiocarbon-dating range (see dates listed for Alaskan Pleistocene bison in chapter 10). Because there was only one phylogenetic line of bison in Alaska during the late Pleistocene (Guthrie 1970, 1980), I argue that vertebrae and other postcranial material should be relatively similar, which is not to say that Pleistocene bison did not evolve. One would expect differences between bison living tens of thousands of years apart, but bison seem to exhibit a continuity of body shape within the late Pleistocene throughout Eurasia. Working on that assumption, I used bison vertebrae from other Alaskan late Pleistocene deposits to reconstruct Blue Babe's hump profile. And although there were sexual and individual variations, the fossil bison vertebrae I examined followed a similar pattern.

Girders and Gallops

Most morphological, evolutionary, and systematic studies of bison are based on male skulls (e.g., Skinner and Kaisen 1947; Guthrie 1970; McDonald 1981). Thoracic vertebrae needed to reconstruct dorsal contours are among the most poorly preserved elements of fossil skeletons. Thoracic vertebrae are not conservative characters, and, taken singly, they are not very diagnostic. As a result, biologists and paleontologists have given hump shape little thought. Dorsal contour is actually the net outcome of a suite of physical features, including manes, bulging muscles, and connective tissue humps. But neural processes, or spines on vertebrae, are the key factor. Relative to body size, these rise higher in bison than in any other mammal, making a bison skeleton appear almost like the Permian reptile *Dimetrodon.*

Functionally, an ungulate's neural spines are comparable to the uprights in a suspension bridge. As in the Golden Gate Bridge in San Francisco Bay, truss members reaching skyward are attached to cables (tendons in the case of neural spines), which carry the long span between piers (legs) that reach to the bottom of the bay (fig. 5.1).

In mammals this structure of uprights and cables does something more than provide a stationary support. Muscles combine with bones and tendons so that half the body can be cantilevered without leg support from the other end, like a cantilever bridge. This is exactly what happens in the gallop; the body is supported alternately from first the pectoral (front) end and then from the pelvic (rear) end. Although little is known about hump function, I can begin with the hypothesis that the most important role of shoulder humps is increasing the efficiency of the gallop. This role can be seen (1) as a cantilever for a phase of the gallop, (2) as a girder to aid in the length of foreleg stride, and, probably most important, (3) as a structure to add leverage, allowing an efficient use of the stretch tendon of the neck.

At this point, to appreciate the function of hump shape we must approach it in a roundabout way, taking a close look at the physical demands of different gaits. The *walk* of a quadruped is analogous to two people walking in tandem, carrying a short pole across their shoulders, "out of step" with each other. A *trot* can be pictured as these same two people running, still out of step with one another. The advantage of a walk or trot is its efficiency. The body is always directly supported on both ends (anterior and posterior) and from alternate sides at the same time, so it does not tend to fall left or

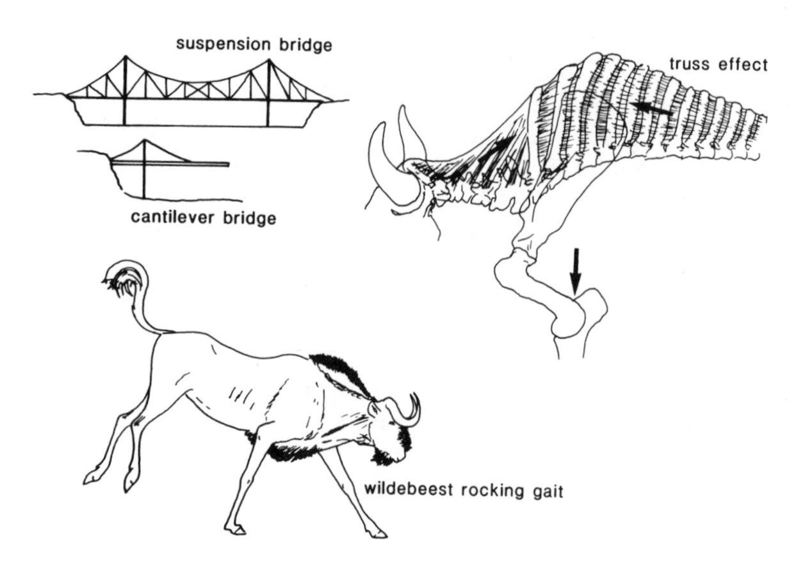

Fig. 5.1. Hump engineering. The construction and action of neural spines have been likened to the upright trusses of a suspension bridge, but during the gallop they are, in fact, part of a cantilever system. Weight is supported from below by forequarters or hindquarters, while the alternate end is held aloft by a cantilever effect. A strongly developed hump thus contributes to an efficient gallop, as seen in black wildebeest (pictured here) and bison.

right. Because of the energetic efficiency of a trot, most quadrupeds use that gait at moderate speeds. Unlike an upright biped, a walking quadruped has to step twice over the same ground, once with the fore feet and once with the hind feet, which makes the gait easy to model with two bipeds walking in tandem. The main disadvantage of a walk or trot is that each stride, hind and front, must be the same length, just as two people carrying a pole on their shoulders should walk in cadence and have the same stride length. A quadruped's hind foot cannot overstep the front foot in a walk or trot.

Camels and giraffes have developed a gait that avoids this problem but produces others. Instead of moving front and hind limbs out of step, they stride in step in a *pace* or *rack*. The problem with this gait is that the entire weight of the body is alternately thrown left and right, which is why camel riding can make a person seasick. The special build of a camel's legs and its broad feet are accommodations to this inherent instability.

Trotting is energetically efficient but has an inherent maximum speed limit. A trot cannot make full use of the flexible forelimb's potentially greater stride length. As a trotting animal gains speed, the stretch of the front leg is limited by the stride length of the hind limb. Unlike the more freely attached forelegs, rear legs have a fixed radius, being securely anchored in the pelvic girdle. To gain greater speed an animal must shift into a *canter* or *gallop*.

While a walk or trot can be modeled by two bipeds with a pole, and we can see that a walking or trotting quadruped does twice the work as two bipeds, it is not easy to model a gallop. To gallop our bipeds would have to alternately throw and pull each other forward by the pole while running. A gallop enables a small quadrupedal squirrel weighing a half kilogram to outrun an 80 kilogram bipedal human. A canter or gallop loses the support efficiency of a walk or a slow trot where the nonsupporting leg has only to move forward to catch the body as it falls. Canters and gallops are energetically more costly, in theory, because they have a greater vertical component, that is, the body is thrown into space, unsuspended, for substantial periods of the gait. However, a gallop has a different kind of efficiency because the hind and fore limbs do not duplicate each others' work.

To understand how a gallop works we look at it in four beats or phases (fig. 5.2), one of which discloses the anatomical importance of humps. A galloping animal hurtles through the air unsupported after it pushes from the hind legs (phase 1) and after it pushes from its fore feet (phase 3). The other phases are the steps on the ground of the forelegs (phase 2) and between both stretched hind legs (phase

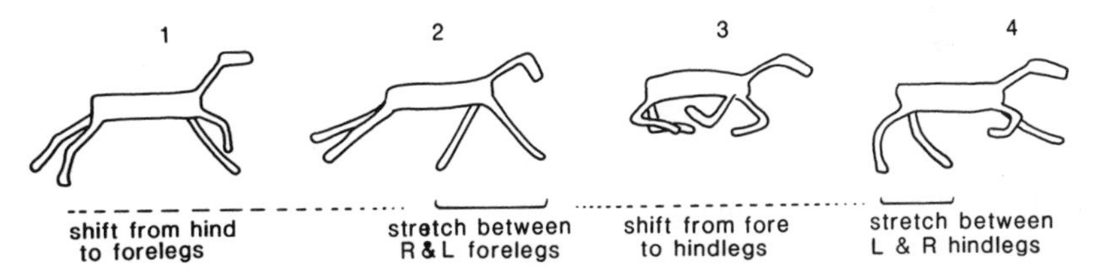

Fig. 5.2. Four phases of the gallop. I divide the gallop into four units relating to suspension and distance traveled. The *first* phase marks the shift from hind legs to forelegs. European and American bison differ, as the latter seems not to leave the ground during this phase. The *second* phase is the distance gained as one fore foot steps ahead of the other. Animals vary considerably in this phase, but bison excel, taking an extremely wide foreleg step. Their tall hump is especially important in phase two. The *third* phase marks the shift from forelegs to hind legs. Most gallopers leave the ground during this phase, even a one-ton bull bison. The *fourth* phase marks the length of step from one hind leg to the other. This is especially short in bison; related to this, bison hindquarters are rather low and their hams relatively small.

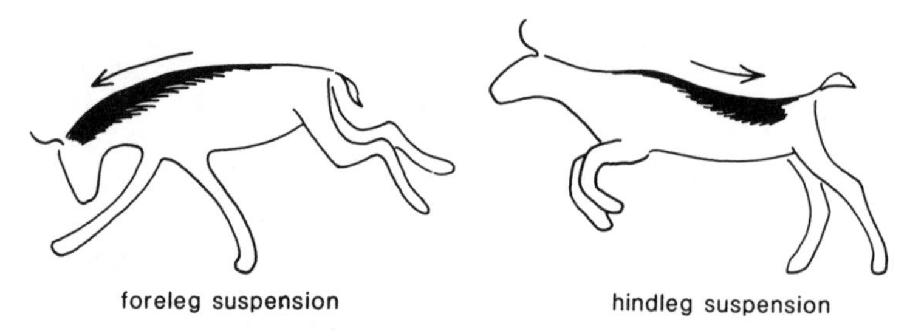

Fig. 5.3. Cantilevers and gallops. During phase two and phase three, one end of the quadruped is suspended by the other via a cantilever system of muscles, tendons, and bony struts. Bison have accentuated phase two more than phase four; consequently, the forequarters used in phase two are more exaggerated than the hindquarters.

4). So the complete four-phase cycle of a gallop can be measured and classified by the footfalls or tracks. Phase 2 is where the hump becomes important. Normally the forelegs do not land simultaneously; rather, one hits the ground, taking the whole weight of the animal, and pushes backward at the same time that the second foreleg is reaching forward and making contact. That second leg then carries the animal's weight and pushes backward, whereupon the entire body is projected forward through the air. The hind legs repeat this left-right sequence, again propelling the animal through space until it lands on its first fore foot (called the lead foot).

This shifting suspension of a galloping animal between anterior and posterior ends emphasizes the importance of the cantilever system of tall neural spines and tendon cables (fig. 5.3). Unlike some other large mammals, bison do not leave the ground in phase 1. Rather, the forelegs carry the center of gravity for an exceptionally long time (extended phase 2). Since the hindquarters are suspended out in space for a longer time than the forequarters, the main cantilever struts would logically be located in the shoulder area. Indeed that is where they are in most ungulates; bison are but an extreme case.

For a large mammal the story of an efficient phase 2 of the gallop includes the shoulder hump. Freed from the restricted arc of the hind limb, the forelimb can swing not from the elbow as it does in a trot, but from the highest crest of the scapula. Thus a galloper needs to maximize the stride length of the forelimb, that is, phase 2. Among African antelope species, the alcelaphines (wildebeest, topi, hartebeest, blesbok, etc.) are all high humped. These species are smooth, efficient canterers and gallopers and seldom, if ever, trot.

Bison are a bovine version of an alcelaphine. Wildebeest and bison can maintain their rocking, distance-eating canter for hours and days. Today they are here in the thousands; tomorrow they are gone to some distant pasture. This mobility is, I believe, the ecological significance of humps.

Fortunately, Swiezynski (1962) has done a careful study of bison musculature that helps us understand how bison humps work. The high neural spines near the anterior portion of the thorax are the major origins for the *splenius* muscle, which inserts on the head and neck. These spines are also the origin of the powerful extrinsic muscle of the pectoral limb, the *rhomboideus.* The *rhomboideus pars thoracalis et pars cervicalis* attach to the dorsal portion of the tall scapula, acting as a powerful lever to pull it forward during the backstroke of the foreleg (fig. 5.4). These muscles are especially important during the gallop. Roskosz and Empel (1961) argue that these specialized muscles of the hump provide agility and flexibility, enabling bison to jump a 2 m obstacle from a standing start, but they also conclude that (p. 69) "the powerful muscles of the withers are decisive factors concerning the growth of the spinous processes. These muscles are connected with the characteristic pose of the head, manner of feeding, the great weight of the body, and the strongly developed anterior part of the European bison. The specific gait and great activity of this animal is not without importance."

As I said earlier, the second phase of the gallop is one of the most important for bison. The tall scapula hinges from its apex, allowing a wide stretch between forelegs, even for such a heavy creature as a bison. But the spines of the anterior thoracic vertebrae are tall in all bison species, and so are those near the posterior end of the thorax, but these latter are proportionally and absolutely much larger in the steppe bison.

Unfortunately, we need the muscles, tendons, and ligaments to say for certain how the steppe bison's unique hump functioned, and we do not have those because they were removed from the Blue Babe specimen by predators and scavengers. We can, however, make a good guess. Perhaps the *rhomboideus* extended more posteriorly; we know from the shape and size of the anterior neural spines that it had a different shape in steppe bison. But another muscle, the *spinalis,* is a series of bellies that originate on the distal ends of each thoracic vertebrae and that travel posteriorly, inserting toward the lumbar vertebrae. These bellies act in synchrony to arch the back convexly (dorsally), and because of their size and origination height they provide a powerful mechanism by which the back can add push to the hind limbs for greater distance during the first beat or phase of the gallop.

This same enlarged *spinalis* muscle originating from a higher angle would also have given the males more power in fighting, as the back comes from a collected convex to an extended concave position when the head is forced upward during pushing bouts. There seems to have been a difference in fighting mode between America bison

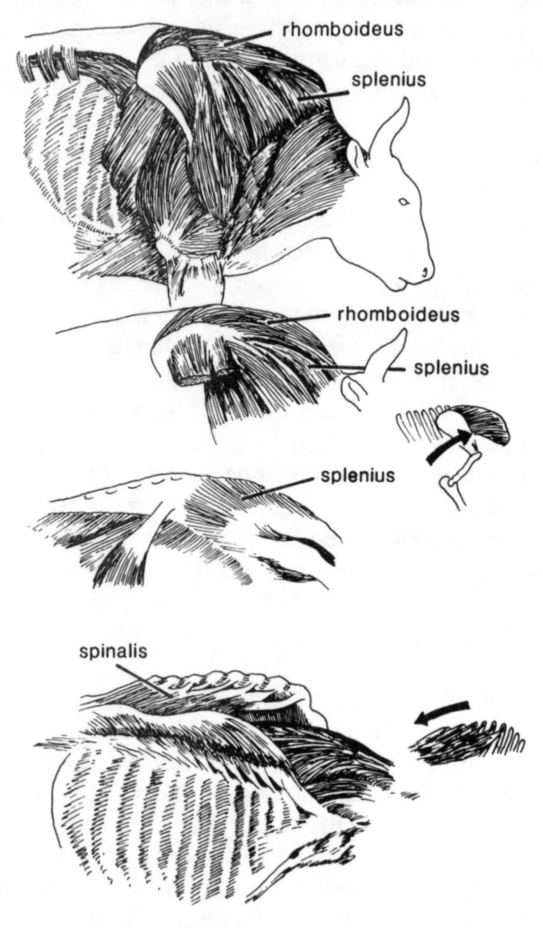

Fig. 5.4. Thoracic muscles of European bison. Some shoulder muscles used in the suspension process are portrayed in this anatomical dissection redrawn from Swiezynski (1962). The rhomboideus and splenius pull the scapula forward, rotating the foreleg backward (the main strength of the bison's gait). The spinalis pulls the neural spines backward, using that leverage to arch the vertebral column in phase one, as well as supporting the rear quarters in phase two.

and the large-horned Pleistocene steppe bison (Guthrie 1980, and reviewed elsewhere in this book). So in addition to its differences in locomotion the steppe bison may have had a muscle-lever system which accompanied a different adaptation in fighting strategy and style.

The long neural spines of a bison's hump are also important as girders or struts for the attachment of the elastic neck ligament, or nucal ligament. This is a thick elastic yellow cable which allows energy to be stored as the head is lowered by gravity (for example, when the animal grazes). Stretch ligaments make it less costly to return the head to its natural position (fig. 5.5). The nucal ligament originates from the tips, or distal ends, of the neural spines and in-

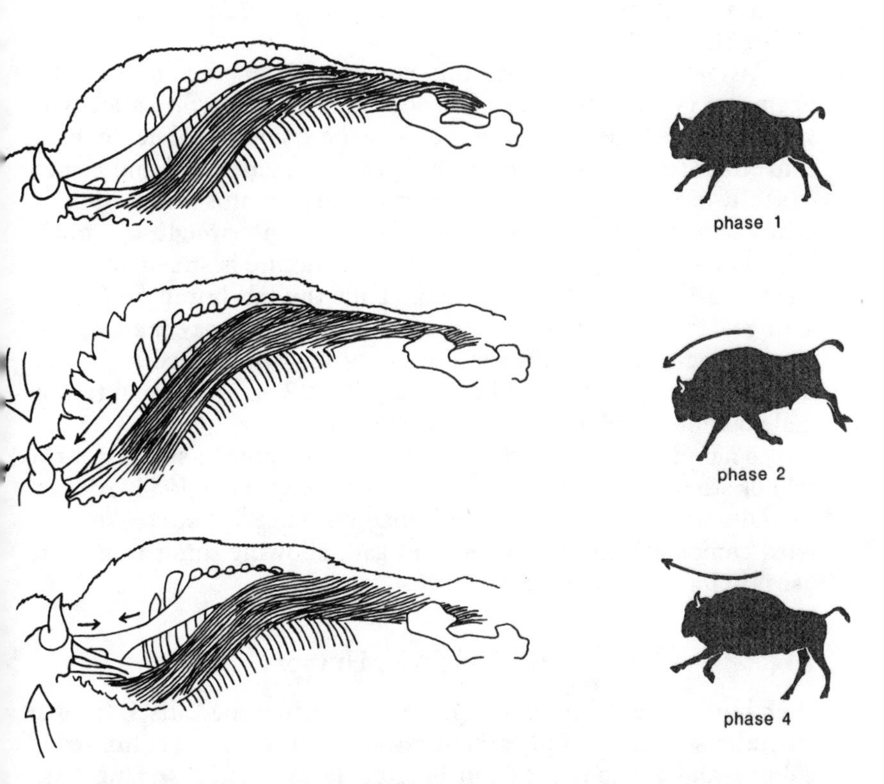

Fig. 5.5. Role of stretch ligament in gallop of European bison. The neural spines in the hump act as girders to which the nucal ligaments attach. The tall spines provide leverage as the head bobs up and down during the gallop. During phase two the head is lowered by gravity and by the pull of the muscles attached to the rearward-moving foreleg, providing a counterweight to the suspended hindlegs. The nucal ligament aids in raising the head, decreasing the expenditure of energy otherwise required.

serts onto the back, or occiput, of the skull. But this ligament may have another function.

To my knowledge, no one has mentioned the nucal ligament's importance in galloping. A galloping animal alternately lifts its head up and down (a novice horse rider must learn to move the taut rein in this rhythm). The down thrust of the galloping animal's head comes during phase 2 (fig. 5.5); it seems to act as a counterweight in the suspension of the cantilevered rear end. The tall trusses of the hump increase that mechanical advantage. The nucal ligament seems to be the main cable of energy transmission in that process. The lowering of the head requires little energy expenditure because it uses gravity. And as we have seen, raising the head is greatly assisted by the energy storage capacity of the nucal stretch ligament. The upward spring of the head aids in the vertical lift of the forelegs during phase 4 (fig. 5.5).

A familiar analogue to this energy storage process in stretch ligaments is the pogo stick, where one bounces along on a stick with springs. The spring stores the energy of the gravitational descent and converts it to a vertical ascent, making movement much more energetically efficient than that performed on an unsprung stick. A heavy window or garage door sprung by a counterweight is similar. The energy of gravity used to lower the structure is stored, allowing one to easily raise the garage door. Thus the tall hump of a bison, acting as the origin of the nucal stretch ligament, may increase galloping efficiency by allowing the heavy low-slung head to be employed as a counterweight lift during phase 2 and recovered to normal position at reduced metabolic expense.

There is an increasing awareness in locomotor studies of the role of stretch tendons in energy storage (Alexander 1984; Alexander, Dimery, and Ker 1985). These energy-storage devices make a gallop a comparatively energy efficient gait, allowing some animals to use it in long-distance travel.

The Ecology of Humps

High humps are the mark of open-plains grazing specialists. In areas where grasses grow thinly and the sward is short, forage is limited in volume and any one area can be quickly exhausted as rangeland. Likewise, quality forage is often scattered across the landscape in a mosaic—rains may come ten kilometers away. The phenological peak of plant quality also occurs at different altitudes and at different aspects of slope at different times of year. A plains-adapted social creature traveling with a herd must keep on the move. An animal adapted to fibrous forage cannot go very long without water, yet the highest quality of range is often quite distant from water and frequent trips must be made between the two. An efficient and fast gait is important on the plains.

Humps are often part of another set of traits. As animals moved away from the forest where thick vegetation afforded protection from predators, many species resorted to sociality as a substitute "cover." The presence of a hundred other animals dilutes a predator's focus on one's self. Forest ungulates are built for great speed over very short distances. Unlike bison, most forest ungulates emphasize phase 1 of the gallop; the rear is large and the front small. Forest antelope accelerate quickly to foil an attack and can disappear in dense habitat. Thus they do not require sustained speed over long distances. Ungulate biologists (Estes 1974; Geist 1974) have referred to this as the "duiker syndrome," named after small forest- or brush-adapted African antelope that have large hams and high rumps, giving the body a downward tilt toward the shoulders, quite the opposite of wildebeest.

One can actually see these morphological traits in horses bred for different purposes. The American quarter horse has been bred to run fast over short distances (a quarter of a mile or about 400 m). It is a breed favored by cattlemen who have to quickly cut-off a cow as it veers from the herd. Quarter horses are built like dragsters; they are large hamed and slope forward like duikers.

Another part of the hump syndrome is an adaptation to grazing out in the open. Diets also changed as animals moved into more open country. Grasses have most living-tissue and energy storage organs beneath the ground. Grasses can treat their seasonal, aboveground photosynthetic tissue as disposable. As such, tissue above ground does not have to be so well defended from herbivores. Woody plants, on the other hand, usually produce toxic chemicals to protect their above-ground woody parts because these parts are more costly to grow. The open sun-drenched plains can have excellent forage part of the year when grasses are abundant and high in nutrient quality and calories. But forage on the plains is more seasonal than in shade-buffered woodlands. Seasonal plenty alternates with lean, inhospitable times, when the only food available to a herbivore is standing dead grasses which offer abundant empty calories, but lack other essential nutrients. A plains-adapted animal weathers this bleak period by having the ability to use poor food (many woodland species cannot survive on such undigestible material). Most northern plains species also rely on fat reserves gained during earlier seasons of plenty. To live year-round on the open plains a species must make full use of available resources. Mobility that is both rapid and efficient is a critical part of survival.

American bison are like wildebeest in their adaptations to grasslands. American bison have large shoulder humps, and they frequently use a rocking canter. But because their body mass is bovine in scale, they have different problems in locomotion. Bison humps are unusually large, the largest of all ungulates in their size range or smaller. When one examines photos of galloping bison, the span between the fore feet in the second phase of the gallop is enormous. A high hump allows bison scapula to be quite tall; it is a feature of

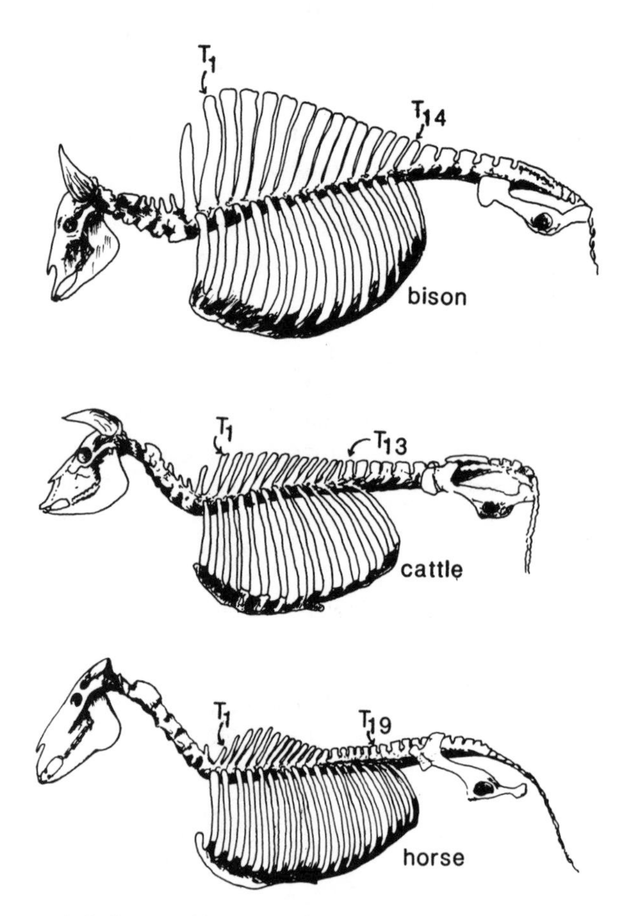

Fig. 5.6. Partial skeletons of bison, cattle, and horse. Note dramatic differences in neural spines.

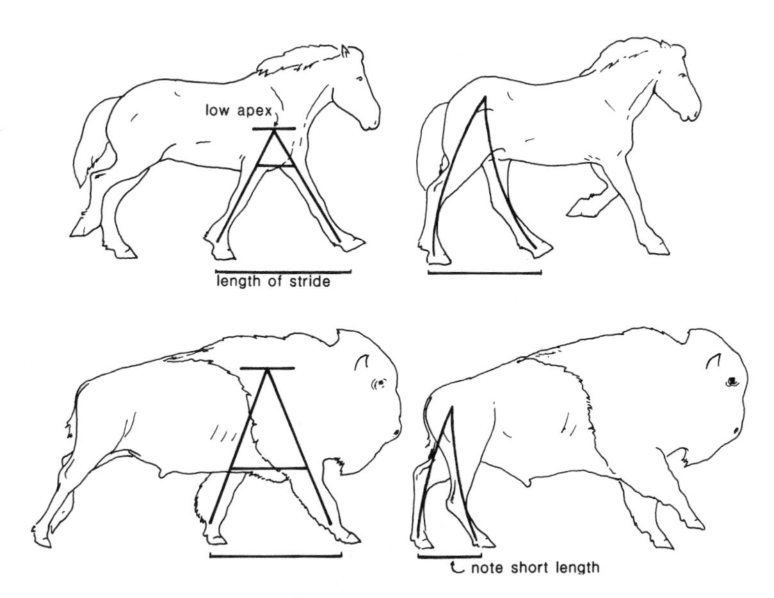

Fig. 5.7. Relative stride lengths of galloping horse and bison. Even though bison are short-legged, they travel about the same distance in phase two (distance between right and left fore feet) as horses, because their high shoulder raises the apex of the angle of foreleg rotation. In phase four (the span of hind feet), bison travel a shorter distance than horses.

bison osteology that easily separates (fig. 5.6) bison from cattle (*Bos*) and buffalo (*Bubalus, Syncerus*) or even equids. As I pointed out earlier, a tall scapula increases the height of the arc through which the forelegs can travel and hence the potential length of stride, making a bison's gallop exceptionally efficient (fig. 5.7).

It was once thought that humps related to heavy heads or big horns (Koch 1932), but there seems to be very little correlation. The open-ground grazing niche is a better predictor of who is humped. Among bovines, bison are the most well adapted to life on open grasslands, and bison have by far the tallest humps. Cattle, in comparison, are small humped; they are woodland edge or parkland creatures. Even among cattle breeds with heavy heads and immense horns, for instance, Watusi cattle and Ankol cattle of Africa, the "withers," as cattlemen call the hump, are low (Roskosz and Empel 1961); that is, the neural spines of the thoracic vertebrae are short. In America we do not think of cattle as adapted to the woodland edge, but associate them with cowboys on the open plains. But cattle on the plains must be fed hay for a big part of the year. Compared to bison, cattle are not well adapted to living wild in these habitats.

Blue Babe's Hump

My first step in reconstructing Blue Babe's hump was to assemble vertebrae from three American bison, in a long band down three pine boards, using one-inch (25mm) strips of clay about four inches (100 mm) wide. The ventral part of the centrum was pushed into the clay, with neural spines pointing dorsally (see fig. 5.8). Of course the board provided a base quite unlike the complex contour of a living bison, and the dorsal outline of neural spines it produced was not the same as that of a living bison. Still, this exercise allowed us to take precise measurements because all vertebrae were aligned on a common baseline (Poplin 1984). Other American and European bison, as discussed later, were also compared using this same baseline.

Vertebral columns of European bison have been thoroughly studied by Polish mammalogists, so I compared illustrations and measurements from their work with fossil vertebrae from the Alaskan steppe bison. In addition, Poplin (1984) made a careful study of the dorsal outline of living bison species as well as the extinct European Pleistocene *Bison priscus.* Although these published studies were excellent treatments of neural spine shapes and sizes as well as vertebral musculature, I chose to examine three skeletons of American bison in order to make direct comparisons with fossil Alaskan specimens and to get a better feel for potential differences. Because Blue Babe was a mature bull, only bones from mature bulls were used. One was a large plains bison (*Bison bison bison*), National Mu-

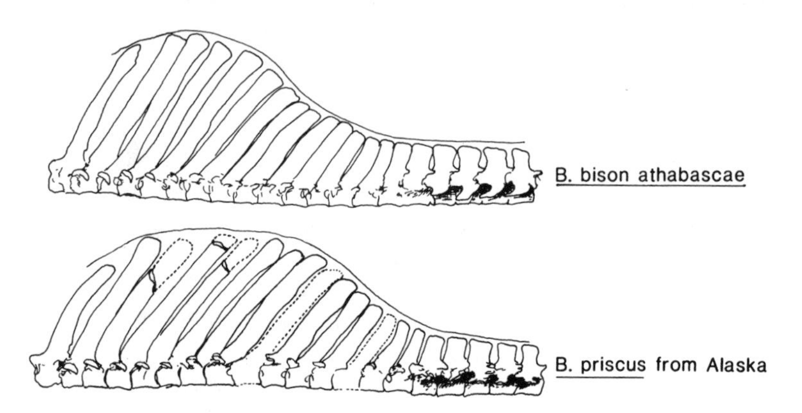

Fig. 5.8. Thoracic and lumbar vertebrae. Vertebrae are assembled on a flat base to illustrate differences in hump contour. The highest point of the steppe bison hump is far posterior to that of plains bison. Also, the neural spines of plains bison fall smoothly to the rear; in steppe bison this contour is more abrupt.

Table 5.1 Bison Neural Spine Length

	B. priscus	*B. b. bison*[a]	*B. b. athabascae*[b]	*B. b. athabascae*[c]
Cervical no. 7	15.5	14.75	15.5	13.0
Thoracic no. 1	16.5	16.0	18.5	18.75
Thoracic no. 2		15.75	18.5	18.75
Thoracic no. 3	17.0	14.75	18.0	17.25
Thoracic no. 4		13.0	18.0	16.0
Thoracic no. 5	18.75	11.25	18.0	14.25
Thoracic no. 6	17.5	10.0	16.25	12.0
Thoracic no. 7	16.25	9.0	13.0	10.5
Thoracic no. 8		8.0	11.50	9.5
Thoracic no. 9	12.0	7.0	9.25	8.25
Thoracic no. 10	11.0	6.25	7.50	7.25
Thoracic no. 11	9.25	5.5	7.25	6.0
Thoracic no. 12	6.25	4.5	6.25	5.25
Thoracic no. 13	4.75	4.0	5.0	4.0
Thoracic no. 14	4.25	3.75	4.5	3.75

Note: All measurements are taken from the posterio-dorsal rim of the neural canal in a straight line to the most distal part of the neural spine, to the nearest 0.25 cm.
a. National Museum of Canada No. 45632 mature male.
b. National Museum of Canada Wood Buffalo Park mature male.
c. National Museum of Canada No. 32628 mature male.

seum of Canada No. 45632, from Montana. The second was a wood bison (*B. b. athabascae*), National Museum of Canada No. 32628, from an introduced bison population in Elk Island Park, Alberta. The third, also a wood bison subspecies, National Museum of Canada was from Wood Buffalo National Park, taken before plains bison were introduced to that area (table 5.1).

Several hundred vertebrae of Pleistocene Alaskan bison at the University of Alaska Museum were sorted for specimens with complete neural spines. These were ordered in categories of cervical, anterior thoracic, posterior thoracic, and lumbar. The individual shapes of these bison vertebrae seemed similar, but they varied greatly in size and robustness. Fortunately the collection contained an assortment of vertebrae from a single animal, University of Alaska Museum No. 1178, collected in 1956 at the Utica Mine on the Imnachuck River. We know this was an old bull, judging from its size and also from its arthritic asymmetry and the pearling around the centra. As an animal grows older the vertebrae exhibit arthritic osteoporosis and roughness at points of muscle origin and insertion and in edges of the vertebral centra. The Utica Mine bison bones were thus the right sex and age to allow for a good comparison with the vertebrae of the three mature bulls from Canada. Using vertebrae from a single steppe bison avoided the problem of assembling vertebrae from individuals of slightly different age and size.

The vertebrae of this steppe bison from the Utica Mine could be ordered in sequence by using the recent bison vertebrae as a comparison. In addition, we were able to cross-check our identifications and vertebral ordering because the articulations showed many unique patterns and asymmetry which had to match exactly. No two articulation facets of the anterior and posterior zygopophyses were alike in this regard, so our identifications were certain.

Noting specific features of these different vertebrae help us reconstruct Blue Babe's hump. Cervical vertebrae of steppe bison and the two subspecies of modern American bison differed mostly in the steppe bisons' greater robustness of muscle and ligament attachments, attributable, I think, to their massive horns which increased the weight of the head. The last cervical vertebra, C7, however, was quite different in shape as well as size. This cervical vertebra is incorporated in the thoracic series and can be thought of as thoracic in shape and function. It is immediately identifiable in any jumble of vertebrae because it has no demifacets for rib attachments, yet it has a very long neural spine, like the thoracic vertebrae.

The shape of steppe bison vertebra C7 is very different from that of a living American plains bison. In steppe bison the C7 is tightly arched posteriorly, that is, it has a convex surface all along its anterior face. The C7 neural spines of plains bison rise from the base in the same posterior bend but reverse themselves halfway in an anterior bend (concave on the anterior face). Also the anterior-posterior width of the neural spine base is greater in steppe bison than in plains bison. From this wide base the spine tapers gradually to a narrow distal end. American plains bisons' C7 spine keeps much of its anterior-posterior width instead of narrowing to a point. Unfortunately, few terminal portions of C7 are present among fossils, as this part is quite edible and is usually scavenged.

In contrast, the C7 neural spine of the Canadian wood bison more closely resembles that of the steppe bison, but is not so wide at the base. Wood bison are in between plains bison and steppe bison, but more akin to the latter. One specimen of wood bison, NMC 32628, had a longer C7 spine (390 mm) than plains bison (355 mm); the other wood bison, NMC 45632, had a much smaller neural spine (330 mm). The C7 of European bison measured by Roskosz (1962) seems smaller than that of both American bison and steppe bison. Roskosz pictured only the C7 of female bison. These were straight in profile and tapered in width to the distal tip, like those of steppe bison, but were not arched posteriorly.

The first thoracic vertebrae, the T1, evidences another difference. Two subspecies of living American bison have a more smoothly curved concave anterior border to their neural spines, whereas the neural spine anterior border of steppe bison is convex on the proximal half and slightly concave only on the distal half. The T1 neural spine of wood bison is longer than any of the other neural spines. The wood bison neural spine lengths of the T1 were 470 mm and 485 mm. The T1 length of the American plains bison was 400 mm and the steppe bison's was 415 mm. These measurements and those that follow are from the dorsal-anterior point on the neural arch to the distal end of the neural spine, along the anterior face.

Thoracic neural spines were more convex on their anterior borders among steppe bison than among either modern American bison subspecies. Both wood bison and plains bison spines almost always started from a concave sweep as they rose from their bases and arched more convexly on their distal half. Comparing neural spine lengths within individuals, both plains bison and wood bison had longer T1 and T2 spines than any other vertebrae. They differed subtly in pattern in that wood bison tended to have longer absolute length than plains bison. I had only one mature male plains bison for comparison, so perhaps some individuals reach lengths comparable to wood bison, but in general a higher forward portion of the hump is characteristic of wood bison (Geist and Karsten 1977).

A striking difference between Alaskan steppe bison and the above specimens is that in steppe bison, the longest neural spine is T5. As can be seen in figure 5.8, the thoracic hump of steppe bison is located farther back than in wood or plains bison. For example, in contrasting the two, the wood bison had longer neural spines from T1 through about T4; the steppe bison had larger spines between T5 and T12.

A comparison of neural spines of vertebrae T4, T5, and T6 (fig. 5.9) highlights thoracic hump placement. Those of the plains bison (T4, T5, and T6) are short relative to other vertebrae. These same neural spines are longer in the wood bison. Among steppe bison these neutral spines are longer than in either living American bison species, and they are set differently, being bent posteriorly in a more acute angle. A similar pattern can be seen in T8, T9, and T10 (fig. 5.10). Neural spines of steppe bison are also broader anterior-posteriorly than the others. Posterior to T10, neural spines among all of the bison are rather similar, including those of the lumbar vertebrae. Among both female and male European bison pictured by Roskosz (1962), the thoracic vertebrae had concave anterior surfaces to their neural spines, more like those of American plains bison.

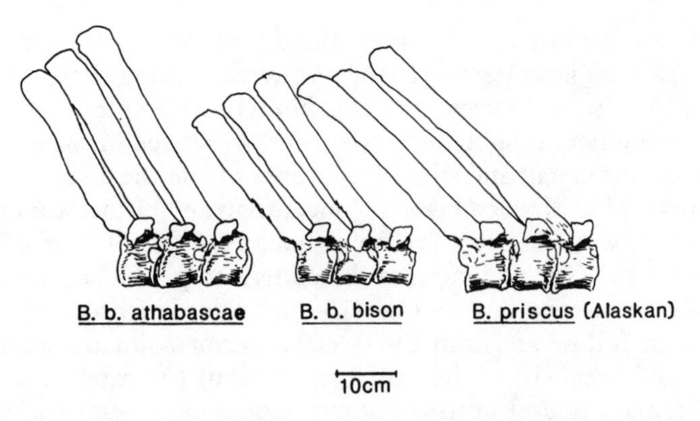

Fig. 5.9. Lengths and angles of the fourth, fifth, and sixth thoracic vertebrae. These are longest in fossil steppe bison, and they emerge from the centrum at a lower angle, bending more posteriorly.

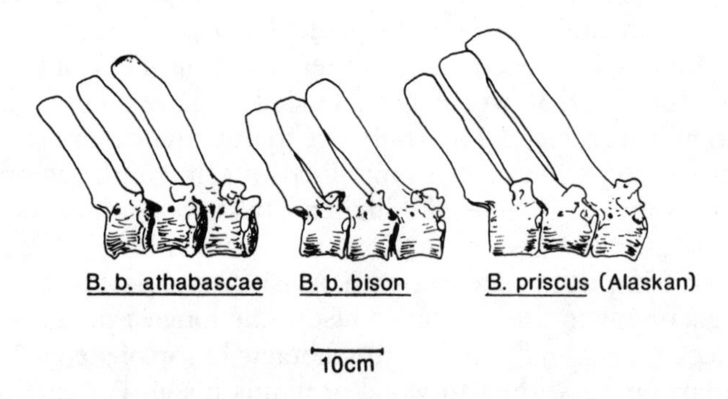

Fig. 5.10. The neural spines of thoracic vertebrae eight, nine, and ten. These are located at the posterior "stop" or drop-off of the hump of steppe bison, *B. priscus*, and, as a consequence, differ from their homologues in wood bison, *B. b. athabascae*, and plains bison, *B. b. bison*. Those of the steppe bison are broader (anterior-posteriorly) and longer.

All this means that the hump shape produced by thoracic vertebrae of Pleistocene Alaskan specimens differed considerably from that of living plains and wood bison. Wood bison have a high hump that is straighter in profile from the saddle of the back (lumbar area) forward to the neck. In wood bison, the forward edge of the hump drops abruptly, almost vertically. In plains bison, the hump is high in front but drops less abruptly to the neck. Overall the plains bison has a slightly concave profile to the highest point of the hump.

As reconstructed from Alaskan fossils, steppe bison had a more gently ascending hump from the front which reached a high point well posterior to that of both American bison subspecies. Posterior to the scapula, the hump arched upward then dropped rather abruptly in a convex arch. The steppe bison's hump was more similar to that of the living European bison, *B. bonasus,* but the former's hump was much more exaggerated in height and contour.

Although he did not have access to Alaskan fossil material, Poplin (1984) examined bones of American *(B. bison)* and European *(B. bonasus)* bison, comparing the dorsal contour of the neural spines. Poplin did not distinguish the two subspecies of American bison in his work. However, I wanted to use the subspecies' differences because I felt the geographic proximity of wood bison to Alaskan Pleistocene bison might be significant. Poplin compared many individual bison, both females and males; a diagrammatic summary of his findings is presented in figure 5.11. Poplin was primarily concerned with testing the idea that European cave artists drew bison with a dorsal contour different from that of now living bison, not

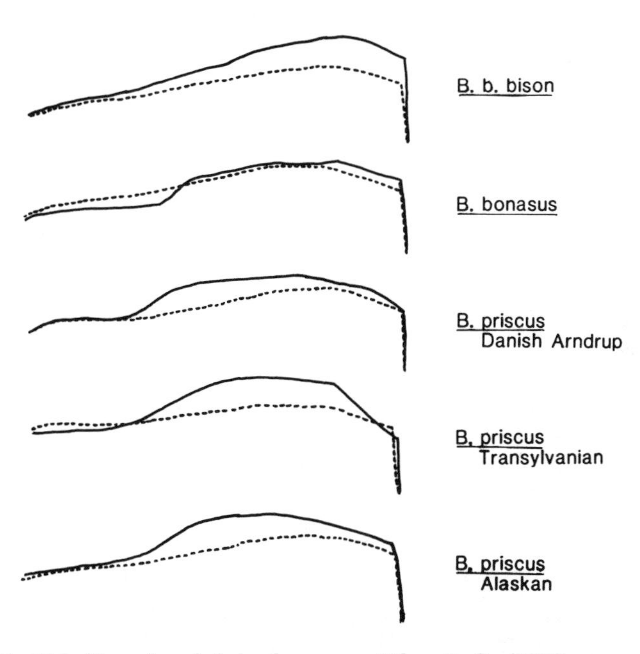

Fig. 5.11. Male bison dorsal skeletal contours. When Poplin (1984) compared dorsal contours of vertebrae from different bison groups, he was able to show there were differences in hump shape. Humps of living bison and extinct European steppe bison (*B . priscus*) have diagnostic shapes. The dashed base-line contour is that of an American female bison; solid-line comparisons are different males. The lower three are fossil steppe bison. (After Poplin 1984)

because of artistic style but because of a biological difference in their model. Fortunately, he was able to find two almost complete specimens of European steppe bison, *B. priscus,* the same species that, on the basis of other evidence, I have argued lived in Alaska, the species to which Blue Babe belonged.

In length of neural spines, Poplin found that among American bison the first thoracic, T1, is normally the longest. Among European bison it is T2, and for European Pleistocene steppe bison it is T3. Remember that Alaskan Pleistocene steppe bison had the longest neural spine of T5, even farther posterior than European steppe bison. More important, Poplin was able to show that, unlike living bison, Pleistocene steppe bison had a dorsal hump placed posterior to the shoulder and, furthermore, that the hump's posterior edge was strikingly convex. As I have shown, this pattern is identical to that of steppe bison from Alaska.

Functional Significance of Steppe Bison Humps

As our comparison showed, the biggest difference in hump shape is between the humps of the steppe bison and the American plains bison. The crest of the American plains bison hump is located in front of the scapula, while the steppe bison's crest is behind the scapula; a possible difference in function may be the humps' mechanical relationship to the nucal ligament. Let me propose a new theory.

The low-slung character of the American bison's head is, I believe, derived from its specialized adaptation to the thin sward of the American shortgrass plains. Not only is the grass normally low in growth form after being grazed, but it readily regrows into a very short "lawn" pattern in an especially thin sward. To get sufficient quantities of grass, American bison must spend many hours grazing with their heads almost touching the ground. There is a dearth of midheight forage.

As a direct product of this adaptation, the entire anatomy of the American plains bison has located the normal position of the head lower than in other bovids (like a lawn mower set low to the ground). This changes the optimum mechanical arrangement of the nucal ligament for head raising and lowering to a more forward emphasis on the hump, its point of origin (fig. 5.12). Additionally, the lower head position changes the optimum mechanical angle for the neck ligament to a more forward or anterior location. Thus the hump of the American plans bison may be part of a network of adaptations to a lowered-head position as a lawn grazer, quite unlike the environ-

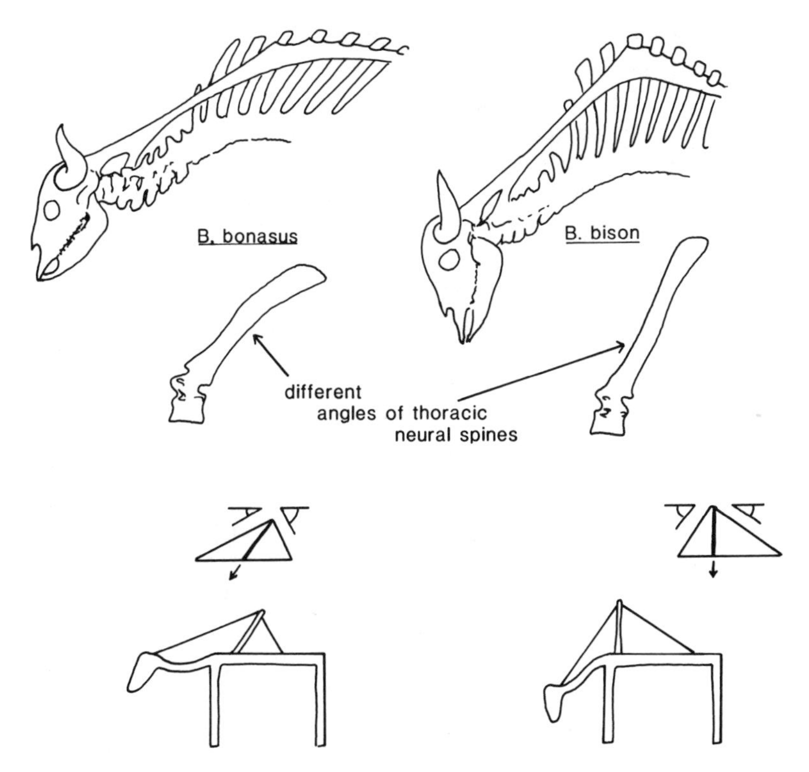

Fig. 5.12. Arrangement of nucal ligaments in living European (*B. bonasus*) and American bison (*B. bison*). Normal head carriage is lower in American bison, necessitating a more vertical alignment of the neural spines and a more forward location of the longest neural spines.

mental adaptations of other large bovines. If that is correct, their hump would be an evolutionarily derived trait from a more steppe bison–like hump.

Steppe bison seem to have held their heads higher, at rest, than do American plains bison—more like fossil and recent European bison. This can be seen in Paleolithic art. Steppe bison probably did not occur in vast enough herds to promote a lawn system of grazophilic grass species, as do wildebeest and plains bison. Rather, modest densities on a relatively cold-arid grassland would have necessitated considerable mobility. The larger shoulder hump apparently enhanced the ability of these large creatures to move rapidly across the landscape. Not only did Pleistocene bison have to move great distances to find fresh ranges and water, but they also had to run away from predators. This necessitates seeing predators before they come dangerously close.

On the generally open Mammoth Steppe landscape, approaching lions could be seen at considerable distances by a bison with head raised to a position of modest height. Among small herds of steppe bison, predation alert would have been a selective pressure countering a permanently lowered-head arrangement. Given some advantage to a higher resting head position, the location of the steppe bison's hump crest behind the scapula may be the optimum pattern to maintain both speed and galloping efficiency of this large and powerful bison.

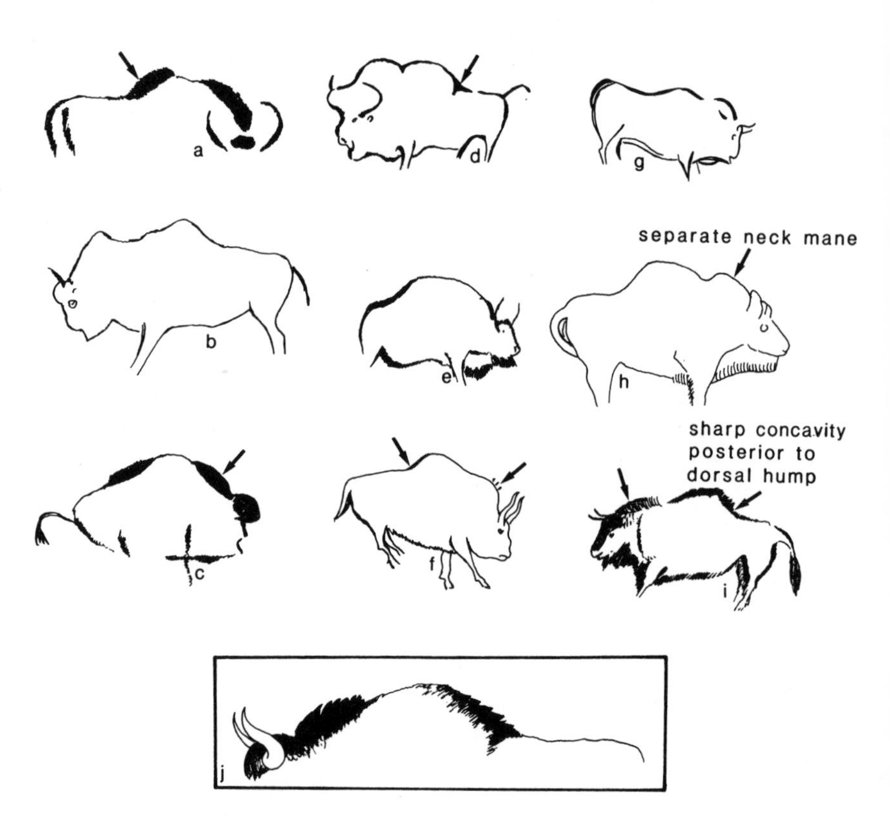

Fig. 5.13. The dorsal contour of steppe bison from Paleolithic art. Numerous drawings of steppe bison in Paleolithic art show an abrupt "stop" just posterior to the hump which seems to have been a darker color as well. A distinct neck mane is shown which is also dark. This mane is not present in American plains bison (*B. bison*), but there are slight traces of it in European bison (*B. bonasus*). These bison are drawn from (*a*) Marcenac, (*b*) Etchberriko-Karbia, (*c*) Santimimine, (*d*) La Pasiega, (*e*) Le Portel, (*f*) Pindel, (*g*) El Castillo, (*h*) Le Tuc d'Audoubert, and (*i*) Niaux.

Despite the absence of Blue Babe's hump, we have successfully reconstructed it indirectly. Given that Blue Babe's hump is similar in structure to the hump of Pleistocene European bison, it is likely that the pelage of Blue Babe's dorsal line was also more similar to European steppe bison than to any other bison, European Paleolithic art provides many drawings of steppe bison (fig. 5.13). In these drawings the stop in the posterior thorax is emphasized by a contrasting, dark-colored hair patch; the dorsal neck hackles also form a hump and are of a contrasting dark color. These appear as two humps.

Working on the ethology of extinct species may strike most ethologists as a strange enterprise, but most ethological studies rely heavily on the literature, that is, on others' earlier observations. In many ways, Paleolithic works of art are very old references, just recently uncovered. While they do not always tell us exactly what we want to know, they provide a wealth of information, some of it obviously incorrect, as indeed in the case of older scientific literature. But we can see by their work that Paleolithic artists were empirically oriented, drawing mainly from observation rather than copying some stylized icon form.

6
STEPPE BISON ETHOLOGY

Social Organs and the Art of Spying on Ice Age Behavior

Many aspects of an animal's appearance are "social" in function; the wattles, stripes, manes, horns, and color patterns shown in an illustrated field guide, for example, are usually social organs. Social organs distinguish the sex, age, and social stature of the bearer; they are flaunted in displays of aggression, submission, sexual attraction, or social bonding. Of course, much of a species' social behavior is shaped by environmental adaptations that also affect anatomy, and thus appearance is an interacting composite of social garb and environmental adaptations. Too brilliant a coat can be dangerous, so while an individual concerns himself with winning stature and females he must also worry about predation. Likewise, features that make an individual stronger among peers (large horns, muscles, and heavy bones) may also make him slow and less agile in defending against predators. Social anatomy and survival have to strike an evolutionary compromise, but the balance of this compromise should theoretically vary with environment.

Estes (1974) outlined the interaction of behavior and environment among African bovids in a gradient from woodland edge to savanna and fully exposed plains. I retailored these (fig. 6.1) to describe evolutionary trends in northern bovines (Guthrie 1980). I think that Pleistocene bison living in an open steppe experienced increases in agonistic behavior, nonlethal fighting apparatuses and techniques, group size, class hierarchies, and gaudy social display organs. In addition to the above changes, homeland fidelity probably declined and was replaced by more migratory or nomadic tendencies. Feeding in a steppe environment became more unselective, and wintertime diets included a high percentage of fiber.

Animals living on the Mammoth Steppe experienced a highly seasonal environment: the quality, abundance, and digestibility of

Fig. 6.1. Woodland versus steppe adaptations. Early bovines originally were adapted to wooded habitats. The expansion of open grasslands created a potentially rich niche, but the move to more seasonally variable grasslands involved new demands and produced different adaptations. Behavioral and morphological traits characteristic of woodland and plains bovines are summarized in this illustration.

their food changed markedly throughout the year. Monocotyledons and nonwoody forbs were relatively rich and abundant during the short summer, but each fall, as growth stopped, these plants moved most nutrients to their roots, leaving poorer quality dead tissue above ground. Pleistocene bison depending on this low-nutrient winter forage required a large rumen to incubate quantities of the high-fiber, poor-quality food. Severe northern winters and attendant mortality would have kept bovine populations well below summer carrying capacity, thus reducing competition for resources during the growth season. Abundance and quality of food in the growing season gave each animal the resources to attain its full genetic potential for horn and body size. I have proposed (Guthrie 1980, 1982, 1984a, 1984b) that such strong seasonality shifts selection optima and favors a larger body size.

The larger group size characteristic of plains ungulates (Estes 1974) generally precludes a linear social hierarchy. Linear hierarchies are common in feral cattle (Schloeth 1961; Frazer 1968) as long as the cattle are in small groups. Hierarchies in large dairy herds of fifty or more often include such subgroupings as "triads" (Brantas 1968).

Both American bison (McHugh 1958; Lott 1974) and European bison (Jaczewski 1958) have rank hierarchies, but the former tend toward a class ranking system, particularly when studied in more natural situations in large parks (Shult 1972; Petersburg 1973). Holocene bison of Europe inhabited parkland openings in a forested landscape and probably experienced selection pressures for homogeneity of color and reduced display organs. The reverse is true of American plains bison on expanding Holocene grasslands. These grasslands had no counterpart in Holocene Eurasia. American plains bison developed extreme display paraphernalia, weapon morphology (short stubby upward-oriented horns), and fighting styles (head clashing). Along with wildebeest, American bison are at the farthest end of Estes' (1974) spectrum of plains-adapted bovines that live in large herds (fig. 6.1).

Migratory species living in highly seasonal plains environments sometimes do not form reproductive territories; instead, males display *themselves,* and outside the brief reproductive period, mature bulls form separate herds. Sexual dimorphism is most exaggerated among these animals. During the breeding season, males join the female bands where they seek out and "tend" females in estrus, fending off an array of contenders, then move on to the next estrus female. Intense selection pressures favor males sporting the most extreme display paraphernalia. These social organs are used for only a brief period, but without them bulls are reproductively neutered (Petersburg 1973).

In other plains ungulate species (e.g., *Oryx*), where both sexes live most or all of the year in mixed herds, there is little marked sexual dimorphism (Estes 1974). Presumably, this sexual similarity is due to both sexes experiencing similar resource availability (fig. 6.2).

In some situations (e.g., wildebeest and plains bison) large herds feed so intensively in one area that they keep grasses clipped short like a lawn. Grasses respond to this "mowing" by ungulates and reshoot laterally. The resulting lawn is highly productive and provides quality grazing, but with keen competition from other herd members. The main advantage to herding has to do with predation, not improved forage, although grass "lawns" can be maintained in highly productive condition by large herds and hence benefit individual animals. The antipredator advantages of herd life are usually accompanied by a decrease in forage quality. Ungulates select their food from an array of plants that are not equally desirable. The "cafeteria" selections in any one area differ in quality and digestibility,

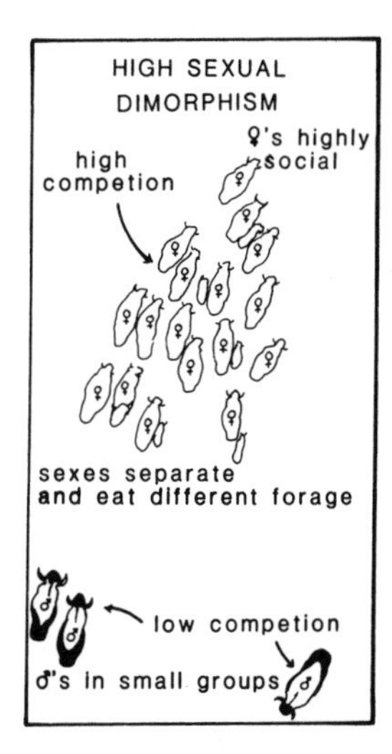

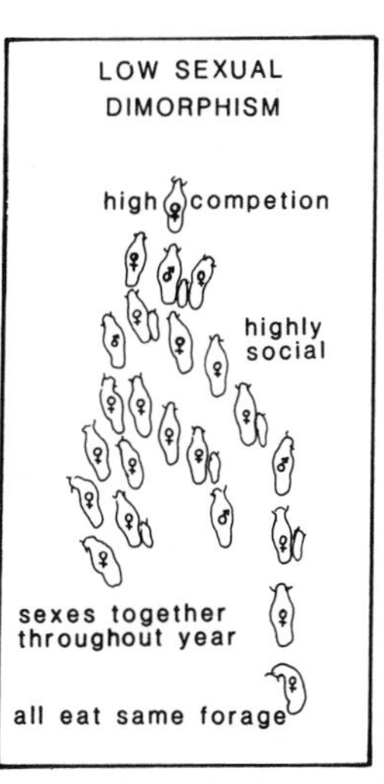

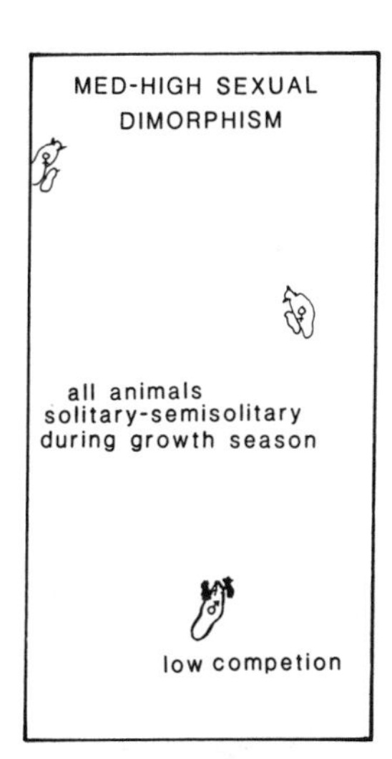

Fig. 6.2. Ethology and sexual dimorphism. Competition for resources is the accepted explanation for radical sexual dimorphism in certain ungulate species. In a polygamous social structure, there is a selective premium for large, powerful bulls; dietary resources become the main limit on their body size and social paraphernalia. If males and females live more solitary lives, as do moose, sexual dimorphism is modest. Sexual dimorphism is most pronounced when females live in herds and males live a solitary life or exist in small herds, as do bison.

and less choice vegetation normally dominates. Single animals can afford to take the "cream," but members of large herds must necessarily duplicate others' routes and feed on second-choice plants or plant parts.

American plains bison are flexible in their response to opportunities offered by different environments. Small bison herds are usually nonmigratory, with marked fidelity to their home range. On such ranges the sexes live apart for most of the year. On the Great Plains, where bison herds built to enormous numbers, the entire herd was migratory for most of the year and sexes remained in proximity if not in the same subgroups. This phenomenon can be seen today in caribou. Small herds of caribou, numbering in the hundreds, are local and move little, whereas the long-distance migratory herds generally number in the tens of thousands.

Thus we can look at sexual differences in social organs to learn something about the ecology and social behavior of extinct steppe bison. We know that sexual dimorphism is greatest among herd ungulates when adult males spend most of the growing season apart from the herd, foraging in small groups or solitarily. Today's American male bison behave in this way. As bulls do not have to share resources with a large herd, they can select better-quality food, thus sustaining large body size.

Sexual dimorphism in body size of *B. priscus* was great, greater in fact than that of any other bison species (McDonald 1981). While the mature bulls had a gigantic skull with long robust horns, the females were very gracile horned, with a skull more like that of cattle. Also steppe bison cows did not have heavy protruding eye orbits like males. The extreme sexual dimorphism of the steppe bison (fig. 6.3) suggests that (1) bulls remained separate from cows throughout the growth season, at least, and (2) herds of steppe bison may not have been so large as bison herds on the Great Plains. The Beringian steppe landscape probably favored moderate-sized groups of bison and a class hierarchy in an "open" social system oriented toward frequent display to many individually unrecognized herd members.

Not only are male-to-female differences important in reconstructing social behavior, but so too are differences between adults and young. American plains bison differ from living European bison with regard to color of the young. The young of American bison (which normally live in large herds) are very light, contrasting with darker adults, while the young of European bison (a species that lives in small groups) are similar in color to their mothers: a medium monotone brown. This pattern conforms to the general rule that highly social species, living in open environments, generally have contrastingly colored young. This is true, for example, of baboons, wildebeest, reindeer, and others. Theoretically, the contrasting color gives the young special social distinction, protecting them from the full brunt of adult aggression. In large groups, individuals are often not known personally and must wear badges of rank, age, and sex.

Paleolithic art is uninformative on the color of steppe bison calves (fig. 6.4), and we have, as yet, no mummified ones. I would expect, based on other plains-adapted features we have been able to reconstruct, that young steppe bison were light colored, in contrast to the dark adults.

The size and shape of Blue Babe's horns can also tell us about steppe bison behavior. Clutton-Brock and Harvey (1976) have proposed an association between larger horns (fighting and display paraphernalia) and environments where bulls can accumulate more females. Since extant bison have much smaller horns than steppe bison, their reproductive behaviors are probably different as well. I have used morphological features of fossil Beringian steppe bison to reconstruct their fighting behavior (Guthrie 1980): (1) Wound damage on the frontals indicates the animals received sharp forceful blows. (2) Moderately developed shock-absorbant front sinuses also indicate the use of frontal concussion blows. (3) The complex oval pattern on the horns of adult bulls makes it experimentally difficult to articulate the horns simply and smoothly with those of an oppo-

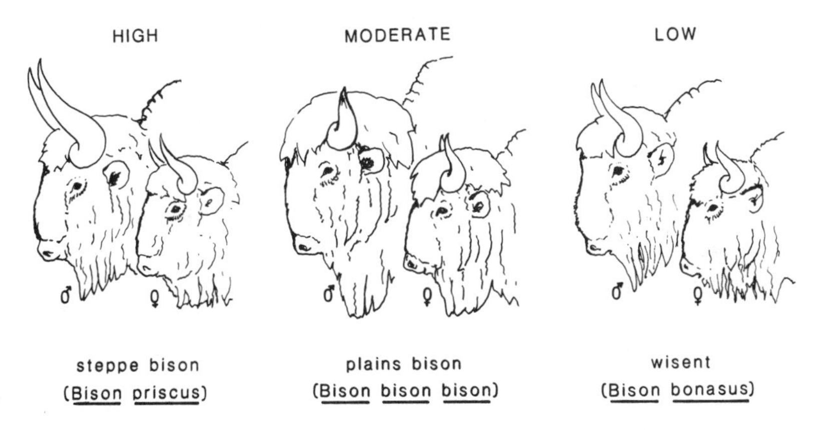

Fig. 6.3. Sexual dimorphism. Size and shape differences between sexes of plains ungulates are associated with the degree of separation of males and females throughout the year. Sexual dimorphism in skull size in steppe bison was extreme, suggesting that male and female steppe bison lived quite separately.

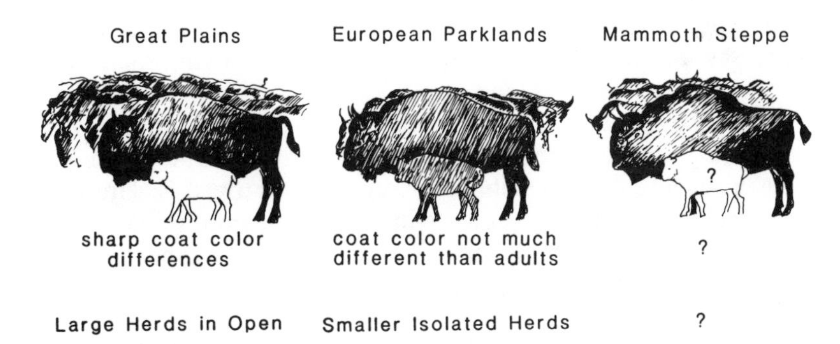

Fig. 6.4. Calf-adult coat color differences. The young of ungulates living in large herds on open plains are usually distinctly colored; for example, plains bison calves are light reddish in contrast to the much darker adults. European bison in less open situations have young that are almost the same shade as adults. The color of young steppe bison is unknown.

nent. (4) The rarity of broken horn tips in older males, compared with this occurrence in living bison, is striking. Most horn tips of older Beringian bison bulls are gracile and sharp. (5) Antitorsion fluting or louvering between horn sheath and core suggests substantial torsion in horn engagement during fighting. And (6) construction of thoracic suspension of the head points to an ability to exert great (lift) force upward.

In early bison the horns were deflected backward or outward; eye orbits were telescoped, and the frontal sinuses were expanded. As bison evolved, these trends continued to increase in an irregular fashion. The element that links these changes in skull and horn morphology and makes them intelligible is a shift in fighting—from a more cattlelike, pushing of foreheads to the sharp impact of forehead clashing (fig. 6.5).

Other ungulates have shifted to a clashing form of fighting; using the horn base as the contact surface. Sheep (*Ovis*) and goats (*Capra*) catch the leading edge of an opponent's horns in the V of the inner horn crotch. In African buffalo (*Syncerus*) and musk-oxen (*Ovibos*) a flattened horn bos evolved as a catching or butting plate. Unlike the above animals, bison use the frontal bone as a clashing platform.

Cattle engage their frontals and then push one another (Haféz 1975), while horns are interlocked. In the bison line this pushing or punching must have graded into violent butting and later into clashing. The horn cores, covered by a relatively thin horny sheath ending in a sharp tip, were adapted for "hooking" in the bovines. They were thus relatively fragile and susceptible to damage. The shift toward a more violent butting-clashing confrontation was apparently accompanied by a backward movement of the horns, moving them out of the way of the clashing plane.

Associated with the above changes, the frontal sinus in bison expanded to a large-domed pneumatic cavity separated from the brain encasement by an elaborate strut complex (fig. 6.6). This cavity extended into both horn cores which are also hollow. The double wall separated by struts seems to have served as a bumper—a device to disperse and dampen the opponent's blow away from the brain tissue (Schaffer and Reed 1972).

In an earlier study (Guthrie 1980), I concluded that steppe bison displayed more and fought less violently than living plains bison and that horn size was a significant aspect of that display. Unlike the horns of living bison, those of steppe bison continued to grow longer even into old age, more like those of wild mountain sheep. Steppe

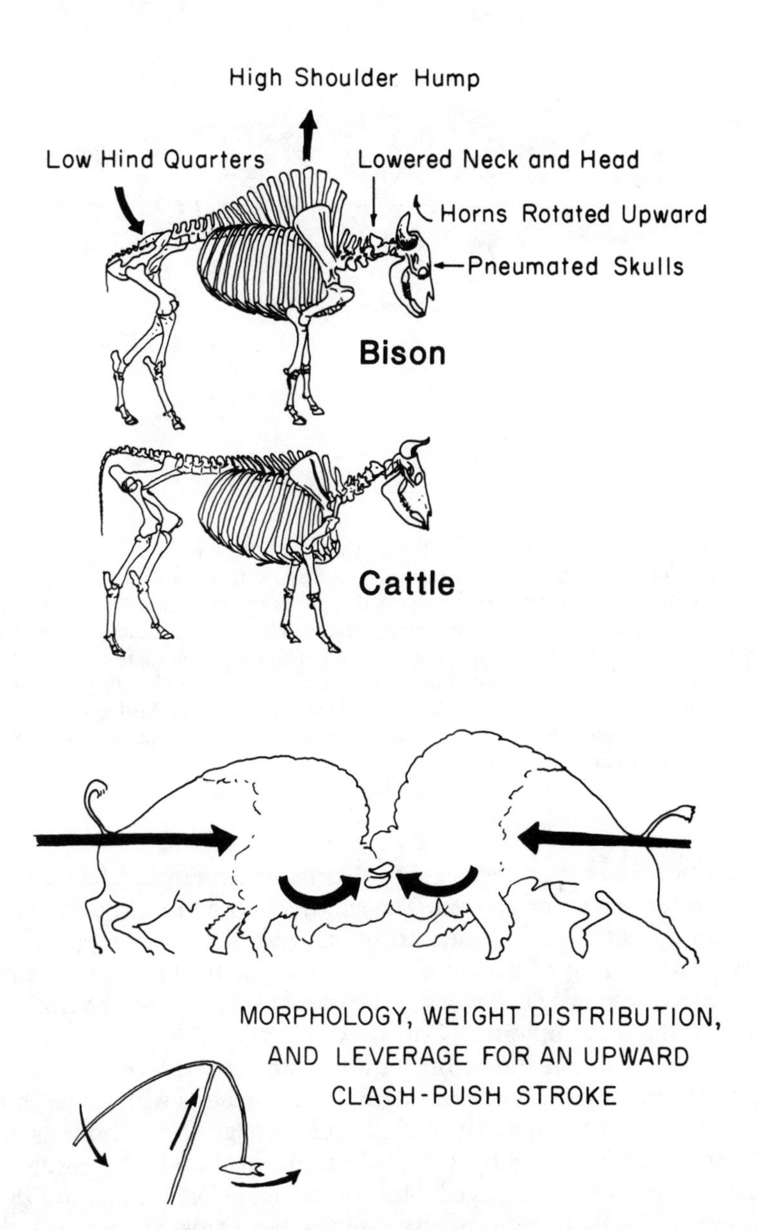

Fig. 6.5. Fighting style and anatomy. Bison and cattle are at opposite ends of bovine environmental adaptations and morphology. They fight in quite different ways: cattle hook and push; bison clash and push upward, lowering their hindquarters as a counterweight and using the shoulders as a fulcrum.

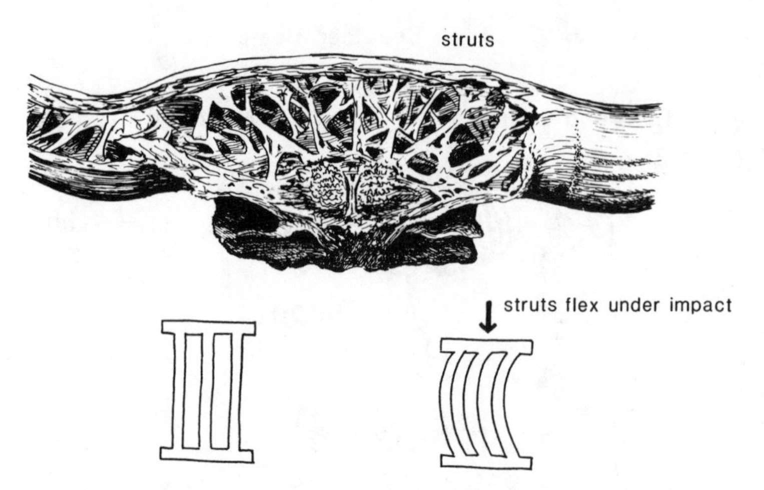

Fig. 6.6. Cranial shock absorbers. Bison fight with their heads, clashing frontals against frontals, in violent concussions which create a dangerous shock to the brain. Bison skulls have a structure that reduces this shock by smoothing the transmission wave and by carrying the blow away from the brain. An intricate network of struts separates the inner and outer walls of the skull. Though appearing to be rigid, these flex on impact, absorbing some of the peak of the shock wave. The above drawing is taken from a naturally broken fossil steppe bison skull from the Fairbanks area. The brain is small, lying behind the convoluted olfactory "chestnut-appearing" structures in the lower medial portion of the skull.

bison horns are oriented forward. Horns of American plains bison are shorter and point upward. The smaller hair bonnet and the more forward-oriented long sharp horns suggest that the steppe bison "must have engaged from a close distance, with the horns slightly occluded diagonally but most of the pressure carried on the frontals, each trying to lift upward" (Guthrie 1980: 60).

Soft tissue on the bison mummy supports information I collected from skeletons of other Beringian bison bulls which suggested that they fought frequently. Ubiquitous damage on the frontals and the antitorsion fluting (fig. 6.7) of the horn sheaths and cores suggest that Beringian bison engaged with considerable force (Guthrie 1980). However, the flattened-oval shape of the horn cross section and the frequency of sharp horn tips indicate a fighting mode unlike American plains bison.

Modern plains bison bulls run full speed toward an opponent, clashing heads, with contact primarily on the parietals. The horns of American plains bison are blunted and broomed, and the large

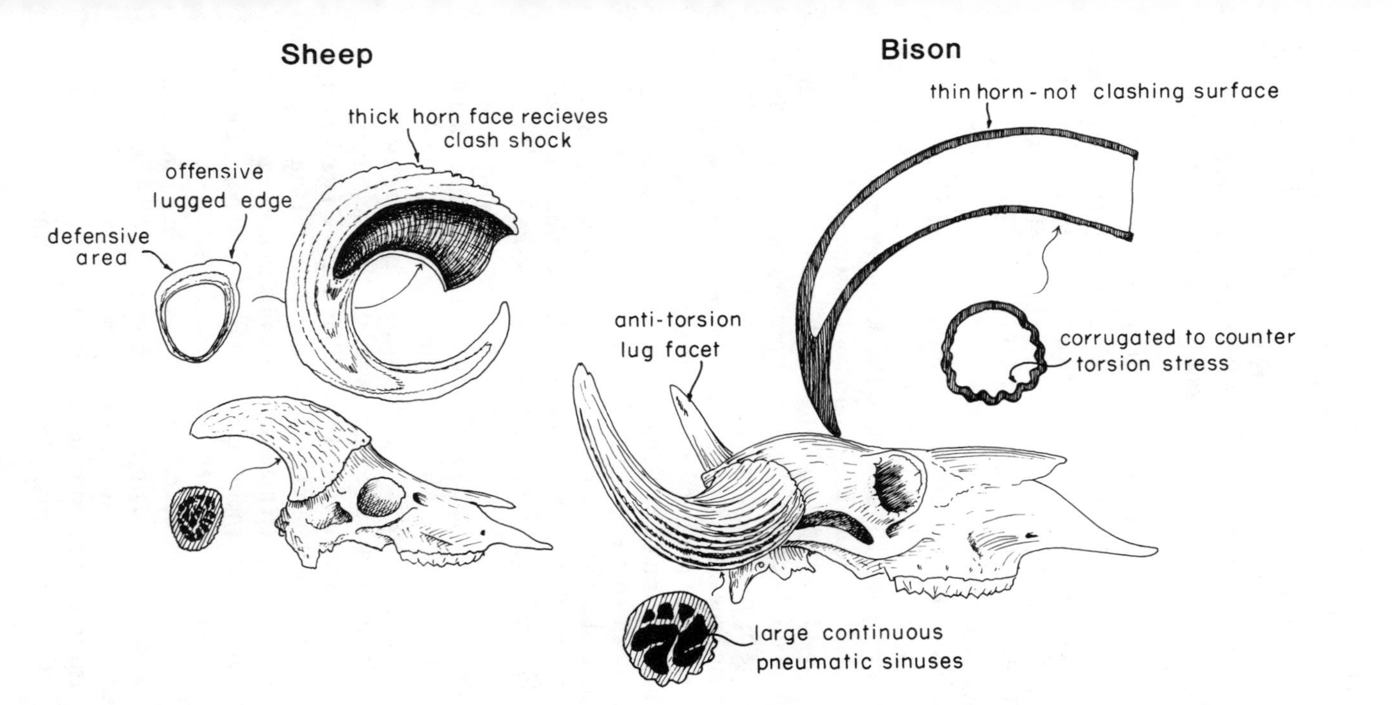

Fig. 6.7. Horns used in pushing versus clashing. The outer horns of steppe bison were supported by long, bony inner cores with deep flutes. This fluting helped bind the horn, locking it against twisting free during combat, and it suggests that a part of the steppe bison's fighting technique was to engage horns, twisting and pushing the opponent in a bout of strength. The thick and heavy horns of sheep, African buffalo, and musk-oxen are used to absorb some of the clashing force of an opponent's charge; in contrast, bison horns are relatively thin.

bonnet of hair serves as a shock-absorbing cushion for the head-on clash, which is followed by a head-to-head pushing contest.

The basic Bovini threat posture is a high head with chin pulled down (Sinclair 1977). Basic fighting technique is a head-to-head pushing contest. Among the Bovini, American plains bison and African buffalo are unique in that they clash. The former uses a clash prior to the pushing contest and the latter just clashes, like sheep; the pushing phase has been eliminated (Sinclair 1977). From clues present on steppe bison fossils, both bones and mummies, we know that they clashed, probably as a prefix to a typical Bovini head-push fight. This behavior has ties to the living bison of the American Great Plains and not to European bison, which have a poorly defined clash at best (Flerov 1977).

The development of a full-blown, noninjurious head clash in American plains bison shifted emphasis from large horns to the upper forehead area (parietals) and led to the development of a padded-hair-bonnet clashing organ. Large horns of Pleistocene bison served to catch and hold an opponent while both pushed. The width of these sweeping horns essentially guaranteed protection from an aggressor slipping by to gore an opponent. By contrast, the short horns of plains bison allow an opponent to slip by, and wounds on the fore quarters after the rut are frequent (Petersburg 1973). I suggest that the forequarters of American plains bison have such a heavy coat of curly hair to act as a protective mantle.

The shock-absorbing hair bonnet (fig. 6.8) in American plains bison is more appropriate for concussion, whereas thick skin on the face of the Alaskan bison mummy acts as a protective shield against horn puncture (fig. 6.9). These traits are correlated with horn size and shape differences in the two species, but they are more a difference of degree than of kind. Although Blue Babe did have facial hair padding, it was not as well developed as that of plains bison; likewise, while plains bison have thick skin on the face (Petersburg 1973), it is not as thick as Blue Babe's. Beneath the facial skin of American bison is a thick layer of elastic connective tissue attached to the skull surface; no such material was found in Blue Babe.

The mummy's extremely thick frontal skin supports the idea that this was the area of primary contact; thick dermis acted as a shock absorber, dispersing energy over a wider area, and limited penetration of an opponent's sharp horns. Another peculiar pattern in Blue Babe's skin over the frontals is a network of ridges and folds (fig. 6.9). These also must have functioned as a unique shock absorber. Yet despite such heavy armor, some horn tips occasionally pene-

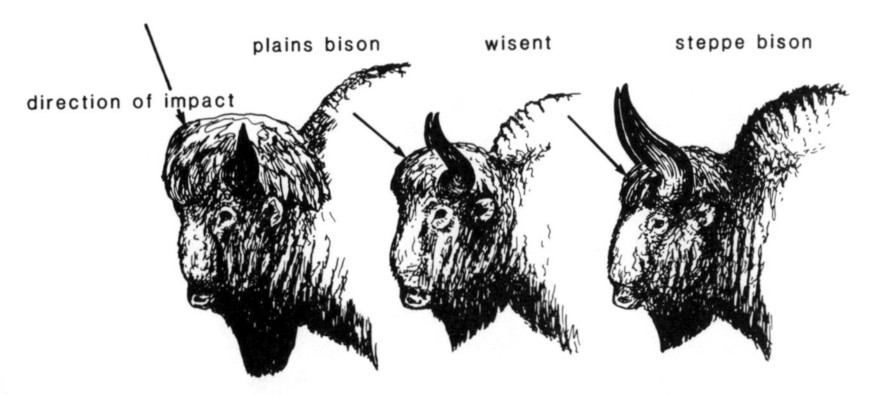

Fig. 6.8. Bonnet and horn variations related to fighting mode. Living bison species differ in their fighting style. Plains bison run at each other and clash head to head with great impact. European wisent are more like cattle, hooking horns and pushing from a fixed position. Steppe bison, like Blue Babe, probably used neither tactic; instead, their cranial anatomy and healed head wounds suggest clashing from a shorter distance, then pushing and wrestling for advantage.

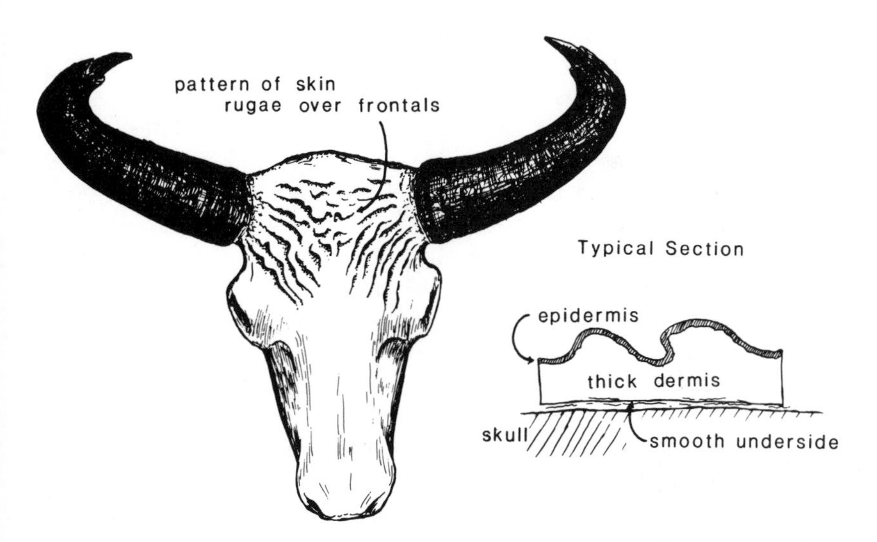

Fig. 6.9. Defensive skin armor over frontals. Blue Babe had strange folds in his thick forehead skin. (*Right*) A diagrammatic section of these folds shows the flat base on the dermis. These folds are not known to occur in living bison. American bison bulls (*B. bison*) have a thick, almost gelatinous, layer of connective tissue beneath the dermis in this area and farther out into the orbits and nasals.

trated the skin and underlying skull roof. We know this because many bison skulls from Pleistocene sediments near Fairbanks have puncture wounds with healed bone growth on the periphery. Fighting between Beringian bison probably began with a hard punch from a standing position, followed by a pushing contest in which each bull tried to reach his opponent's most vulnerable spots with his sharp horns.

In the course of working with Blue Babe and other steppe bison fossils I have developed what I believe is a significant new insight about antler and horn use; this new theory applies not only to steppe bison but to the majority of antlered and horned ungulates. Our current understanding of horn and antler use has developed from models first proposed by Walther (1966) and Geist (1966), which seem fundamentally correct. Walther and Geist focused on the ritualized behavior that occurs prior to fighting, tracking through comparative ethology, and the evolutionary origins and forms of horn and antler displays. This emphasis on ritualization and displays influences their models of actual combat activity by deemphasizing its bloody nature and the damage-producing design of antler and horn weaponry.

Walther and Geist proposed that early ungulate groups had stiletto horns and antlers and mainly attacked the opponent's flanks, while standing facing opposite directions. The goal of fighting among these ungulates was apparently inflicting lethal abdominal punctures or debilitating damage of locomotor muscles. From this early stage, ungulates began to limit injury by catching and holding an opponent's horns with their own, preventing access to more vulnerable body parts. Horns and antlers took on a more defensive design: lugs, hooks, and forks developed to more easily lock onto the other's offensive weaponry. Geist (1966) proposed that the long arching hook of a bison's horn is one of the first steps in this stage; the shape evolved to hook an opponent as it tried to get by the defensively padded head.

Once the level of defense reaches the point where rivals can seldom get by the more elaborately shielding horns or antlers to strike, fighting takes on a new form. Geist, Walther, and others argued that very social ungulates, instead of attempting to injure an opponent, first use ritualized displays to convince a rival of its lower rank; then, if equally matched, noninjurious but sometimes painful tests of skill and strength ensue. These formalized fights involve pushing or wrestling contests with locked horns or antlers; other phylogenetic lines have evolved a head-clashing technique. But all these fights accomplish the same thing: nondamaging combat which settles hierarchy disputes. Until recently this has also been my view, but now I think the emphasis on conventionalized combat formality is somewhat askew.

The skew is consistent with the pervading mood of the late 1960s when our understanding of ritualized behavior in ethology first blossomed. Social signals, or expressive behaviors, as Walther called them, were seen as a means of ritualizing or conventionalizing agonistic behavior. This new awareness of formalized conventions for aggression largely supplanted the previous image of "nature red in tooth and claw." Although theoretically independent from it, this ethological understanding emerged concurrently with the now discredited, but then popular, "epidectic display" (Wynne-Edwards 1962; proposal that many behaviors consisted of a rechanneling of violent, individually oriented competition toward ritualized resolution, for the best interests of the group). It is not too farfetched also to see ethological thinking of that time in the context of the counterculture peace movement in the United States. Remember, ecologists and ethologists (such as Paul Erhlich and Desmond Morris) played a prominent role in the newly developing generational Zeitgeist.

Whatever its exact history, the trend of the time was to envision social evolution as generally culminating in noninjurious agonistic forms. We saw the horn and antler structures of more social ungulate species as devices to hold an opponent, thus converting potentially deadly fights into noninjurious tests of strength. Evolution had transferred the spearheads of early horns and antlers into a multitude of shielding shapes that were now used in ritualized dances of aggression and dominance.

I realize now that the spearheads are still there, mounted on the shield itself. And they are there because they are in active use. I propose that the horns and antlers of most social species consist of two main portions: a defensive apparatus, as defined by Geist (1966), and other structures specifically formed to breach this same defense—to inflict as much damage as possible given the effective defenses. The offensive aspects of horn and antler form are structured to injure a rival by reaching (1) *through*, such as with the oryx's long, sharp horns, (2) *under*, for example, with the long, upward-hooking brow and bez tines of the wapiti, (3) *over*, such as with the high, upward-reaching antlers of reindeer and caribou, and (4) *around*, as in the widely hooked horns of steppe bison.

During my sabbatical year in Africa, I watched a fighting gemsbok (*Oryx*) reach through the horns of a rival and thrust his long rapiers with total energy, obviously intent on skewering the other male, who responded with (sometimes incompletely successful) parries. I examined freshly shot older impala males, just after rut, with deep gouges in their head and neck, their ears in fringes from violent tears. Living in Alaska, I have watched dozens of caribou rutting battles, both in the flesh and on film; these are not just formalized pushing battles but are in deadly earnest. Those long beams reach over and catch an opponent's strained shoulder and neck muscles between inwardly pointing tines and, like giant claws, dig and scratch, hair tufts flying, with the full weight of the body thrown into every gouge. I think the majority of horned and antlered ungulates that lock head to head have in their armory specially adapted offensive structures used to injure their rivals by breaching his defenses.

This behavior and horn structure I had watched in African antelope and saw afresh in Alaskan ungulates caused me to rethink the image of fighting behavior I had reconstructed for steppe bison. Two important pieces of evidence from Alaskan fossil steppe bison show that the male's long, widely arching, hooked horn had an offensive role—to cause injury. One was the frequent skull damage seen in Pleistocene bison bulls from a sharp-pointed "instrument," undoubtedly another bull's horns, and the second was the unusually thick neck skin noted in Blue Babe, which dispersed the pressure of the pointed horn tip and thus decreased its damage. Steppe bison horns were not horns for clean wrestling, but were used to protect the animal while at the same time allowing him to reach around the opponent's defenses and stab the dickens out of him.

I propose that steppe bison horns work this way: The broad forward-facing cross section of the horn's trunk, discussed previously, acted as a defensive blocking device, engaging the corresponding portion of an opponent's horns, above or below his horns, or in an offset rotation with one above and one below (fig. 6.10). The slightly flattened trunk of the horn reached well out, away from the skull, and would have added leverage to keep an opponent from swinging around to the side. The rough annuli on the forward surface probably assisted the defensive holding process. Occasionally, the broad, telescoped orbits of these bison may also have helped hold an opponent's head from rotating. The occiput of steppe bison is especially broad from left to right (fig. 6.10), providing more leverage to hold the skull from a lateral twist. The bison's fighting strategy must have been to keep parallel to his own body the sharp forward-pointing horntips of an opponent. As long as the animals were standing horn to horn on a line, the match was a "draw," that is, the fight was stalemated. In a serious fight, one bison would try to push the

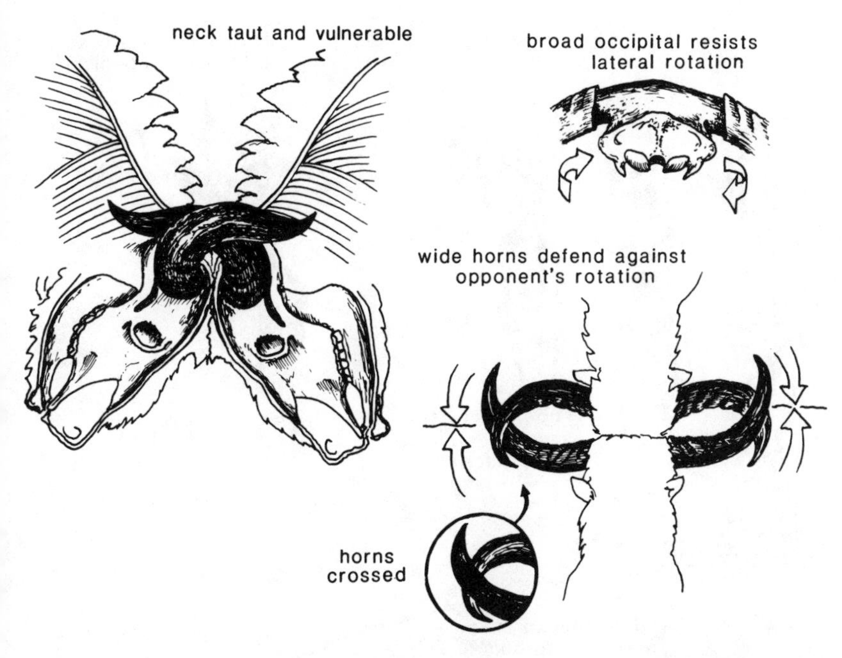

Fig. 6.10. Pattern of horn engagement. The horn and head structures of Blue Babe and other steppe bison show that their horns engaged well away from the face. This holding structure would have added leverage to keep an opponent from swinging around to the side.

other backward faster than it could retreat and still keep its body in alignment. The greater strength of one bison would thus cause the body of the other to rotate while their heads remained locked and held rigid. This would have protected the body of the advancing bison, while the twisted neck and shoulders of the retreating bison would be within range of the advancing bison's sharp horns (fig. 6.11).

The inward curve of horn tips found among many fossil specimens can be better explained by this strategy than as a hooking device to keep opponents from slipping by. An inward-pointing hook decreases the angle an opponent's body must rotate before horns can be dug into his neck. When bison had their heads lowered and horns locked in place, the taut neck and shoulder muscles would not be far from the inwardly pointing horn tips (fig. 6.10).

When two bulls with the same-sized and -shaped horns twist symmetrically, their horns contact each other's neck at the same time and to the same degree. If one animal has a significantly longer outward reach of the horn trunk, or a longer forward-pointing horn

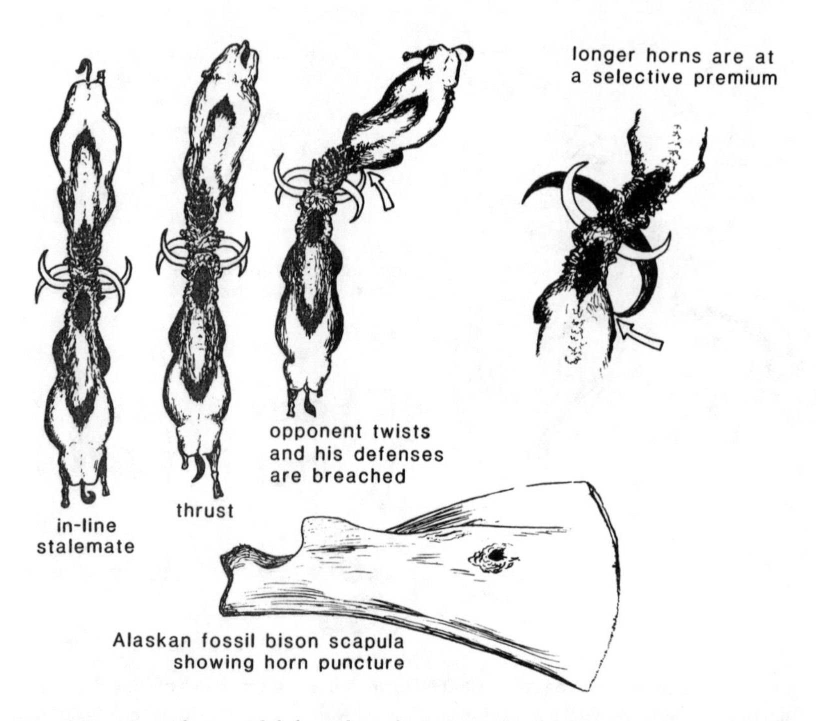

Fig. 6.11. Horn shape and defense breaching strategy. The widely arching sharp horns of steppe bison were curved so that they could reach through an opponent's defense and puncture the opponent's neck or shoulder. The scapula shown, UA-V-54803, was punctured all the way through by a horn stab and has healed.

tip, it will cause proportionally greater damage (fig. 6.11). This is probably one of the selective forces that created the long and sharply pointed hooked horns in steppe bison.

I propose that evidence of horn shape, preserved skeletal wounds on fossil steppe bison skulls, and Blue Babe's thick protective head and neck skin suggest that the hooked shape of steppe bison horns functioned as a breaching device. Once locked with the opponent's horns, this form allowed the stronger bison with the larger horns to ultimately breach or circumvent his opponent's holding defenses by reaching around and stabbing into his body.

For steppe bison the initial display of social stature would have been concentrated more toward the anterior body than it is now in European bison. The length of the tail characterizes this difference: Blue Babe's tail is only half the length of the European bison's, both living (*B. bonasus*) and extinct (judging from cave art). A long tail can be used to express emotion if the opponent is watching the en-

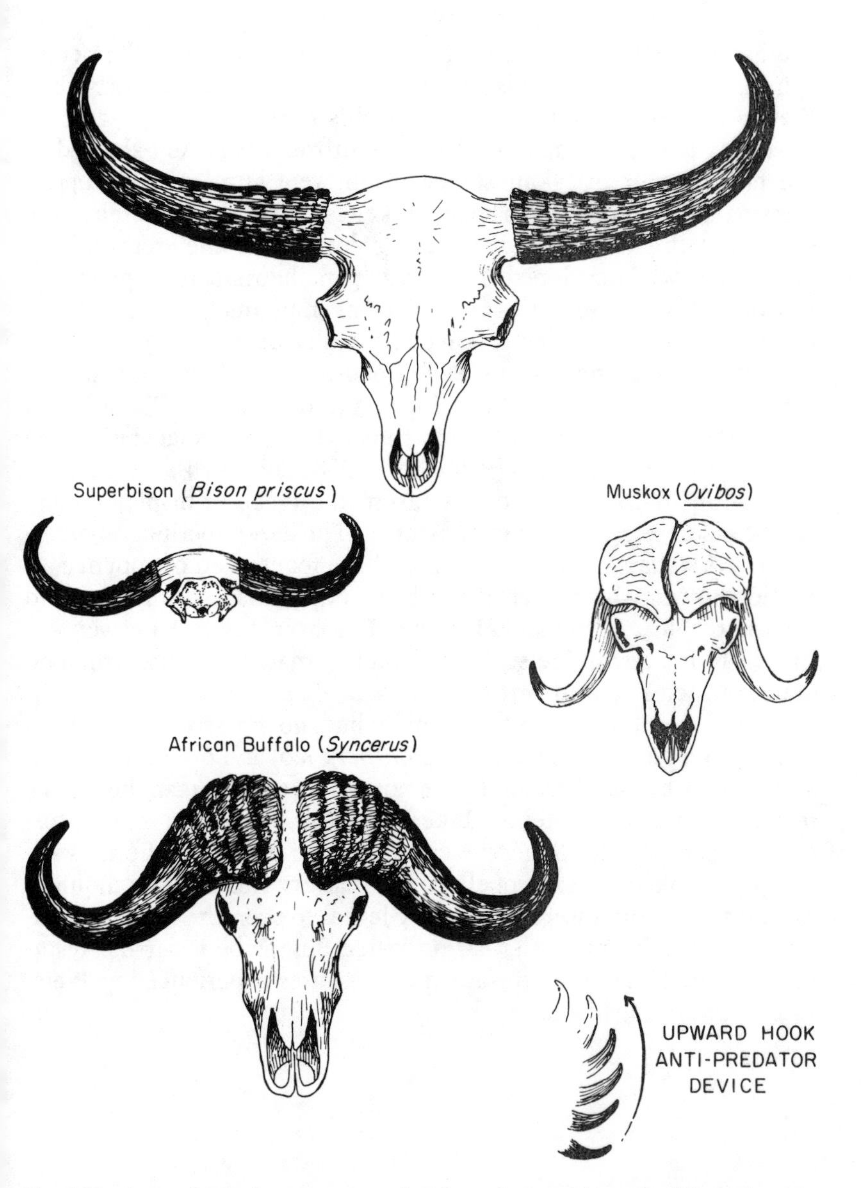

Fig. 6.12. Steppe bison horns as an antipredator device. African buffalo and musk-oxen have a sharp hook on their horns which seems to be used primarily against predators and has little or no role in interactions within the species, unlike bison.

tire body. Among the European *B. priscus,* the height of the neural spines, hence hump, was larger and extended more posteriorly than that in *B. bison* (Poplin 1984). Also, this posterior hump had contrastingly colored hair (Geist 1971b; Guthrie 1980). As I showed in the last chapter, Alaskan steppe bison were similar to European Pleistocene bison in hump shape. Likewise, the beard was relatively short, but the ventral mane running down onto the sternum was moderately well developed. Yet display paraphernalia of steppe bison in Paleolithic art are not as cephalized as in plains bison. The latter has an enlarged head bonnet and elongated beard. Foreleg pantaloons also add mass to the anterior end of plains bison, *B. bison bison,* as part of their exceptionally developed anterior display. Plains bison swing these pantaloons in their threat displays, exaggerating the movements of the forelegs (Petersburg 1973; Lott 1974).

Compared to living plains bison, the focus of display among steppe bison was subtly more posterior. The darker peripheral coloration of Alaskan steppe bison would have focused an opponent's attention toward the center of the body, but perhaps farther forward than with Pleistocene and Holocene European bison. However, unlike all living bison, Pleistocene bison had massive horns, which no doubt formed a central part of their social display.

Unlike Holocene bison, which had no predator larger than wolves, steppe bison faced lions (*Panthera leo*) and other large predators. Their horns, therefore, were not only a social organ, but also a major antipredator defense. Like musk-oxen (*Ovibos moschatus*) and African buffalo (*Syncerus cafer),* steppe bison had horns with sharply hooked tips, propelled by massive neck and shoulder muscles; the horns were potentially lethal to even the largest predator (fig. 6.12). The fact that a lion killed Blue Babe is further testament to the different environmental pressures experienced by Pleistocene and living bison.

7
STEPPE BISON ECOLOGY AND PHYLOGENY

The Character of Steppe Bison

Steppe bison such as Blue Babe were one of the most common large mammals across the north during the Pleistocene. From western Europe to the Yukon Territory, for over a hundred thousand years, bison, mammoths, horses, reindeer, and others occupied a biome of locally diverse habitats. Soviet scientists refer to these animals as the "mammoth fauna" (Flerov 1977; Vereshchagin 1959; Sher 1971). I have expanded this designation and used the name *Mammoth Steppe* (Guthrie 1980, 1982, 1985), including both plants and animals in my discussions. I think *Mammoth Steppe* fairly describes the continued presence of these key grazing species over a truly vast geographic range and chronologic depth.

Mammoths (*Mammuthus*), horses (*Equus*), and bison (*Bison*) were the "big three" of the Mammoth Steppe (Guthrie 1982). We know that each of these species was a grazing specialist, but how and why they were such a successful team is not completely clear. Vereshchagin and Baryshnikov (1982) and I (Guthrie 1982) have proposed theoretical specializations that would have accommodated this association. We know from contemporary studies that large bovines, equids, and proboscidians tolerate different degrees of fiber in their diets. This was undoubtedly true of Pleistocene bison, horses, and mammoths, and it is likely that other specializations existed as well. Perhaps each species retained specializations that let them tolerate the others' presence despite changes with time and geography. This would not necessarily exclude evolutionary change in themselves or in other species in the community. If fact, all underwent marked evolutionary changes, particularly in the latter part of the Pleistocene. Horses increased and then diminished in size, from giants to petite ponies (Sher 1971). Mammoths also changed, particularly with regard to cheek-tooth complexity (Sher 1971), to the extent that some paleontologists have wanted to give the later mammoth separate generic stature. Bison changed too. Bison fossils record changes in horn core size and shape and subsequent cranial effects (Guthrie 1980).

The large-horned steppe bison found in western Beringia (Flerov 1977, 1979; Sher 1971; Vereshchagin and Baryshnikov 1982) and eastern Beringia (Guthrie 1970, 1980) are interpreted as chronological representatives of the same species. McDonald (1981) separated Beringian fossil bison into two phylogenetically separate species and proposed that these alternated in appearance during interglacials and glacials throughout the Pleistocene. But this seems to be a misinterpretation of the data (Walker and Boyce 1984). Rather, Beringian fossil bison seem to be one variable species; such variation within a species is comparable to that found today among reindeer-caribou (*Rangifer tarandus*). Morphological differences between northern steppe bison and bison farther south, as well as changes in steppe bison over time, probably represent adaptive tuning within the context of the far north and events there. At least this would be a logical start in accounting for changes within *Bison priscus*—its similarity to and differences from other bison. I use these different specific names for bison groups, knowing that in reality they are all closely related and are better thought of as one species complex.

We know that during much of the last hundred thousand years Beringia has been herbaceous steppe (Hopkins et al. 1982). Such conditions would have favored grazing species. During most of this time we can document the presence of bison in Alaska and, in fact, know that bison were the dominant large mammal (Guthrie 1968). A comparison of equids, proboscidians, and bison remains from central Alaska showed that bison were overall the most numerus and especially predominated in the uplands (Guthrie 1968). My observations since that time have supported those earlier studies. Among the large numbers of Pleistocene bones collected at Pearl Creek, where Blue Babe was found, over 80% are from bison. These include stratigraphically dated bison bones from the interstadial and undated fossils emerging from all other parts of the section.

In Alaska, dozens of male steppe bison skulls have been collected which still have the entire horny sheath attached to the horn cores. At least the proximal annuli are very marked, unlike many living bison, and some sense can be gained as to life expectancy and perhaps maximum life span. I have plotted age against length around the outside horn curve (fig. 7.1). Although the sample is too small for a precise look at survivorship, we can conclude that steppe bison had a life expectancy similar to that of American plains bison.

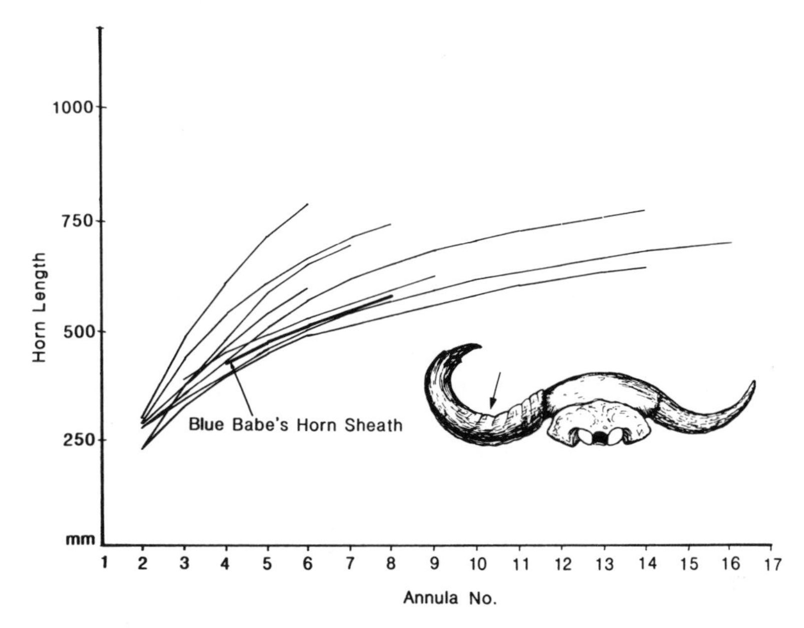

Fig. 7.1. Cumulative length of male Alaskan steppe bison horns. Bison horns grow only during the summer, leaving an annulus, a construction between the segments of summer growth. Among bovids the first summer's segment is often worn away, so this was not measured. The cumulative lengths of horns from eight Alaskan fossil steppe bison are shown for comparison with Blue Babe. The first three annular rings are not visible in Blue Babe's horns, but the plots of later annuli show its growth rate is in the lower range of this sample. These plots show that maximum horn length is reached at 8 or 9 years (Blue Babe's age). Variations in individual horn lengths seem to depend mainly on growth rate of the early segments.

Diet of Steppe Bison

We can determine what Blue Babe ate, or more generally, what steppe bison ate, by looking at diets of extant bison and examining the anatomy of fossil bison. Extant bison are eclectic foragers. Yet despite this apparent versatility, bison are actually adaptive specialists and prefer low-growth herbs, particularly grasses (Guthrie 1980, 1982). I think this was also true for steppe bison (*Bison priscus*). Vereshchagin and Baryshnikov (1982) have argued on the basis of hypsodont molars and cochleariform incisors that the steppe bison was a grazer. Gut contents of a steppe bison mummy from the Siberian Kolyma Basin were analyzed, along with pollen from surrounding sediments, by Korobkov and Filin (1982). All plant macrofossils they found in the gut were grasses, and pollen showed a predominance of Graminae, Compositae, Chenopodiaceae, and Cruciferae. Anatomically and physiologically modern bison are grazing specialists (Hofman and Stewart 1972). They consume some forbs and woody browse, but browse seems to be their lowest choice (Soper 1941). Grass comprises 80% to 90% of the diet of American plains bison studied under natural conditions (Meagher 1971, Peden et al. 1974; Hansen 1976; Hubbard and Hansen 1976; Hansen, Clark, and Lawhorn 1977; Olsen and Hansen 1977; Vavra and Sneva 1978; Van Vuren 1984). However, at the margins of their range or on poor range they can subsist on other plants. For example, bison in Wood Buffalo Park rely on sedges. European bison now confined to atypical dense woodland habitats consume a lot of browse (Borowski, Kra, and Milkowski 1967).

Bison, which were introduced to Alaska in the 1930s, provide an interesting sample. There are several populations; all are confined to windswept riverine areas within or adjacent to mountain passes where wind keeps snow cover to a minimum depth, allowing the bison to reach winter food. Campbell and Hinkes (1983) found that the *Fairwell* bison eat grasses and sedges through the winter (browse contributed to just over 1% of the bison's diet). Bison in the *Delta herd* take some browse and *Equisetum* but mainly eat riverbar grasses and farmer's barley (Gipson and McKendrick 1981). Two smaller herds at *Chitna* and the *Copper River* are on very poor range (Miquelle 1985). There are about fifty bison in the Chitna herd, and they consume considerable summer browse, about 50% willow and 50% graminoids (based on five fecal droppings that were analyzed). The winter diet was about 75% browse and the rest graminoids (from a sample of four droppings).

I use this dietary information (see table 1) from living northern bison to address two issues. The first is an argument used by some ecologists—that if bison can survive today in Alaskan boreal forests on browse, their presence in the fossil record cannot be taken to indicate a steppe environment (Colinvaux and West 1984). However, steppe bison are only part of a large community of grazers. The inability of reintroduced bison to spread from small special habitats argues, if anything, against their adaptive ability to cope in the north without widespread graminoids. I return to this point later in the context of arguments against the Mammoth Steppe.

During the time I've spent thinking about Blue Babe and writing this book, I've also been working on several other projects which turned out to be relevant to the Blue Babe study. In one of these I investigated the diets of Mammoth Steppe fauna by picking small plant fragments from the infundibula of teeth in fossil skulls (fig.

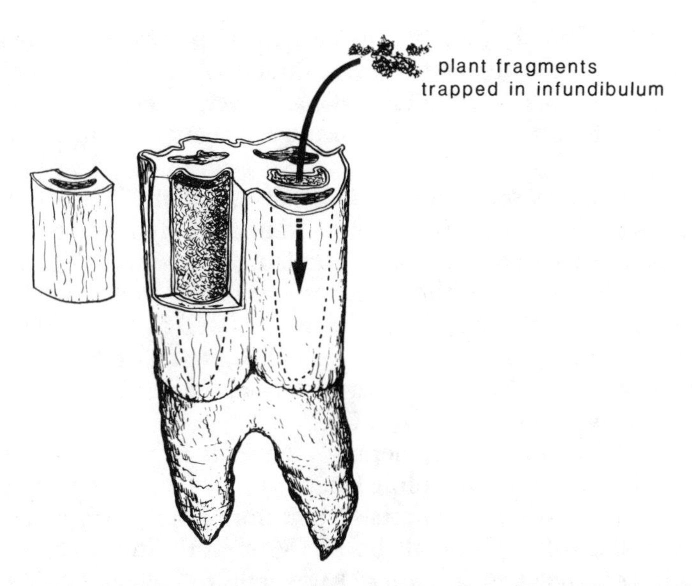

Fig. 7.2. Plant fragments compacted in molar infundibula. Histological structure allows even tiny pieces to be identified to plant group. Decay is so limited in frozen soils that plant fragments taken from Pleistocene-aged Alaskan fossil teeth can be identified.

7.2). In part this work was a response to proposals that the dominant Mammoth Steppe animals—bison, horses, mammoths, saigas, and others—were not grazers at all but simply ate sedges or willow as moose, caribou, and musk-oxen do today. Fossil steppe bison skulls were especially numerous from the Fairbanks mining district, and I was able to get plant fragments from these bison as well as from other fossil herbivore teeth. Plant fragments taken from the molars were sent to one of two laboratories that now do most histological plant identifications for range managers and wildlife biologists. Biologists and range managers rely on these two labs to identify plant fragments in fecal samples. The waxy cuticles covering the plant epidermis are especially resistant to digestive enzymes and acids, and these cuticles pass through the gut undamaged. Except for a slight etching, the cuticles can be readily identified in the lab. Plant cuticles carry an imprint of the underlying epidermal cell pattern. This pattern is characteristic of each plant group, even more so than pollen coats. In fact, pollen is undiagnostic for many plant groups, especially grasses and dicot forbs, whereas cuticles are quite diagnostic. An animal's diet can thus be approximated by this kind of laboratory look at plant remains in their feces. This approach is less laborious than trying to follow an animal at close range, observing every bite; often observations are simply impossible. It is difficult to get close enough to wild animals, and even when rather tame they can be difficult to study. For example, many deer feed mainly at night.

There are some frozen feces of Alaskan Pleistocenen large mammals, but these are very uncommon. Studying diets via plants in fossil teeth was much more promising, and I was especially fortunate to be working with fossils preserved in frozen ground because plant fragments in the teeth were usually well preserved. This is not the case among most fossils from temperate and tropical climates, where it is usual for plant parts trapped between the cusps to be thoroughly decomposed.

Relying on fossil epidermal fragments in teeth was, however, quite different than sampling feces; I needed a modern analogue to tell me how well the material I took from fossil teeth reflected the actual diets of the animals on the Mammoth Steppe. In 1985–86, while on sabbatical in Africa, I was able to collect tooth samples from a number of large mammal skulls, including animals that had died naturally and ones that had been culled by wildlife managers.

I sampled strict grazers, like the Alcelaphines (wildebeest, blesbok, etc.), mixed feeders (for example, springbok and impala), and browsers (such as eland and kudu). Fortunately, I was able to sample in areas where the diets of these species and the plant fragments found in the molars had already been studied. Correspondence between the natural diets of these species and the plant fragments found in the molars was almost complete. This information from African analogues allows us to say that the plants we find in the teeth of Pleistocene large mammal fossils probably indicate their usual diets. I report on this study in greater detail elsewhere, but the general conclusion is that grass fragments were the predominant food found in the teeth of steppe bison, horses, and woolly rhinos. Data summarized in table 7.1, based on plant material taken from 44 steppe bison, are especially relevant to questions about Blue Babe's diet.

Histological studies of bison known to be eating a grass diet show that their feces contain only 80–90% grass fragments. There are two reasons for this. First, a big-bite grazer is not selective enough to eat only grass; dicot forbs and moss grow in between grass stems and are ingested with the grass. Although more toxic than grass parts, small amounts of these dicot forbs from diverse species actually cause little metabolic harm. In fact, small doses of them can add greatly to overall nutrition because dicots normally are

Table 7.1 Plant Cuticle Fragments Taken from First Molar of 44 Fossil Steppe Bison from Interior Alaska

Gramindoids	Number	Percentage
Agropryon	224	5
Bromus	153	3
Carex	7	—
Poa	480	11
Glume	107	2
Unidentified grass	2540	58
		80%
Dicot forbs		
Dicot seed	335	8
Dicot stem	68	2
		10%
Woody plants		
Wood	124	3
Bark	200	4
Vaccinium	17	—
		7%
Cryptogams		
Moss	123	3
		3%

higher in protein than are grasses, which is why dicots are so well protected with pharmacologically active defense compounds. Second, woody plants such as willows and some other shrubs can be quite digestible, even for grazers, if they are eaten in the spring. These shrubs' new growth tissue is fairly high in protein and low in toxic defense compounds.

A second dietary issue is rate and pattern of toothwear. Bison vary tremendously from area to area and through time in the rate at which they wear their teeth (Reher 1974, 1977; Frison and Reher 1970; Haynes 1984). The lower M1 crown height (enamel height of the metaconid) has been chosen to quantify differences in wear. It is the first permanent cheek tooth to erupt and thus wears out first, giving the most dramatic index of tooth wear.

Frison and Reher (1970) calculated wear rate for Holocene bison from the late Pleistocene Vore Site to be 3.5–3.8 mm per year. They point out that tooth wear rate changes with geological time among plains bison. Haynes (1984), however, observed that plains bison teeth always wear faster than those of extant bison in Wood Buffalo

National Park located at the border of Alberta and the Northwest Territories.

Haynes found Wood Buffalo Park bison have unusually low rates of tooth wear, almost half that of plains bison (Haynes 1984). The northern Canadian bison averaged 1.7 mm per year of enamel attrition on the lower M1, using the same measurement of metaconid height as Frison and Reher did. Additionally, the occlusal surface of their teeth is different. It wears in a deeply furrowed pattern, creating a "steepling" effect. Most young bovines have this pattern, but with age it is smoothed into low ridges. This steepling is characteristic of an unabrasive diet and is common among browsers such as moose (*Alces*), which eat vegetation that is very low in phytoliths.

Haynes (1984) attributes this pattern and slow rate of tooth wear to the diet of the Wood Buffalo National Park animals, which consists mainly of sedges growing there in dry pans of Pleistocene lake beds. Unlike grasses, sedges have less biomass in opaline phytoliths. These bison therefore consume fewer abrasive particles per volume of forage. Haynes proposed that amounts of phytoliths in the diet and the amount of dust on leaf surfaces are the main factors in tooth wear. The differences between the teeth of bison Haynes studied and those of the Great Plains bison offer the opportunity to assess teeth of fossil bison in light of these ecological differences.

There has been considerable controversy over the nature of Pleistocene vegetation in Beringia (which I discuss in detail in chapter 9). Several botanists (Cwynar and Ritchie 1980; Colinvaux 1980) have argued that Beringian vegetation was more alpinelike mesic tundra, not very different from today's high-arctic tundras. These present-day tundras are dominated by sedges with few grasses, and the grasses that do grow there tend to be more mesic adapted. If Beringian vegetation was more tundralike, one would expect fossil Alaskan bison, particularly those living during the interstadials when conditions were even more mesic, to be mainly sedge eaters and would show tooth wear rates and patterns similar to those studied by Haynes.

We can figure how rapidly Blue Babe was wearing his teeth. The metaconid height was 16.2 mm. Haynes uses 41.71 mm as the crown height for a newly erupted lower M1 in a 3-year-old animal. Assuming Blue Babe was 8 years old at death, we can calculate rate of tooth wear to be 5.1 mm per year. This rate is much more rapid than Haynes found among his Canadian bison; in fact it is among the more rapid rates of tooth wear observed in bison anywhere. Reher (1974) calculated that the average enamel height of an 8.5-year-old

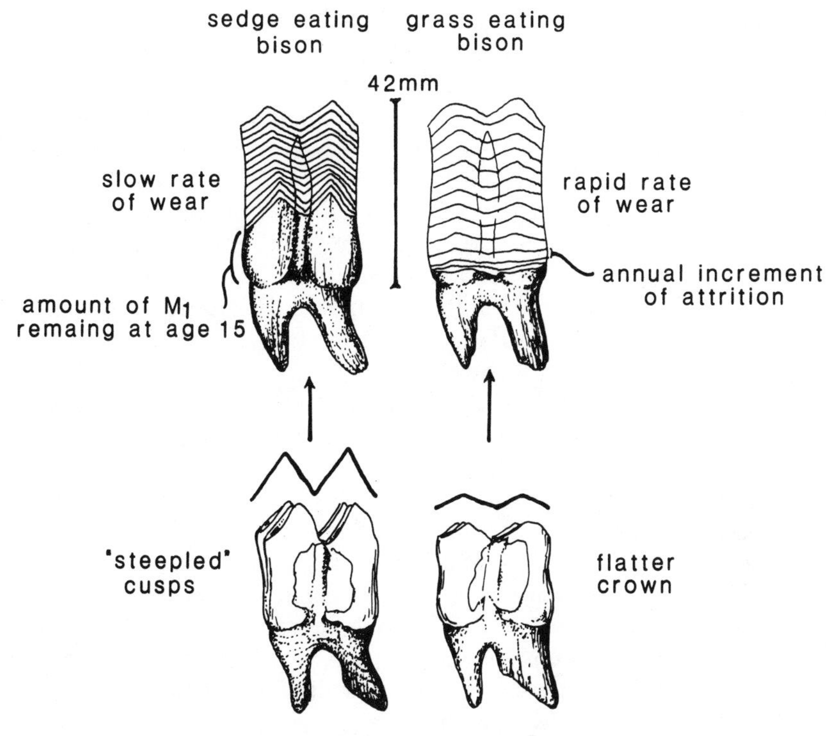

Fig. 7.3. Bison diets as indicated by tooth wear. Teeth are a rough indicator of diet. Sedge-eating bison in Wood Buffalo Park have peculiar teeth: they wear down at less than half the normal rate and show a pattern of high steepled cusps like teeth of a browser. Steppe bison teeth are worn more like those of grass-eating bison on the Great Plains.

bison, at the 10,000-year-old Casper Site in Wyoming where the tooth wear rate was considered rapid, to be 20.5 mm (in contrast to the 16.2 mm of Blue Babe).

The wear pattern on the mummy's cheek-tooth occlusal surface is more similar to that of living plains bison and quite unlike that of the Wood Buffalo Park bison (fig. 7.3). This suggests Blue Babe's diet was more like that of open plains bison, which are obligate grazers.

Age tables of Pleistocene Alaskan bison made by Skinner and Kaisen (1947) on the basis of mandibular tooth wear clearly show wear classes and attritional survivorship curves similar to those from plains bison and dramatically unlike the sedge-eating Wood Buffalo Park bison. From a sample of 1,322 individual Alaskan Pleistocene bison, Skinner and Kaisen showed that the majority, that is, those dying in the senility mode of the curve, had almost exhausted or had exhausted the entire crown of M1. In their sample, 14–16 years represents a very old animal (Skinner and Kaisen 1947), judged by horn sheath annuli; this corresponds well with the rate of tooth wear in Blue Babe (8 years for half-worn M1). Skinner and Kaisen's first age-class (S-4) identified an extreme degree of crown wear with an almost completely worn M1. This age class contained more individuals than any other (40% of the 1,322 mandibles).

Examining the same hundreds of mandibles studied by Skinner and Kaisen, I also found flatter patterns of tooth wear than those Haynes observed among his nongrazing bison; rather, the steppe bison molars were similar to those of Blue Babe and Holocene Great Plains bison. I did not find teeth with extreme furrowing or steepling characteristic of the Wood Buffalo Park bison. These data suggest a grassy arid habitat in Beringia.

In addition to studying bison cheek teeth to determine patterns and rates of wear, the shape and size of the nipping incisors and the premaxillary plate against which they "occlude" can tell us something about steppe bison ecology. There are considerable differences in these characters among different bison groups. Like cattle, European bison have narrow premaxillary widths and American plains bison have the broadest, while steppe bison have intermediate widths. The question is why.

Generally speaking, grazers have very broad incisor batteries and premaxillary widths, and those of browsers are quite narrow. Actually, the story behind these differences is that selective feeders must have a more refined "forcep" because they choose to eat very specific foliage, high in energy and nutrients, but often adjacent to other plants or plant parts which are to be avoided because of their antiherbivory defenses. Some grazers are also quite selective, like roan and sable antelopes (*Hippotragus*); these have pointed snouts and comparatively narrow premaxillary widths. So only certain kinds of grazers have these traits exaggerated. When one examines grazers more carefully, the anatomically wide nippers are the ones that graze on a sward of shortgrasses. More important, these species are large herd animals which often regraze the same area several times during the same growth season. Many grasses are adapted to this kind of repeated grazing and respond by shifting away from a tall-growth form and sexual reproduction to a more prostrate growth form and vegetative reproduction by tillers. This low-growth vegetation makes grazing not only difficult for a large grazer but reduces the "take" of each bite. In response, the "lawn grazers" have expanded the width of their biting apparatus—the premaxillary plate and incisor battery—not to be less selective but to increase the volume intake of each bite such that these regrazed lawn growth forms can be exploited. This is not to say that wildebeest or plains bison always or regularly needed this broad nipper, just that they have met with those requirements sufficiently frequently in the past to be able to exploit and survive those situations when they occur.

Whatever the character of the Mammoth Steppe rangeland across northern Eurasia, the mammoth fauna probably never were so numerous that they resorted to lawn-forming kinds of grazing behavior. This is not true of American plains bison. We know from early records that herds of millions of bison did regraze. Likewise, the kinds of grasses growing on the shortgrass plains were adapted to lawn formation. Some of the grasses, like the ones eaten by wildebeest in Africa, are cultivated today as commercial lawn species because these grasses are already adapted to "mowing."

Although the premaxillary and incisor width of Blue Babe and other northern steppe bison, *B. priscus,* is not as wide as that of the American plains bison, *B. bison,* it is quite wide (figs. 7.4 and 7.5). These bison are very unlike the living European wisent, *B. bonasus,* which has the narrow nipping arrangement of a more selective grazer indicative of either thicker grasses or a more eclectic diet, and a diet higher in dicots.

We can say that Blue Babe and steppe bison in general were grazers; that these bison chose bites from a thin to modest sward but were not well adapted to using and reusing law-forming grass species to the extent of modern bison on the American Great Plains. We have assumed that the steppe bison did little browsing because for tens of thousands of years during the full glacials there was little browse. There were few trees on the mammoth steppe. The anatomy of steppe bison incisors and premaxillaries supports this assumption. During the interstade, when some woody vegetation reinvaded the Mammoth Steppe, bison probably utilized the leafy parts of dicots, especially during the summer, as a dietary supplement.

I found compacted plant matter in Blue Babe's first molars. Cheek teeth of large herbivores have small infundibular pits between the cusps which become permanently filled with plant fragments from the food being chewed. Leaf and stem cuticles are very resistant to both chemical destruction and decomposition, as they form the outer protective coating or seal of the plant against parasites or infection. Because of this durability they preserve quite well. Also, they

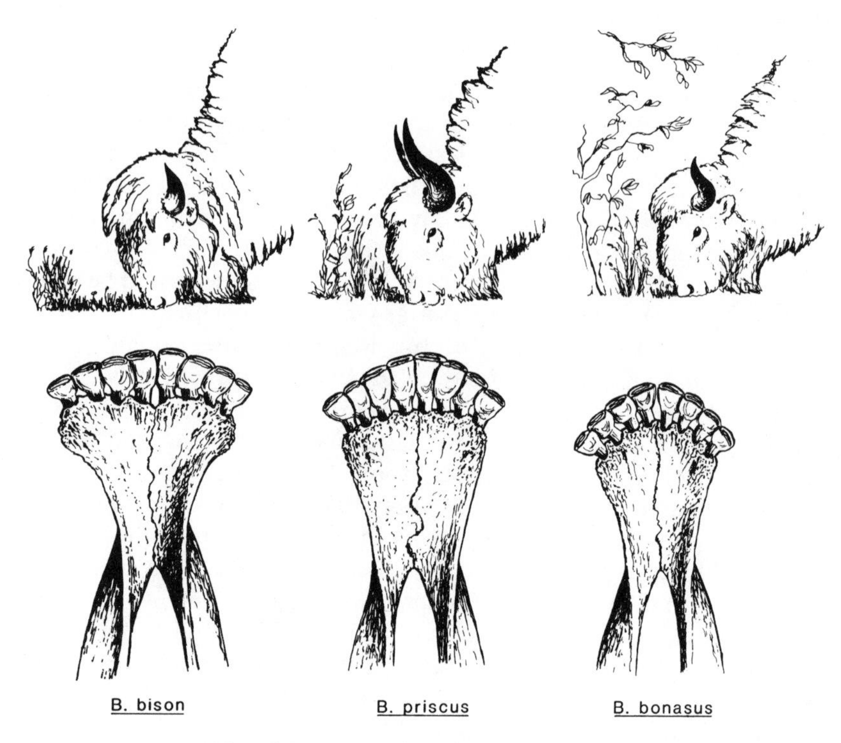

Fig. 7.4. Incisor width and diet. Among grazers the width of lower incisors (the lower canine is incorporated into this row in incisiform pattern) is associated with adaptation to grazing on short grasses. The incisor row is widest in American plains bison (*left*) and intermediate in steppe bison (*middle*). European bison have the narrowest incisor row (*right*).

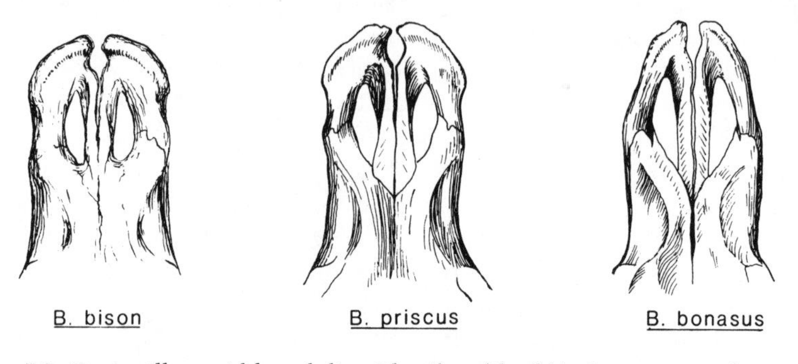

Fig. 7.5. Premaxillary width and diet. The "breadth of bite" among ungulates is a general indication of dietary selectivity; species with narrow mouths can be very selective about each bite. Browsers are normally more selective and hence have narrower biting structures than grazers. Bite width among grazers also corresponds to the volume of sward regularly available in one bite. Animals that feed on very short, previously grazed "lawns" tend to have wider mouths. Premaxillary bones which underly the biting pad are widest in American bison (*left*), intermediate in steppe bison (middle), and narrowest in European bison (*right*).

form an imprint or "fingerprint" of the underlying epidermal cells which tend to be very characteristic for each plant group. So, not only do epidermal fragments preserve well, but they are very informative as to which plants were eaten. Many studies have used epidermal fragments from fecal pellets or stomach contents to reconstruct diets of grazing mammalian herbivores.

Blue Babe's tooth contents were sent to a commercial histological laboratory (at the Department of Range Management, Colorado State University, Fort Collins, Colorado) for identification, where the cuticles mentioned earlier in this chapter were analyzed. Only about 20% of the cuticle fragments were identifiable, and then only to genus. Two grasses were the most common. Wheatgrass or *Agropyron* was one of them. There are several species of wheatgrass in Alaska today. They are somewhat "weedy," that is, they are found on disturbed gravel bars and riverbanks or high alpine meadows. They all occur in rather dry habitats (Hultén 1968).

The other kind of grass cuticle in Blue Babe's teeth was *Danthonia,* a grass that is moderately common at mid latitudes in the plains states, but does not now occur in central Alaska. There are some specimens collected from the Matanuska Valley and Seward Peninsula, and it is possible that these are introduced from farther south. From Hultén's (1968) map this genus seems to be especially widespread in the southern Yukon Territory.

In addition to these grasses, a willow (*Salix*) twig tip was found among the cuticle remains, so the presence of grass and some dicots is consistent with the dietary portrayal of steppe bison predicted by its morphology. This would be especially true of the interstadial time, in which Blue Babe lived, when more willows would have been available.

Interstadial Bison in Alaska

Discussions about the ecology of large mammal species on the Mammoth Steppe have more or less been limited to glacial episodes (see articles and references in Hopkins et al. 1982). We know much less about Beringian mammals during interglacials, primarily because the depositional and preservational environment was better during glacials (Guthrie 1985a). More acid soils and complete vegetational mat characteristic of warmer and wetter peak interglacials left a peaty deposit relatively devoid of bones. However, some specimens, including Blue Babe, date from the last interstadial (Hopkins's Boutellier Interval); these can tell us something about interstadial conditions.

Strangely enough, most of the Mammoth Steppe fauna continued throughout the last interstadial (from about 60,000 to 28,000 years ago), in roughly the same proportions. Three species (bison, horse, and mammoth) dominate most faunal assemblages reported from that time (as an example see the radiocarbon dates listed from the articles in Hopkins et al. 1982). What is odd about the persistence of bison, horse, and mammoth during the interstadial is that this was a time when trees recolonized much of Beringia, approximately reaching their present distribution. Hamilton (1979) described spruce stumps 41,000 years old found on the Chandalar River, near the present northern limit of spruce. However, the presence of Mammoth Steppe fauna during the last interstadial suggests that the climate and landscape must have been rather dry, unlike that of today. The continued presence of bison, horse, and mammoth indicates an arid, grassy environment, with facies not too different from those occurring during glacials over a wide area in all kinds of topographic terrain. Yet tree species that now characterize the boreal forest were present (Schweger and Janssens 1980). Such a combination is strange to consider because it is not comparable to our experience with the present boreal forest or with an arid and treeless steppe. This picture reflects what Andersen found in the pollen around Blue Babe (Appendix B).

Today, both the northern and altitudinal limits to the boreal forest are apparently controlled by warmth, that is, by degree days. Since the present amount of moisture throughout the north is so meager, we might assume it represents a minimum quantity, but in fact it can be reduced. The southern border of the boreal forest is controlled by aridity, at least in the southwestern part. Picture for a moment a combination of these factors: aridity controlling treeline from below, so to speak, and summer warmth controlling the upper reaches of treeline. We can better imagine the interstadial as a combination of warmth allowing trees to expand northward to their present altitudinal and latitudinal extent, but under more arid conditions. Trees would then grow only in well-watered locations, particularly along streams, like the way trees penetrated more arid southern grasslands (Wells 1970). To empty Beringia of trees, as occurred during glacial peaks, by cold alone would have required very acute changes. But increased aridity in conjunction with cold would be a severe challenge to trees. The interstadial climate then was an unusual mixture of warmth, with a little less than the same degree days as present, and considerably less moisture. This allowed woodlands to grow along valley bottoms, with restricted accumulation of peats; at the same time extensive grasslands marked better-drained environments and probably much of the broad lowlands away from stream arteries.

Several fragments of information point to this interpretation. One is the presence of saiga antelope (*Saiga tatarica*) on the north slope of Alaska and in the Yukon Territory around 37,000 years ago, about the same time that Blue Babe lived in the Fairbanks region (Harington 1981). Today, these areas have over twice as much snow as saigas can tolerate. And in summer our ubiquitous tussock meadows would be totally unnegotiable. Nelson (1982) identified an assortment of insect fossils from interstadial deposits (Boutellier Interval) on the north slope. He found a number of species that do not occur there today and are found only in very dry conditions elsewhere. Yet these insects from arid habitats occur in deposits with other species still found in the north slope that are now limited to moist habitats. Pollen from this same Boutellier Interval found in Siberia and parts of eastern Beringia suggests that although spruce, larch, birch, cottonwood, and aspen were present, they grew much more sparsely than today (Hopkins et al. 1982). Unfortunately we have no pollen cores from interior Alaskan lakes which reach so far back into the past.

As in Alaska and Siberia, Europe was without trees during the full glacial episodes. However, the European interstadial (isotope stage 3, in the deep-sea cores) was not characterized by a reinvasion of oak hardwood forests as occurred in the early Holocene. European interstadial pollen continues a pattern of herbaceous dominance with some pine (Bowen 1981). In that regard Europe was probably more like interior Alaska during the interstadial, as indeed it was similar to Alaska during the peak glacial.

How did these glacial and interstadial changes affect steppe bison? Does Blue Babe, our interstadial (Boutellier) bison, differ from his glacial counterparts? We know that the Holocene profoundly affected bison all across the Holarctic, markedly reducing body size and changing horn size and shape (Sher 1971; Wilson 1975, 1978; Guthrie 1970, 1980). Did a similar thing occur during the last interstadial? Unfortunately our data are not good enough to make a detailed comparison, but we can gather some ideas from Blue Babe.

Natural Trap Cave in northern Wyoming contains many bones that provide a long record of large mammals during the late Pleistocene. Gilbert and Martin (1984), who have worked with these fossils, found no reduction in body size during the last interglacial among sheep (*Ovis*), pronghorns (*Antilocapra*), bison (*Bison*), wolves

(*Canis*), and wolverines (*Gulo*). They argue that this is a general pattern and that the dramatic dwarfing at the beginning of the Holocene is unique to that time. I agree with Gilbert and Martin and have argued (Guthrie 1984b) that Holocene changes indeed appear unique when viewed from biotic evidence, even though physical climatic data are not striking. From this we would expect the interstadial bison to be as large as a full glacial bison, but was it?

Blue Babe and other interstadial bison remains collected from Pearl Creek are much larger horned than living bison; one would never mistake these large thick horns for those of a modern bison. However, the interstadial bison are somewhat smaller than most full

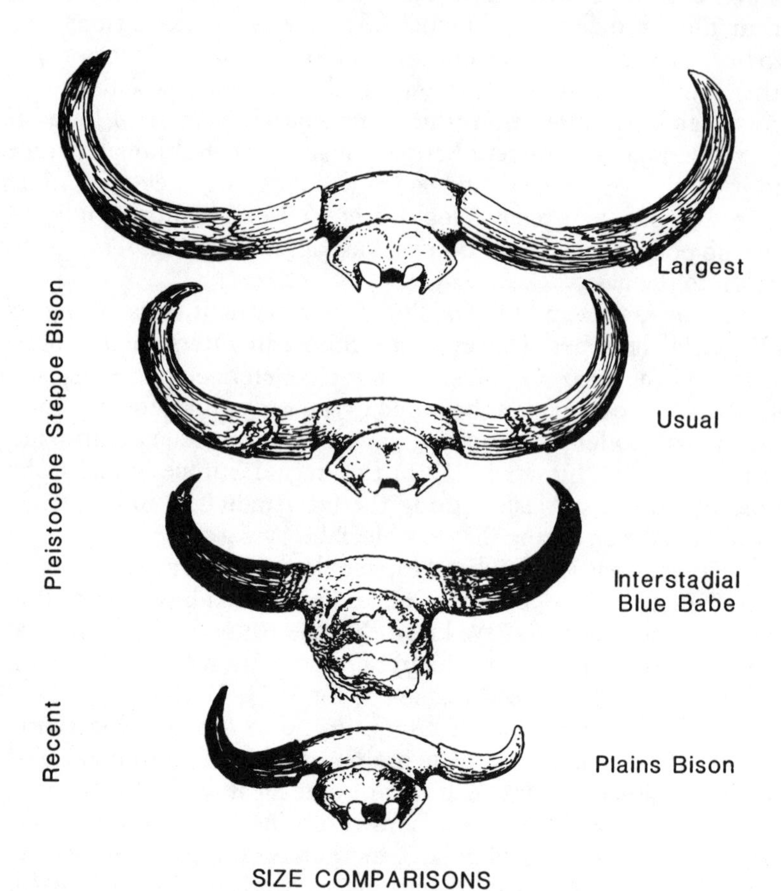

Fig. 7.6. Male skulls. The horn size of Alaskan Pleistocene steppe bison is quite variable. Intrapopulational differences probably account for much of this. Blue Babe's horns are smaller than most steppe bison but larger than horns of living bison.

Table 7.2 Size Comparisons of Blue Babe with Fossil Male Bison from Interior Alaska

	Blue Babe	Mini-mums	Mean	Maximum
Width of cranium between cores and orbits	273	255	288	332
Width of skull at masseteric processes at M^1	182	177	196	213
Greatest postorbital width	332	307	349	408
Horn sheath length on upper curve	565	410	584	810
Greatest spread of horn cores on outside curve	922	790	986	1322
M^1-M^3 alveolar length	92	72	92	115
P^2-M^3 alveolar length	144	131	147	166
Total length of skull	556	558	598	642
Occipital crest of tip of nasals	451	473	491	503

Sources: Skinner and Kaisen 1947.
Note: Measured in mm.

glacial bison (see comparisons in figs. 7.1 and 7.6, and table 7.2). Blue Babe and the other interstadial bison from Pearl Creek fall within the lower range of Skinner and Kaisen's (1947) size distributions of Alaskan Pleistocene bison (fig. 7.7).

I have discussed body size among Alaskan Pleistocene large mammals in detail elsewhere (Guthrie 1984a). In that context, contrary to Gilbert and Martin's (1984) finding, the same processes that caused Holocene dwarfing in Beringian bison were at work to a lesser extent during the interstadial. During the Holocene, the grassland environment was reduced and bison range in Alaska was limited to small islands of habitat. Usually this habitat was adjacent to mountains where windswept river outwash plains provided early succession herbs. Winter winds kept these herbs exposed. These would have been survival habitats but not dependably rich ranges with generous peaks of protein availability. Judging from the smaller size of interstadial bison, invading trees and shrubs as well as changes in the herbaceous community, reduced range quality, and poorer quality range led to smaller bison.

We can make other comparisons between Blue Babe and extant bison. We can measure hoof phalanges to see if foot structure changed commensurate with the more moist, soft substrate we now have in the north. Today ungulates that use interior Alaskan lowlands (e.g., moose and caribou) have specially adapted feet to spread

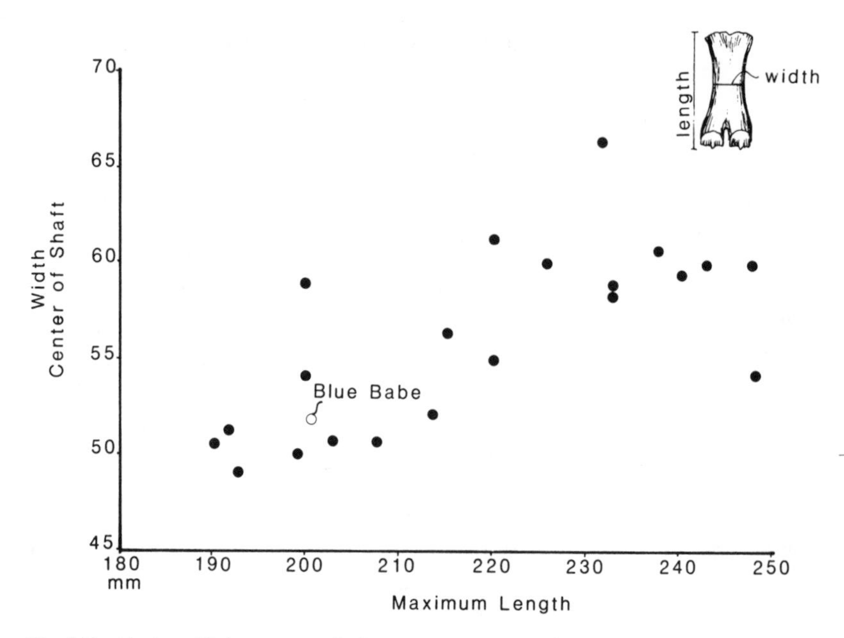

Fig. 7.7. Alaskan Pleistocene male bison metacarpi. In this plot of metacarpal size of Alaskan bison, Blue Babe is near the lower end of the range. Any bone could be used in such a comparison, but metacarpi are particularly good because weight loading more or less determines the size of this particular bone.

their weight, thus decreasing weight loading (kg/cm[2]). Sinclair (1977) observed that water buffalo (*Bubalus*) have broad hoofs which allow them to traverse swampy country, whereas bison have comparatively small feet for their size, being adapted to dry grasslands. Vereshchagin and Baryshnikov (1982) have measured the hooves of Beringian bison and concluded that they were much the same size as living bison. The hooves of Blue Babe were also similar to those of living bison, so the interstadial seems not to have affected hoof size significantly.

Fossil evidence indicates that the main characters of the Mammoth Steppe fauna (bison, horse, and mammoth) continued throughout interstadials. Yet at the end of the last glacial, most of the Mammoth Steppe fauna became extinct or severely reduced in their distribution and were supplanted by the boreal forest fauna of today—black bear, moose, and so forth. By this fact alone one would say that the interstadial large mammal community had a different climate than today exists in those same areas. Bison cannot live in Pearl Creek or the Fairbanks area now. In addition to forage insufficient for winter survival, forage that is there is frequently made inaccessible by deep snows a meter in depth undrifted, and occasionally well over a meter, far too deep for bison to survive the winter.

The paleoenvironment of the last interstadial in interior Alaska is ripe for more study. It is an interesting time: a mixture of the Mammoth Steppe and the boreal forest. It must have been a rather unique period, with many internal variations over the several tens of thousands of years it lasted. Matthews (1982) proposes that the interstadial might have been the critical period when eastern and western steppe faunas and floras intermingled.

Phylogenetic Kinship

Some studies of bison phylogeny and biogeography propose multiple coexisting species across the north (Skinner and Kaisen 1947; Geist 1971b; McDonald 1981). Others see only a central line with considerable geographic and chronological diversity (Flerov 1967, 1977; Guthrie 1970, 1980; Wilson 1975, 1978). If the former is correct, bison are unique among artiodactyls in the far north in having had more than one species per genus at any one time and place in the last half of the Pleistocene. One characteristic of far northern artiodactyls (*Ovibos, Alces, Rangifer, Ovis,* and *Saiga*), proboscidians (*Mammuthus*), equids (*Equus,* caballids, and hemionids), rhinos (*Coelodonata*), or even most larger carnivores—brown bears (*Ursus*), wolves (*Canis*), and lions (*Panthera*)—is that they lived in a community with no other closely related species. This is understood if we see that in the continuous habitat of northern environments there is (and was) little home fidelity, which encourages high vagility and social systems that foster large-scale genetic mixing. Thus, I think it unlikely that two or more species of bison coexisted for any length of time. Although placed in the same genus, hemionids and true horses (caballids) are quite distantly related and may have occupied different habitats in Beringia (hemionids the uplands and horses the lowlands).

The ungulate community in the north is comparatively simple and species are widespread, although individuals exhibit considerable variation in body size and social paraphernalia throughout their geographic ranges. Most if not all northern species show dramatic evolutionary changes when lineages are traced through time (Guthrie 1984a). Social paraphernalia can change rapidly, in either direction. Although Geist (1971b) implies a unidirectional colonization process and McDonald (1981) proposes a semipermanent "stasis" picture of artiodactyl populations moving through time and space, we can document rapid evolution in these populations even since the beginning of the Holocene. Wilson (1975, 1978) showed that Great Plains bison change from a large-horned form to the modern small-horned species in 9,000 years. Contrary to colonization or stasis models, most northern species have decreased in body size and in horn and antler size during the Holocene (Guthrie 1984a). Moose, *Alces alces,* for example, seem to have colonized the New World late in the last glacial or early Holocene, yet southern moose populations are now morphologically different from those that colonized the far north.

Like other artiodactyl lines in the north, bison are a rather plastic and broadly adapted species, readily responding to selection pressures of local environments, both geographically or chronologically. In fact, bison seem especially sensitive in their evolutionary refinement to new environmental conditions and attendant changes in social organization. Wilson's (1975) documentation of body size reductions in Holocene bison on expanding grasslands in the Great Plains is a case in point.

Although Gentry (1967), and Groves (1980) link all the Bovini (cattle, bison, and buffalo) within a narrow evolutionary radiation (they wish to submerge bison within the genus *Bos*), the variety of social behaviors and anatomy within the group shows that similar environments can quickly override old phylogenetic ties, strongly modifying these characters. The use of rutting wallows by African buffalo, *Syncerus,* and American plains bison, *B. bison,* extreme sexual dimorphism in body size and horn size, and the clashing form of combat used by these two species are not an indication of their phylogenetic proximity but rather a product of a parallel behavioral adaptation to more open plains situations.

The use of evolutionarily more plastic, external characters for predicting phylogenetic kinship has strengths and weaknesses. Its strength lies in the high variability of these traits, both between and within species, which make it rather easy to delineate and specify differences. The weakness lies in the swiftness with which these characters can change geographically or chronologically—and the ease with which they can reverse themselves. For example, American bison were widely dispersed in variety of habitats: from Mexico and Florida in the south to Alaska in the north, and there were a number of local forms with distinct external appearances. Most of these bison groups were killed before they were thoroughly described, but descriptions by early explorers portray animals quite different than the bison of the shortgrass plains. Judging from dramatic geographic variations in other North American ungulates, such as white-tailed deer (*Odocoileus virginianus*) and mule deer (*Odocoileus hemionus*), we would expect the same range in appearance among American bison. Although island and coastal populations of mule deer in the northwest are easily distinguished from their counterparts in the Rocky Mountains, phylogenetically both forms are undoubtedly quite close. The point is that differences between steppe bison of western Europe and of Alaska are within the range one would expect in a species of such wide distribution and different habitats. These differences are less than those of mule deer subspecies referred to above, and for that matter are in the same range as differences between wood bison (*B. b. athabascae*) and plains bison (*B. b. bison*) which were contiguously distributed and integrated on the Continent.

Strangely enough the confusing variety of fossil bison names is mainly an American phenomenon. Eurasians have placed most of the large, middle-to-late Pleistocene bison into a catchall species called the steppe bison (*Bison priscus*), although they have acknowledged that it varied chronologically and geographically. European systematists prefer a broad identity, much the same as in the brown bear, *Ursus arctos*, or reindeer, *Rangifer tarandus*, which shows analogous temporal and spatial variations. They have given the various forms of *B. priscus* subspecific status, for example, *B. priscus dimenutus*, *B. priscus longicornis*, and so forth.

There are several reasons for Eurasian conservatism in naming late Pleistocene bison; one is philosophical, but another must be the comparative paucity of bison in the Old World. Although bison are not an uncommon fossil in the Old World, they occur in far greater quantities in the American Great Plains sites. It is relatively safe to say that herds of millions of bison were unique to the Holocene Great Plains.

Unfortunately, in North America there is still no unanimous agreement as to which bison species were present when and where, nor the place of their origins. However, a general pattern is emerging of the overall features of the zoogeography of various "forms." Their exact systematic placement is yet to be agreed on, but that is not my main concern at present. My choice of nomenclature does not necessarily imply agreement with a particular taxonomic status, but is rather a reference to a commonly recognized form.

Bison origins remain obscure. There are a number of bison or bisonlike skulls from the late Pliocene–early Pleistocene of southern Asia—a relatively important center of bovine evolution and diversification. At first they were quite cattlelike, with none of the swollen frontals or orbital protrusion of later bison. The horns began even this early to swing laterally instead of forward as in most cattle groups. *Probison dehmi* and *Bison sivalensis* have been described from India, and *B. paleosinensis* from China. (See Wilson 1975 and McDonald 1981 for further discussions involving the controversy of bison origins.) Very early in the Quaternary these small, somewhat lightly built bison spread throughout Eurasia and were seen in Europe as forms variously called *B. tamanensis, B. voigtstedtensis, B. langenocornis,* and *B. schotensacki.* There may be problems with too fragmentary remains and sexual identification; however, this is currently a focus of interest to several paleontologists and a clearer picture may soon emerge.

Whatever the exact lineage, the generalization of a more cattle-like ancestor seems to hold, as does a stem form that was small bodied and small horned in comparison to some later bison. This early diversification probably still centered around a woodland-parkland environment through temperate Eurasia. However, early in the Pleistocene another form arose that was considerably larger and had long stout horns—the steppe bison, *B. priscus;* it persisted throughout the rest of the Pleistocene, disappearing only at the beginning of the Holocene. *B. priscus* existed as far eastward as England and westward into North America. It was even found northward as far as Novaya Zemlya, 75° N latitude, and southward into Spain and the Caucasus, 40° N latitude. *B. priscus* was the dominant form throughout Eurasia during most of the Pleistocene, and it seems to have been virtually ubiquitous in semiopen and open country. During the glacials it can be found everywhere in the Mammoth Steppe (the Arctic grasslands spreading across northern Eurasia and Beringia) and during the interglacials along the central belt of temperate Eurasian steppes. Interestingly, *B. priscus,* although common, seldom dominates any Eurasian Pleistocene faunal assemblage.

Because of the chronological persistence and ubiquity of *B. priscus,* an understanding of its paleobiology is perhaps a key to an understanding of bison evolution. This is particularly true as it becomes apparent that *B. priscus* probably gave rise to the North American lineages.

The colonization of central North America by bison is also a controversial issue. However, debate has narrowed the time of first arrival from late Illinoian to early Sangamon. (It is central North America which is the area critical to our interest, as Alaska and the Yukon Territory are zoogeographically linked more closely to northeastern Asia.)

Whatever the exact date of colonization, it is agreed that these already large Old World bison, *B. priscus,* became even larger in central North America. In fact, they became enormous, with horn cores reaching 2 m from tip to tip; this New World giant has been given a separate species category, *B. latifrons.* Their bones are found throughout much of unglaciated North America from California to Florida, but the greatest concentrations seem to be along a line from Alberta to Texas, just east of the Rocky Mountains, and in the intermontane basins just to the west.

As with its origins, the disappearance of *B. latifrons* is also controversial. It seems to be the Sangamon bison and probably decreased in size at the onset of the Wisconsinan or soon after, throughout all or most of its range.

There is a smaller form of bison occurring after *B. latifrons* (the chronology of which is again disputed). The main point is that there are enough transitional forms to conclude that the gigantic *B. latifrons* was not a "terminal" line but graded into a comparatively smaller (but still quite large) Wisconsinan form, called *B. antiquus.* This line continued to decrease somewhat in body size throughout the Wisconsinan. Like *B. latifrons* its distributional abundance falls along the line from Alberta to Texas, although it too has a scattered occurrence from Florida to California.

This "*B. latifrons–B. antiquus*" line exhibits decreased horn size and increased doming of the frontals. Also the orbitals are less telescoped outward. The horn cores extend outward in more of a straight line from the central axis of the skull. There is less hook in the horns, and the bases are thick in comparison to northern bison. Wilson (1975) separates these northern and southern "types" in relation to the Holocene bison, a point to which I return later.

In northern Eurasia, in Alaska, and in the Yukon Territory, the large *B. priscus* decreases in size as it continues through to the Holocene. It passes through forms Sher (1971) has called *B. priscus diminutus–B. bison athabascae* in Siberia; in Alaska, Skinner and Kaisen (1947) have called the forms *B. crassicornis–B. preoccidentalis–B. occidentalis–B. bison athabascae.* Whatever names one uses to label the change, there seems to be a diminution of body size in the far north, starting about or slightly earlier than 13,000–12,000 B.P.

Europe is more of a puzzle. Vereshchagin (1959) proposes that in the Caucasus, *B. priscus* decreased in size toward the small living *B. bonasus* of Europe well before the Holocene. The fate of central and northern European *B. priscus* is unclear; whether it was replaced by a smaller bison from elsewhere (Kowalski 1967a) or whether it experienced a size decrease (Degerbol and Iverson 1945; Gromova 1965) is unknown. The living European bison, *B. bonasus,* seems to be an evolutionary regression toward the woodland-edge cattle adaptation. Like cattle, *B. bonasus* even marks its territory (Zablocki 1967).

Thus during the Holocene (and perhaps earlier in some areas), we have a picture of Eurasian bison rapidly decreasing in body size. There was also a decrease in numbers and in distributional extent (Vereshchagin 1959).

The picture in North America, however, is different. There Holocene bison were becoming more abundant while also decreasing in body size. At the end of the Wisconsinan, evidence shows that the northern small form of *B. priscus* (called *B. occidentalis*) reinvaded the Great Plains and came in contact with the line of *B. latifrons* (called *B. antiquus*) from the south (Wilson 1975). The northern forms, having decreased in size independent from the more southern

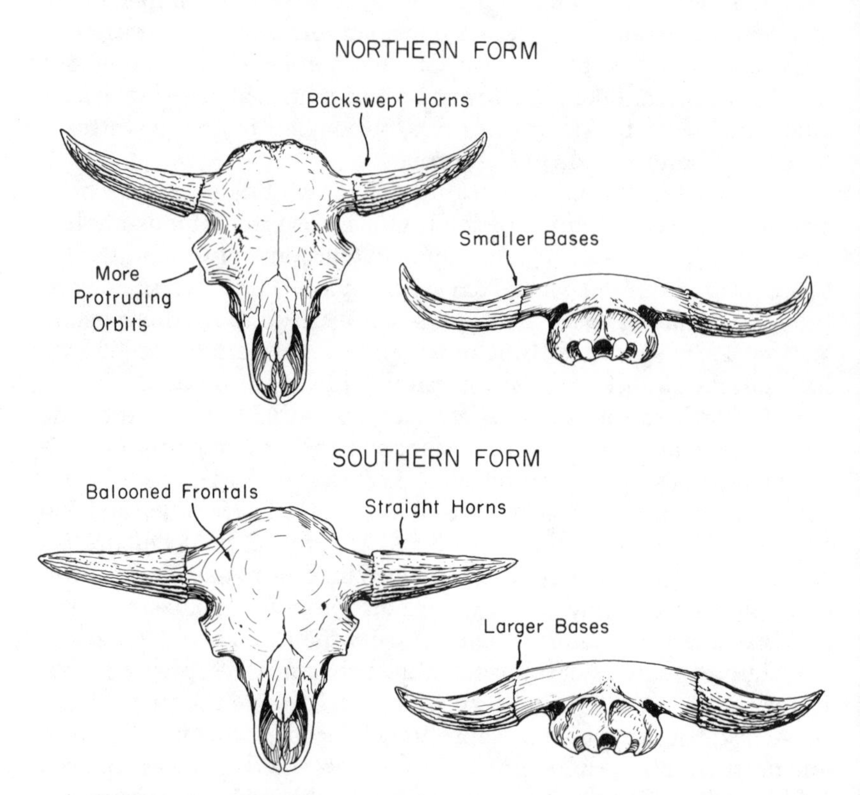

Fig. 7.8. Ancestors of plains bison. Two late Pleistocene-Holocene bison groups are given a variety of names because their biological relationships are as yet unclear. The northern form is referred to as *B. occidentalis* and the southern form *B. antiquus*. One, or both, of these gave rise to the living plains bison, *B. bison*.

forms, were then not very dissimilar in body size at the time of their early Holocene contact, although they differed somewhat in cranial and horn characters (fig. 7.8).

Wilson (1975, and elsewhere), in a study of bison from well-dated archaeological sites, has presented a convincing model of the southern movement of the northern *Bison occidentalis* displacing *B. antiquus.* Additionally, he found several sites in which there were mixtures of the *occidentalis* and *antiquus* skull patterns (fig. 7.8) in various combinations. He interpreted this as strong evidence for genetic remixing of the two lines.

The only thread of information running contrary to this interpretation of *B. priscus–B. occidentalis* is the data about hump shape (chapter 5). All of the fossil bison humps from Alaska and the Yukon Territory have a definite *B. priscus* dorsal arc, which shows a definite concavity in the dorsal contour in the region of T12–T13 neural spines; *B. bonasus* shows a similar contour. However, both *B. bison bison* and *B. bison athabascae* show a rather straight dorsal contour (van Zyll de Jong 1986). The nominal species, *B. antiquus* and *B. occidentalis,* show a slight concavity like *B. priscus,* but the anterior thoracic neural spines are very similar to *B. bison* (fig. 7.9), so the most appropriate specific designation is unclear.

All this controversy over bison nomenclature seems unnecessary. If we had referred all late Pleistocene and living bison to one species as we have done with caribou or African buffalo, as some have suggested (Bohlken 1967; van Zyll de Jong 1986), bison phylogeny would not have clouded the important issues of bison paleobiology and biogeography. Many things, such as horn-size reduction, are obviously parallel adaptations, and multivariate techniques used to show nearest neighbor groups must incorporate biological-paleontological information into their models to give us much real insight.

In figure 7.10 I have summarized this brief history in relation to body size, geography, and chronology. I have emphasized body size because it is conveniently pictured, but size is correlated with a number of other traits such as frontal sinus development, horn size, and many horn characteristics, such as ventral rugae or flutes on the horn core. I have tried to be both abstract and conservative. There are a few "outliers" which do not fit this portrayal and several controversies involving placement of the hatched lines. I think, overall, it is an accurate summary.

Ancestral interaction of these groupings is, however, a different matter. The most logical route is to derive a later from an earlier one in the same region; we know this does not always happen, but it is

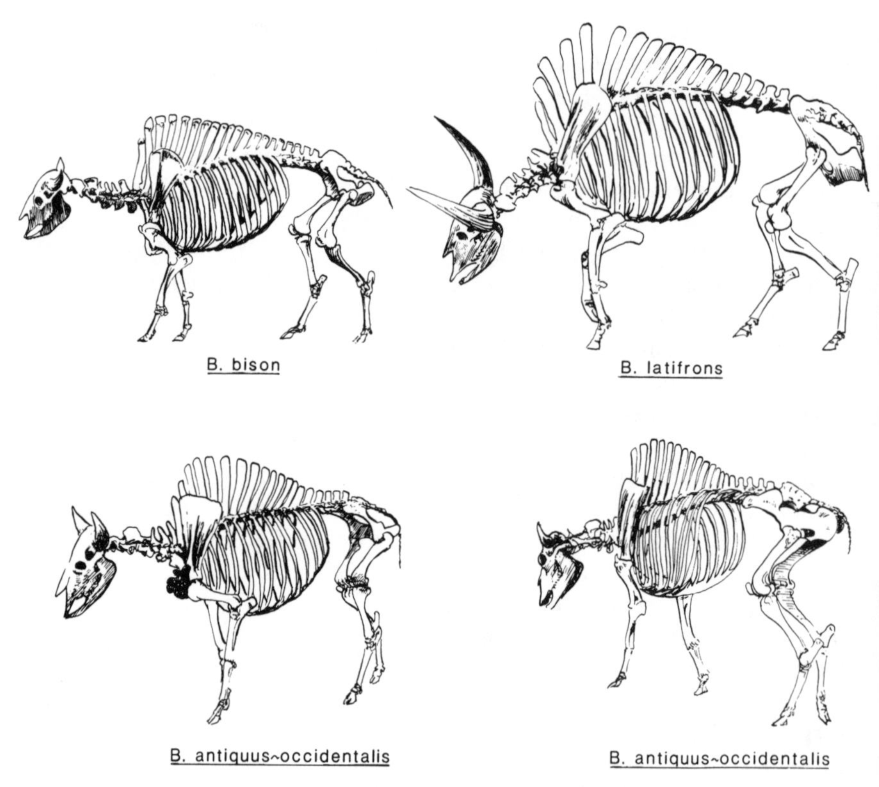

Fig. 7.9. Hump shape in American bison. Bison humps can be seen in these mounted skeletons of *B. bison* (*upper left*), Smithsonian, National Museum of Natural History; *B. latifrons* (*upper right*), Idaho State Museum; *B. antiquus* ~ *occidentalis* (*lower left*), Nebraska State Museum; *B. antiquus* ~ *occidentalis* (lower right), American Museum of Natural History. Though the fossil specimens are not identical to those of recent *B. bison*, they all have a more forward hump than the Eurasian steppe bison, *B. priscus* (compare with fig. 5.8). (Drawn from photos)

the usual case. Therefore my phylogenetic connections are vertical unless there is some evidence to believe otherwise (fig. 7.10).

In previous chapters I have shown that Blue Babe is virtually identical to Pleistocene bison in Siberia, known as steppe bison, *B. priscus*. The Alaskan Pleistocene mummy is also closely related to European Pleistocene steppe bison. Steppe bison ranged from Alaska to western Europe, where they were important figures in Paleolithic art. The Blue Babe mummy allows us to compare the easternmost steppe bison with its counterpart two continents away in Europe. Although *B. priscus* seems rather conservative compared to many large mammals, the species did change some, both geographically

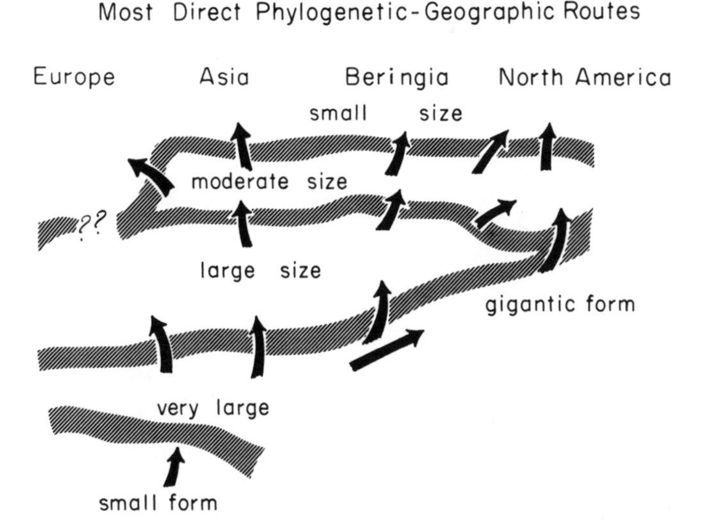

Fig. 7.10. Schematic pattern of bison body size changes throughout the Quaternary. The course of bison evolution is not simple, but it appears that earlier forms were small (there are some exceptions) and that bison increased in size in the mid-Pleistocene. There was a late Pleistocene reduction in size among all bison, and this accelerated during the Holocene. Horn size followed the same pattern.

and over time. Blue Babe also allows us to better compare northern steppe bison with their more southern relatives in the rest of North America.

While different from his Pleistocene contemporaries in Europe and his later relatives in North America, Blue Babe resembles both. In basic color, in pelage length and in hump shape he seems more closely connected to European *B. priscus,* yet, his tail length is more like living American bison. In almost all traits he is more similar to *B. bison athabascae,* the northern subspecies of American bison, than to the more southern subspecies, *B. bison bison.*

These characteristics, both in time and from west to east, seem to follow general social-anatomical patterns first seen by Geist (1971a, 1971b) in ungulates. They include shortening of the tail, concentration of social display into the forward part of the body, and a richer, more contrasting pelage pattern. These trends apparently reversed during the European Holocene, as the earlier European *B. priscus* is much more dramatically adorned than its Holocene descendant, European *B. bonasus.*

8
THE MAMMOTH STEPPE

The Boreal Forest Today

Today, by any judgment, the Fairbanks area is not good bison habitat; yet in the rich Pleistocene fossil deposits of interior Alaska, bison are the most common large mammal. As I write these lines, over a meter of snow covers our yard, and until April this undrifted snow will blanket most of the interior of Alaska. In record years snow may reach a meter and a half, or more. With effort, moose can extract their long legs and wade through such deep snow, but movement in these circumstances is very difficult for bison. Moose now feed easily on willow trees near our house. Moose are browsers, and they have a wide selection of twigs above the snow. Bison have less choice; they have to get under the snow to reach grasses. The energetics of reaching grass buried under a meter of snow would not balance the investment, and bison would not survive such a winter.

Bison can tolerate deep snows for a short time, but their limit for snow depth for a sustained period is 60 to 70 cm (depending on the size of individual bison) (fig. 8.1). Across interior Alaska, ridges above treeline, where snow is windblown and re-sorted into hard pack, provide winter range for dall sheep. Like bison, sheep cannot regularly walk or feed in deep snow. But bison cannot survive on winter ranges of mountain sheep because the ridgetops lack the proper kinds and quantity of vegetation to sustain bison.

Bison were reintroduced to interior Alaska over forty years ago. These modern Alaskan bison are able to live in special habitats where winds, sweeping out of mountain passes, strip snow from broad river floodplains. Frequent summer winds and braiding streams produce grassy river bars and meadows, thus providing the

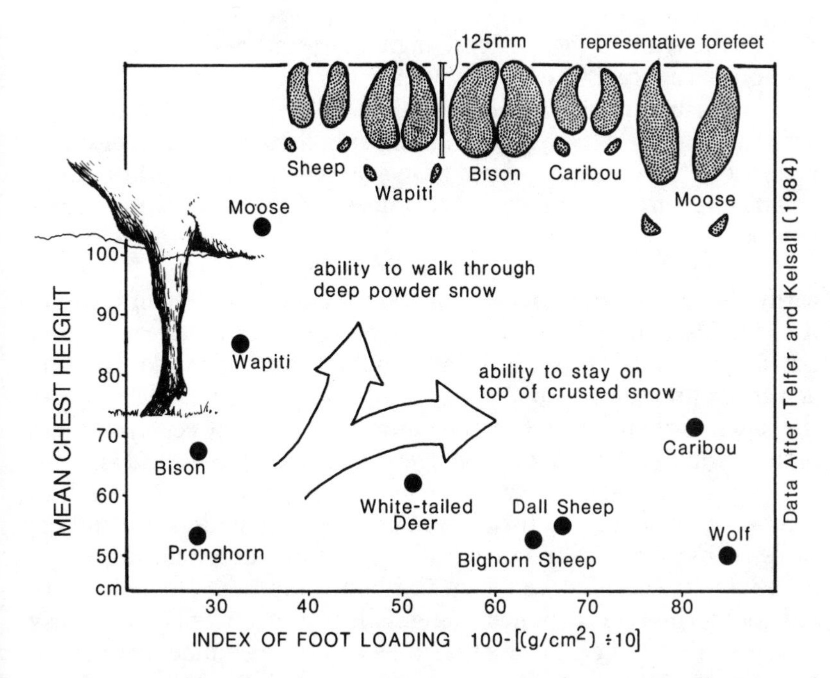

Fig. 8.1. Foot and leg adaptations to snow. Large mammals are not equally restricted by deep snow. When snow reaches the chest (this depends on leg length), an animal can no longer lift its legs above the snow surface, and this consumes a great deal of energy. Foot loading, the ratio of weight to surface area of the foot, is also important. Some animals are able to walk on the snow surface. Both leg length and foot loading are important in the degree of snow tolerance characteristic of each ungulate group. This plot shows that bison excel at neither; they are poorly adapted to deep snow. (Data after Telfer and Kelsall 1984.)

combination of grasses and winter access that bison require. Such appropriate habitat is, however, quite limited, and bison have not colonized other habitats. Bison at Healy are fed supplemental brome hay during the winter; Delta Junction bison forage in the local barley fields, which are frequently left to overwinter unharvested, due to bad fall weather. Although Healy and Delta Junction bison now use some agricultural resources, the habitats in which they live can overwinter a few head of bison naturally. Smaller groups of bison overwinter without assistance on outwash plains of the Fairwell, Copper, and Chitna rivers.

Away from the mountains, the interior has very little winter wind, and as a consequence, large snowdrifts are never produced. Snow accumulates on fence posts in tall top hats that gradually bow toward the sun, as the low sun angle recrystallizes and sublimates the southern face of the snow column.

Most ungulates in the far north are adapted to winter cold. Their major stress is dietary: finding sufficient quantities of digestible food. Northern ungulates have adapted to meager winter diets and usually survive so long as snow does not severely restrict access to food.

Unlike bison, caribou are true snow deer. Broad fore feet allow caribou to dig down to lichen if the snow is not too deep or too hard packed. Bison lack adaptations to deep snow; they have relatively short legs (unlike moose) and very high foot-loading (unlike caribou), and bison hooves are not shaped for digging snow. Instead, bison use the sides of their heads to sweep away snow in a sweep-sweep-bite pattern which is not effective in deep soft snow or in packed snow.

Access through snow is not their only problem; in much of interior Alaska there is little bison can eat. The understory and occasional meadows of the taiga forest and tundra contain little food for a large grazer like the bison. Caribou shun most plants beneath the snow; they select lichen which is low in calories but very easy to digest. Caribou can locate lichen by smell, even under deep snow. One might think a long-legged browser such as a moose would be able to eat anything in sight, but palatable and digestible twigs are not common in a mature boreal forest, and moose must roam about, looking for the right sorts of willows in stream bottoms or old burns. They drift through our yard all winter, clipping and reclipping the willows, garden trees, and shrubs.

Many plants in the boreal forest and tundra are simply unpalatable or even toxic to large mammalian herbivores. Biologists have just begun to understand that that is part of a cycle peculiar to the north (fig. 8.2). Permanently frozen ground and cold soils tie up nutrients. Plants that can grow in these low-nutrient conditions tend to be tolerant of nutrient stress. They are adapted to rather slow growth and very slow rates of nutrient removal from the soil (Chapin 1980); they are conservative and cannot afford to keep reserves stored in underground roots. Thus, most of their biomass is above ground. Plants such as Labrador tea (*Ledum*) and spruce (*Picea*) cannot survive heavy browsing by herbivores because they lack sufficient underground resources to recover quickly. They use nutrients conservatively, allocating some resources to growth and others to toxic chemical defenses. Thus, we may find a six-inch diameter black spruce tree that is a hundred years old. This spruce survived because it produced toxic defenses—terpines and phenoloics that made it

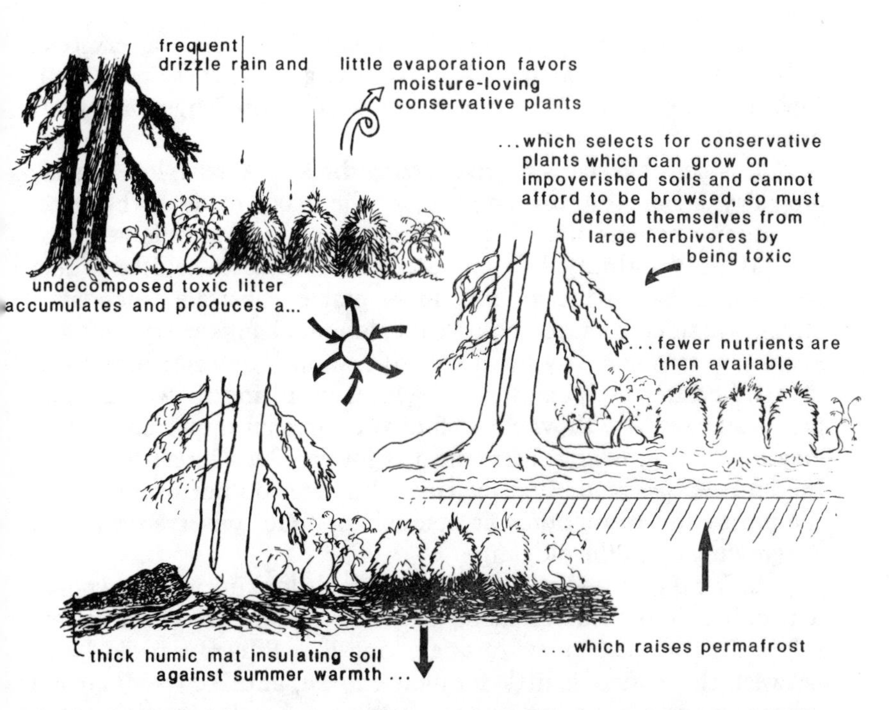

Fig. 8.2. The boreal forest cycle. Present vegetation in the far north is maintained by a cycle of summer moisture, low evaporation, and insulating mats of toxic vegetation that decompose slowly. This insulating layer raises the permafrost zone and limits the rate of annual frost removal. Restricted access to soil nutrients favors conservative plants with few subsurface reserves. These plants are well defended against herbivory by toxic compounds that also inhibit decomposition. Fires rarely burn deep enough to remove this insulating toxic mat; fires tend to interrupt rather than break the cycle.

taste terrible to any passing herbivore venturing a bite. Producing the toxins was effective but costly.

Even dead needles and leaves of these conservative plants are so toxic that decomposers leave them for decades until physical processes begin their breakdown. In a spruce forest or sedge meadow, this toxic plant litter accumulates, forming a thick, spongy mat that insulates the soil and inhibits summer thaw. Gradually permafrost creeps upward until it lies just under the mat; summer thaw does not drive very deep. With deeper soils frozen, only shallow soil nutrients are available and nutrients from dead plants recycle slowly because of slow decomposition. Plants living in these conditions must be able to extract nutrients from a shallow, nutrient-poor zone, and for this reason they have a shallow root system. This is why a fallen spruce tree's roots look like a large suction cup or saucer popped off just beneath the surface; there are no tap roots. The conservative, rather toxic plants that are able to live and grow in these poor, shallow soils are generally not the kinds of plants large mammals can eat.

It is possible that a tourist driving through Alaska, looking day after day at boreal forest, covering all the major roads in the state, and scanning thousands of square miles, may never see a large mammal. Moose are thinly distributed and are limited to habitats where some disturbance breaks this inhospitable substrate, such as a stream that deposits fresh nutrients along its banks every year and keeps permafrost at bay by the relative warmth of water running at nearly freezing temperatures all winter, resulting in well-drained banks and edible willow cover. Fire also plays an important role in creating moose habitat. Permafrost is lowered by a forest fire when the burn strips insulating tree cover. But fires do not usually burn the mossy soil insulation, and before long more conservative plants are growing up in the old burn.

In winter, moose cruise the countryside looking for these polka dots of hospitable edibles: some tasty willows on a gravel bar or, two ridges over, a patch of willows growing in a fifteen-year-old burn. Between these there is little for moose to eat, but likewise there are no niches open for bison, elk, or horses, especially in the winter. Bison do not live here because there is poor to nonexistent summer range and no winter range at all.

A large percentage of the plant mass in the north is poisonous to most large mammals. Even in the vast green summer landscape, there is little to eat for most large herbivores. Far to the south, say in the Mississippi Valley, vegetable resources are more substantial: tubers, nuts, thimble-sized or larger fruits, cereal seeds, and so on, but these are almost nonexistent in the far north. Before Europeans came, northern natives ate mostly fish and mammals. Plants served as a garnish; plant resources supplied few energy or growth nutrients.

The scattered habitats for bison, caribou, and sheep in the far north seem to be a relatively recent phenomenon, dating since the early Holocene. We find bones of these species in Pleistocene deposits many kilometers away from their present habitats. Pleistocene soils seem to have been more fertile over a broad area, favoring the kinds of plants bison and other grazers need. What would happen if northern soils were more fertile—if we were to dump a mixture of nitrogen, phosphorus, and potassium on today's northern vegetation? Chapin (1980) performed this experiment and found that mosses and sedges were quickly replaced by grasses. Grasses and some classes of forbs take up nutrients quickly; they outpace and displace more slow-growing conservative plants. Increased soil nutrients changed the competitive balance, favoring plants with a different life strategy.

These plants (mainly grasses but some forbs as well) can escalate their rate of nutrient uptake in accordance with the amount of nutrients available. Every year they generate new leaf tissue and renew root endings. Root systems of these less conservative plants are elaborate, enabling them to tap richer below-ground resources. Because they can rapidly extract soil resources, these plants keep large reserves in their extensive root systems. Changes above the soil surface thus affect them quite differently. Fire or grazing, for instance, can actually increase their competitiveness. Fire removes insulating litter and quickly recycles mineral nutrients. Unselective clipping of green tissue can eliminate adjacent plants that lack comparable recovery reserves, allowing the grazophilic plant to expand laterally by stolens and rhizomes.

How the Mammoth Steppe Differed from Tundra and Boreal Forest

We can use the above discussion to explain why Pleistocene mammals in Alaska were so different. The subsurface reserves of grasses growing on richer soils allow them to get by with less protection against large-mammal herbivores. Such grass can tolerate being grazed. Large grazing mammals are adapted to eating these kinds of plants; leaf tissue with few toxins and comparatively high digestibility is readily converted into large-mammal tissue. Even the dead winter tissue of these plants is energy rich and can be used by horses and bison as winter range.

It is important to discuss biological adaptations of grass in order to understand the biology of northern Pleistocene bison; the two go together. One characteristic of monocotyledonary plants (grasses and some grasslike plants) is that they grow not from the tip, but from the base, that is, their meristematic tissue is near the ground. These meristems are precious; they contain the highest concentrations of nutrients and are major investments for each plant. A herbivore can bite off a mature stem and leaves without damaging the grass's meristem. This is why the remaining blade of grass that has been mown or grazed keeps its flat top; new growth is from below. Dicotyledonary plants (forbs and broad-leaf woody plants) do not

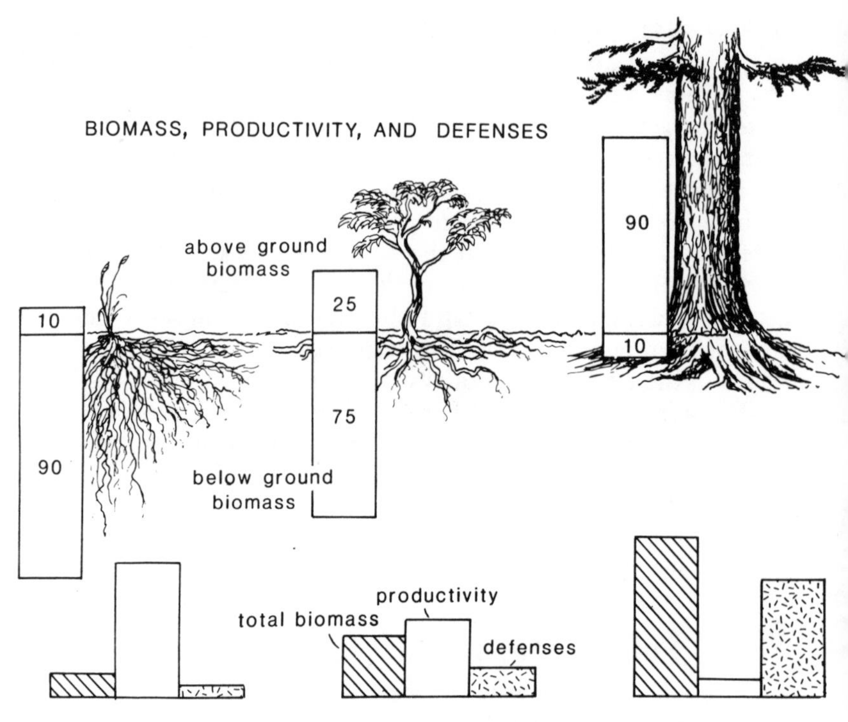

Fig. 8.3. Below-ground stores, defense strategy, productivity, and biomass. Plant groups vary in the percentage of above-ground and below-ground biomass. Because grasses have extensive root reserves, they do not have to have such extreme antiherbivory defenses. Woody plants are much more exposed, having a considerable portion of their tissue above ground, and are, therefore, usually better defended.

work this way. Their meristematic tissue is on the twig tip. It is difficult for them to tolerate herbivore grazing as a herbivore almost always eats new growth on the periphery of the shrub or forb. Dicots must either grow so high as to be out of a browser's reach or they must be rich in toxins, pharmacologically active compounds that disrupt the herbivore's metabolism. Alkaloids such as caffeine, nicotine, and opium are some familiar examples of toxins plants make to protect themselves from insect and mammalian herbivores.

Extensive subsurface reserves (fig. 8.3) enable grasses to grow rapidly and replace damaged above-ground tissue, so they do not require elaborate chemical defenses. Grasses sometimes do use silica from the soil as abrasive phytoliths to discourage grazers. Phytoliths produce the sharp roughness felt when running one's hand the wrong way on a blade of grass; silica is in all soils and is a relatively cheap compound for grasses to produce. Silica deters most plant-eating insects because their chitonous jaws are torn apart by phytoliths. The tough, grass-nipping jaws of grasshoppers are one exception.

Grasses take the chance of exposing their photosynthetic tissue to many potential herbivores during the short summer season. This evolutionary daring-do occurs mainly because grasses thrive in a seasonal environment; nutrients are exposed above ground for only a short time. In the dry season or winter the plant pulls back nutrients to the roots. Tissue left above ground is dehydrated and of poor quality or is abandoned altogether as standing dead material. In the former case, being grazed is not much of a loss, and in the latter it is a benefit. Removing dead tissue before the next growing season allows the soil to warm quicker and speeds recycling of minerals and nutrients left above ground.

Many grasses have evolved concurrently with large herbivores, such as bison, and actually use herbivores as their seed dispersers. Grass seeds are produced in profusion on a seed head. Individual seeds are never very large, as anyone knows who has ever planted lawn grass or hay, but grass seeds do contain energy-laden starches and protein-rich germ cells. They are sought after; however, because the seed coat is so resistant, a herbivore must crush it with its teeth or the seed will not otherwise be digested. Invariably the herbivore misses some seeds. Those that survive are deposited in a warm pile of fertilizer, ready to sprout in a choice microhabitat. Thus grasses give something to large herbivores, and the large herbivores, in turn, aid seed dispersal. Some grass species even have strains adapted to intense grazing (lawn forming), which switch to vegetative reproduction, and others that do well with light grazing. Steppe bison lived in a habitat characterized by plants adapted to large grazers, and vice versa. Today vegetation in these same regions is comprised of plants that virtually exclude large herbivores.

The tundra and boreal forest landscape is thus not simply a product of average annual rainfall and degree days. Vegetation itself affects soil character. The largely toxic insulating plant mat, shielded from high evaporation, promotes permafrost, or at least very cool soils, and limits available nutrients (fig. 8.2). This, in turn, favors the same plants that created those soil conditions. The cycle propels itself; conservative plants on low-nutrient soils must defend themselves against herbivory by large mammals. This largely toxic vegetation limits the species diversity and biomass of the large mammal community.

Aridity: Key to the Mammoth Steppe

The factor that could break this cycle of toxic litter, cold summer soils, and conservative plants is moisture. The Mammoth Steppe was more arid. Aridity, like cold or poor soils, is harder on some plants than others. Aridity is hardest on the very plants that dominate Alaska today—conservative forms with most of their mass above ground. Today, few Alaskan plants are stressed by lack of moisture during the growth season. Aridity is a factor only on some steep south slopes, one of the few places grasses and aridity-tolerant forbs now grow.

As it is, interior Alaska receives very little moisture, around 11 inches (275 mm), technically near the definition of a desert (fig. 8.4). Interior Alaska receives the same annual moisture as west Texas, or the Kalahari Desert, yet one cannot hike comfortably during the summer without rain gear or without getting wet feet. And Alaska has many lakes, muskegs, and rivers in spite of its annual rainfall. The main reason west Texas and the Kalahari are much drier is moisture loss. Desert areas receive tremendous solar input on clear summer days; this combines with wind to dry the surface and pump heat into the soil. In fact, most of these areas have a net loss of moisture over the year, losing more moisture than that gained by rainfall.

Interior Alaska, like most tundra and boreal forest around the globe, loses little moisture in comparison to that received. Some of this retention of moisture can be explained by the long Fairbanks winter (fig. 8.5) which ties up water as ice and reduces both evaporation and runoff. But long winters are not all the story. In the far north, light summer showers are frequent and skies are often overcast. When skies are clear, thunderheads soon return evaporated moisture to the landscape. Also there are few drying winds. A more proximate factor is shallow permafrost, which restricts percolation and fosters the accumulation of litter and mosses that catch and hold moisture. Evapotranspiration by plants in the present environment of the far north is comparatively low (Chapin 1980).

A gravel or dirt road that breaks this damp mat dries easily between showers and soon becomes a ribbon of powder dust. It is strange to walk down the dirt road leading to our house, padding in deep, soft dust, and see in the woods on either side the soggy understory of horsetail rush and moss.

Throughout the far north occasional summer dry spells create ideal situations for forest fires. When woody vegetation above the moss dries out, it is subject to easy ignition. Most fires are started by lightning. Big thunderheads, produced by moist updrafts of warm air,

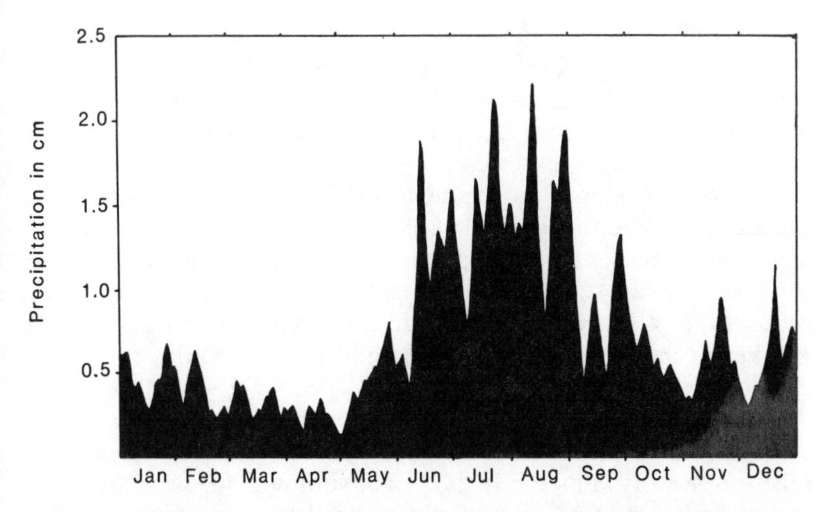

Fig. 8.4. Precipitation curve for Fairbanks, Alaska. At present, and probably throughout the Holocene, precipitation peaks during July and August, while most snow falls before January. (Data prepared by S. Bowling, Geophysical Institute, for years 1949–80.)

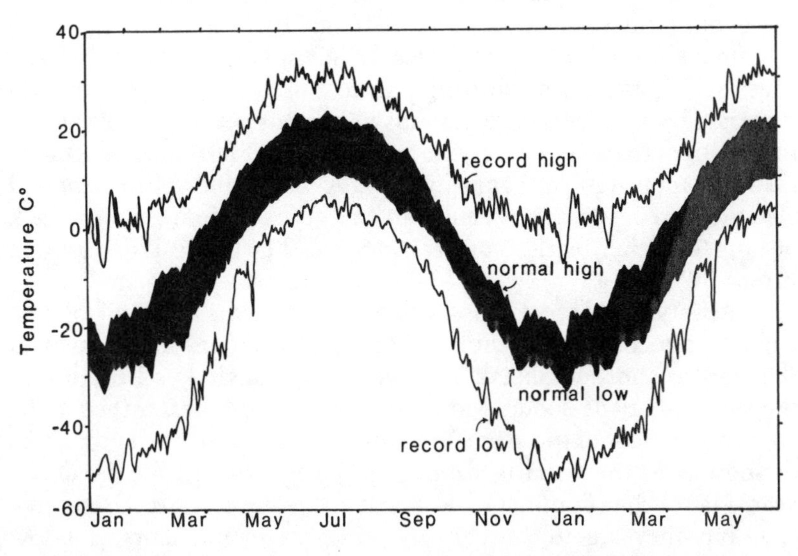

Fig. 8.5. Fairbanks area temperatures. Temperature varies markedly with altitude and aspect in the Fairbanks, Alaska region; the official record from the weather station at just under 130 m provides this seasonal pattern. Present-day temperatures at Pearl Creek, where Blue Babe was found, would be similar. Glacial temperatures were probably only a few degrees different, on average, but the windier glacial climate would have created exceptionally severe winter wind-chill factors. (Data prepared by S. Bowling, Geophysical Institute, for years 1949–80.)

crackle and boom, igniting the toxic resins and terpine hydrocarbons these plants produced to defend themselves against herbivores. Undone by fire, acres and acres of boreal forest and tundra burn when weather is dry and sunny, often giving the best summer days a smoky flavor.

Mountain slopes are frequently drier than the lowlands, not because they receive less rainfall or snow, usually they receive more, but because steeper slopes improve drainage even when underlain by shallow permafrost. Mountain slopes have more wind and barren patches that allow evaporation, but there is almost no tradition in Alaska of carrying extra water when hiking. Most rolling ridges and meadowlands are wet.

Unlike the grassland localities of Johannesburg, South Africa, or Salina, Kansas, rain in Fairbanks, Alaska, usually comes as fine mist droplets; it rains many days each summer, but without much accumulation. Locals call it "dry rain." Pelting rainstorms of the midcontinent grasslands are seldom seen in interior Alaska. One Fairbanks Alaskan tour company had a "money-back guarantee" against rainy weather. Rainy weather was defined as at least one-half inch of rain a day. It sounded good to tourists who did not realize that few of our rainy days are wet enough to qualify.

In addition to absolute decreases in rainfall, aridity can also be effectively increased by shifting rain to a different time of year and increasing evaporation. Most plains grasslands get their moisture in the spring; virtually none has late summer and fall rains. August is the rainy month in Fairbanks (fig. 8.4). A more windy climate would also promote aridity. Wind and sun can dry soil more than the lack of rain. Together, wind, sun, and dry weather create a dry ground litter.

Aridity could break this cycle of low nutrients and toxic plants, and dryness could result in a richer soil. Picture interior Alaska at the start of another glacial. The average sun angle is a little lower, the weather a little cooler. Mountains to the south nearer the Pacific coast catch most of the year's moisture. In the interior it hardly rains or snows and the wind is stronger, probably throughout the entire year. Plants that dominate the north today are stress-tolerant species, but they are tolerant of low amounts of nutrients, not low amounts of water. They are not summer-aridity tolerant. As such, mesic- and hydric-adapted plants would begin to die and cease reproduction, while plant species tolerant of aridity would become more common. At first these plants would have to manage on nutrients from the dead plants they are replacing, but because their aboveground dead tissue is more easily decomposed and easier to burn to the soil surface, the insulating plant litter would eventually thin. The sun, shining all day long in the summer, is making its rounds in normally cloudless skies. As the more exposed soils dried out, summertime soil temperatures would also rise slightly.

Thus a new cycle begins, one keyed to high evaporation and a deeper summer thaw. Plants now have access to nutrients that were previously frozen in lower soil levels, which favors species with much of their biomass beneath the soil surface. The thinner aboveground biomass of these plants allows the sun to reach the soil; they do not generate such an undigestible, nonburnable litter and humus mat. When fire occurs, and it is more likely to occur and spread widely in such arid conditions, it burns down to the soil. But the new plants have most of their biomass beneath the ground and most can recover from fire. For the same reason they are not so vulnerable to herbivores. Unlike spruce and Labrador tea, these plants are quite edible.

Grasses of the Mammoth Steppe probably recycled nutrients more rapidly. Nutrients were not tied up for decades or centuries in woody above-ground tissue or undecomposable litter. Despite less above-ground biomass the productivity of such a grassy landscape would be greater. In addition to access to deep nutrients, without the acidic mantle of undecomposed litter, soils could be more basic, shifting ionic availability. Also, nutrients sprinkled over the landscape—loess deposition itself—would enrich soils with freshly ground minerals. Combined with wind, eolian erosion and substrate redistribution would, in turn, have created "new" soils, that is, soils with unleached nutrients, basic in pH, and readily available cations.

As glaciers flow out away from the mountains, they actively abrade valley walls and bottom, carrying the mountains with them as they come. This newly ground stone is dumped at the terminis where it is washed by summer meltwater into broad river flats stretching for many miles. Strong winds produced by the ice masses whip over these outwash flats, picking up silt-sized particles and carrying them up to hundreds of miles away in thick dust clouds.

Without cataclysmic change, that is, with the same or slightly more oblique sun angles, with temperatures only a few degrees cooler, and with even less standing biomass but vegetation of a different kind, it is possible to have a much more productive natural rangeland. Such a northern rangeland apparently was usable by a much larger biomass of large mammals and by many different kinds of large mammals, allowing grazers such as mammoths, horses, and bison to become widespread.

Today, one can see some effects of aridity on well-drained,

Fig. 8.6. South-facing slope near Fairbanks, Alaska. The scattered grass, sage, forbs, and exposed soil found on many south-facing buffs in interior Alaska are similar to arid grasslands much farther south. The steep slope increases insolation and allows water to drain rapidly, making the soil warmer and drier.

south-facing slopes of river bluffs, where evaporation is high. (Water moving along the river flats creates a relatively windy condition.) These steep bluffs are often grassy with a peppering of sage, *Artemesia,* or other arid-tolerant species, which are often uncommon or nonexistent away from these special habitats (fig. 8.6). If the south-facing slopes and windswept grassy river flats were more extensive, they could support elk (*Cervus*) and bison today. Interestingly, the last sign of the bison and elk in interior Alaska comes from such areas (Guthrie, unpub.). Decreasing summer and autumn rainfall, increasing wind, and clearer summer skies (all interrelated) would allow the vegetation of these south-facing river slopes to expand back up into the hills and across the flats, because it would break the toxic vegetation mat and the shallow annual thaw cycle.

I alluded earlier to the different evolutionary strategy of these plant groups. A spruce tree, which stands exposed for a hundred years, is very likely to be eaten unless it can deter herbivores with poisons. More ephemeral plants face a different balance of pressures, a different game of probabilities. A grass with an extensive root system can more easily extract nutrients and can use the same roots to store reserves out of reach of a bison or horse. Once grazed, it can recover relatively rapidly. Grasses have no woody tissue to care for over winter. Growth starts from the ground surface each year. In a sea of similar grasses, it is likely that any particular grass may reach seed stage without being grazed. Grasses can afford a more "free-wheeling" gambling strategy (fig. 8.3). Natural selection has apparently struck a different balance with grasses, favoring species that do well on fertile soils; grasses devote little energy to defensive toxins and instead push for rapid growth and seed production.

It is because there are such gambler strategies among plants that large herbivorous mammals have evolved. Grazers and ungulates take advantage of plants with fewer toxic defenses, plants that can recover and even benefit from grazing. Like wildfire, bison remove the standing dead sward, allowing sun to reach new plant tissue in the spring. In many cases large grazers also do the plant a favor by dispersing seeds in their feces.

Pleistocene reconstruction has suffered from an inadvertent misapplication of the "uniformitarianism principle" of vegetational control. As one climbs a mountain, one passes through vegetation zones. These differ from place to place, but in interior Alaska they go something like this: spruce forest and its successional facies at the base, grading into a deciduous zone of alder, on into dwarf birch, then into taller tundra, and on to dwarf tundra plants on the highest felfields. In general, this gradation is controlled by temperature, that is, by degree days. As such, this same altitudinal zonation is repeated on a latitudinal scale. There is a northern treeline of spruce, after which one finds mainly shrubs, and farther along northern tall tundra, and into the high Arctic a felfield cold desert roughly comparable to the highest alpine vegetation.

If one sees the Pleistocene differing from the present mainly in degree of cold, or degree days, it is easy to imagine the effects wrought by the reduced insolation during full glacials as simply shifts in altitude and vegetation zones (Ritchie 1984). This is a misconception, however, because it ignores aridity. If temperature alone controls vegetation zones, one must imagine North America as a ladder (fig. 8.7) of temperature isobars, each of which determines vegetational zones as one passes critical thresholds from tundra in the north to tropical forest in the south. However, the actual vegetation map of North America shows major deviations from such a horizontal patterning. Grasslands throw a wrench into this neat system because they run at right angles to this ladder (fig. 8.7). This occurs because grasslands are controlled mainly by aridity rather than temperature. Their north-south alignment corresponds to the rain shadow of the Rocky Mountains, with the shortgrasses abutted

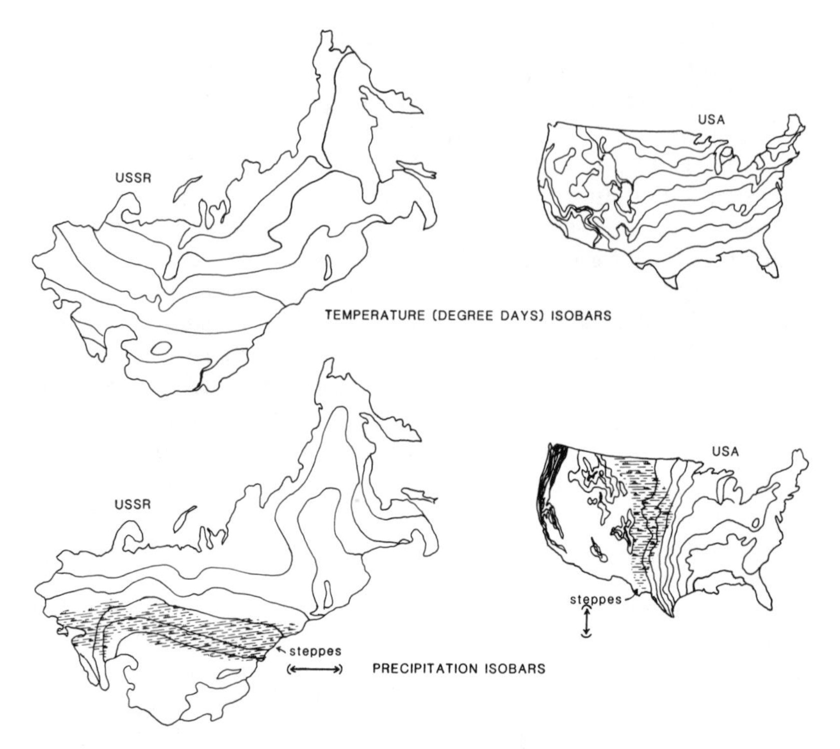

Fig. 8.7. Steppe orientation in Asia and North America. Vegetation zones are not only affected by temperature; moisture is equally important. These diagrammatic maps of the Soviet Union and the United States show temperature isobars (*above*) and moisture isobars (*below*). One can see from these comparisons that the distribution of steppe vegetation is better explained by precipitation than by temperature. Because of the Rocky Mountains, moisture isobars are almost north-south in North America. The east-west tending mountains in southern Asia create the opposite pattern.

against the mountains on the High Plains, with midgrasses farther east and tallgrasses even farther away, breaking up a temperature isobar pattern.

Part of the confusion in thinking about temperature zones is that aridity is complex; it is not a simple matter of rainfall. There are two factors in the aridity-controlled component of plant communities—moisture input and moisture loss—and a lot of subfactors go into that equation: cloud cover, wind, fire, season of moisture input, season of moisture loss, soil permeability, ground cover, and, of course, seasonal patterns of temperature. Aridity is not so neat an equation as mean annual rainfall versus vegetation type, so it is more difficult to visualize. Degree days versus vegetation type is simple and straightforward, and today in the far north it makes a lot of sense. But this sense becomes misaligned in reconstructions of Pleistocene vegetation.

Patches of grasslands on some steep south slopes in interior Alaska are sometimes cited as an example of the temperature-controlled vegetation patterns. But it is a mistake to see these small patches of grass and sage as created by temperature; they are the result of aridity. Their upturned aspect makes them warmer, but it also means that they are always well drained, and these factors combine to produce aridity. Aridity, not warmth, excludes the surrounding boreal forest species.

The second thing that is confusing, especially to North American botanists, is that today we have very few grassy steppes in the far north to serve as an analogue to the Mammoth Steppe. Small grassy patches on south slopes and the distant steppes of central Asia are incomplete and remote images. Furthermore, the relatively rich Aleutian grasslands confuse the issue. Woody plants are unable to grow on the Aleutians because of cold, wet summers (degree days are inadequate for woody growth). Aleutian herbs, on the other hand, thrive in slightly above-freezing temperatures. These unusual grasslands are hard to classify and are sometimes referred to as tundra. Such misconceptions have provided little assistance in reconstructions of the past.

Missing Pica on the Mammoth Steppe

In addition to offering more for grazers to eat, Mammoth Steppe vegetation may also have provided more minerals and other nutrients for growth than is available today. We know that the higher rate of evapotranspiration (fig. 8.8) makes plants in arid regions usually higher in water-soluble minerals than their counterparts in moist conditions (Chapin 1980). Also, the present humic acid-rich soils of Alaska restrict cation cycling. Theoretically, at least, herbivore diets on the Mammoth Steppe would have been richer in minerals. Indirect evidence supports this proposition. Hiking in the Alaskan bush, one occasionally finds bones from an old kill and, even more frequently, shed antlers from moose and caribou. These remains usually show gnaw marks made by rodents (mice, squirrels, or porcupines), but in addition, antlers are eaten by the ungulates that once grew them. Sometimes one finds an entire antler that has been reduced to its main beam, with tines and palms completely eaten.

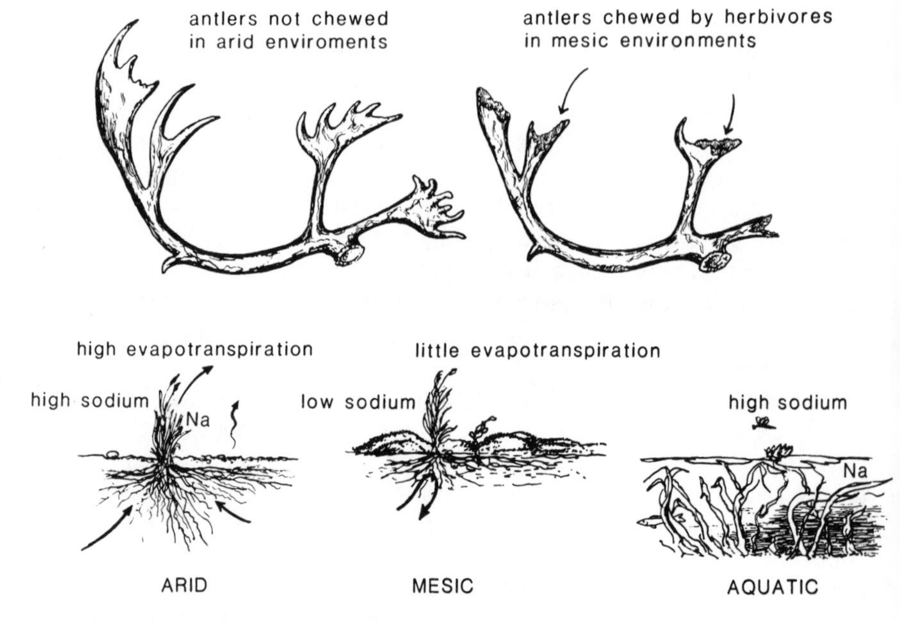

Fig. 8.8. Evapotranspiration, mineral cycling, and pica. Terrestrial plants in arid regions have high evapotranspiration rates and hence generally have high concentrations of sodium in their tissues. Herbivores in these regions have no trouble getting sufficient minerals from plants. Herbivores eating plants with less minerals must supplement their diet, turning to mineral licks or "pica," eating shed antlers or bones to obtain critical minerals. Pleistocene bones in Alaska only rarely show pica chew marks, whereas Holocene bones are regularly chewed.

Antler consumption is so frequent today that it is rare to find old antlers without gnaw marks.

This kind of antler chewing should not be confused with the bone gnawing of carnivores, as seen on Blue Babe's bones. Herbivore-chewed bone looks different, and it is eaten for a different purpose. Carnivores are after fat in bone cavities and protein collagen among the mineral crystals. Carnivores are seldom mineral deficient.

Herbivores chew on antler and bone because they are after the minerals. This kind of bone chewing is well known and is caused by an ungulate being mineral deficient. Nutritionists call this sort of craving for unnatural food *pica* (a Latin term we also use for the magpie, *Pica pica*). But pica (chew) marks are rare on the tens of thousands of Pleistocene bones I have examined. They do occur, but very infrequently. I propose that the higher mineral content of Mammoth Steppe grasses, and probably soils as well, meant that bison and other Pleistocene grazers were seldom lacking essential minerals (fig. 8.8).

It has always been assumed (e.g., Sutcliffe 1970) that bone-eating herbivores are deficient in phosphorus, but more recent data suggest that the critical mineral is sodium (and perhaps calcium and magnesium as well). The flush of broad-leaf growth in the herbivores' spring diet has high ratios of potassium to other minerals and seems to cause sodium, calcium, and magnesium ions to be bumped from their systems and excreted in the urine. Plants have a much lower concentration of minerals than do animals, but potassium is an exception. Plants have over twice as much potassium as animals (Batzli and Juna 1980), which means that herbivores must work physiologically to excrete it from their systems (fig. 8.9). But with the spring flush of high potassium, herbivores on lower sodium diets are thrown into a pathological sodium deficiency. Whatever its cause, spring is a time during which Alaskan ungulates become

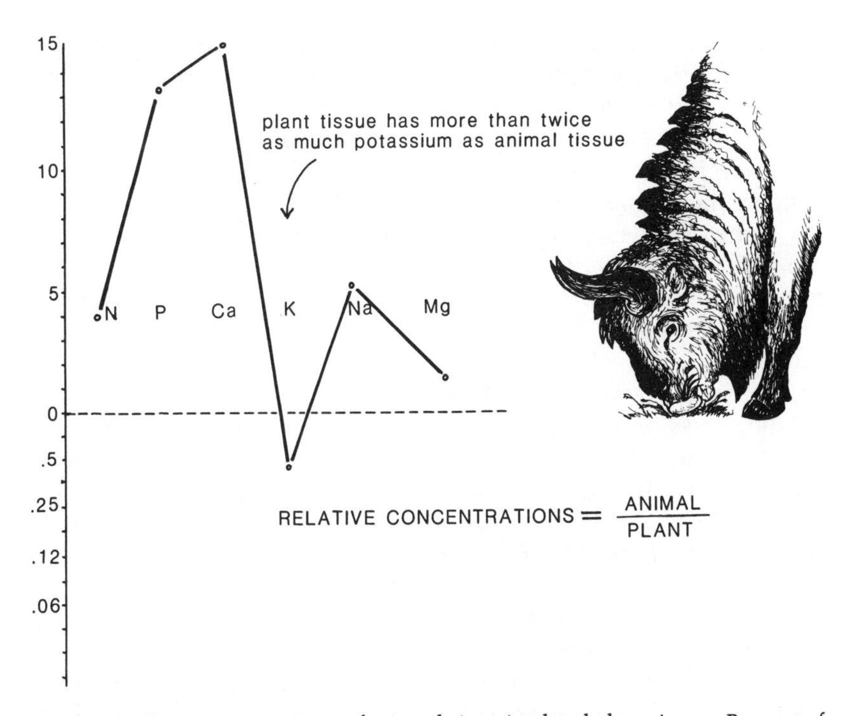

Fig. 8.9. Relative concentrations of minerals in animal and plant tissues. Because of their drastically different physiologies, animals have more minerals in their tissue than do plants. The one major exception is potassium (K).

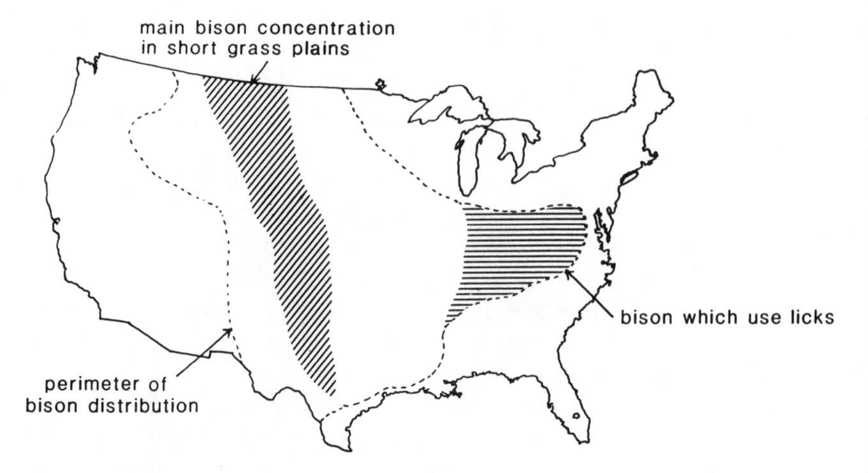

Fig. 8.10. Areas of lick use by bison. Within the contiguous United States, recent bison living in the West seemed to have made little use of licks, whereas bison in the eastern deciduous forest relied heavily on licks (horizontal stripes).

mineral deficient and sometimes travel great distances to mineral licks. This also seems to be the time when bone and antler chewing occurs. Since the new plant growth is rich in potassium, I suspect the comparatively high sodium, calcium, and magnesium in animal tissue (including antler and bone) are within the taste range of a salt-deficient herbivore.

Jones and Hansen (1985) compared eastern parkland bison of the United States with those in the Great Plains, showing how soil minerals affect bison distribution and movement. Eastern bison living on more arid soils were restricted in their movements because they were tied to mineral licks (fig. 8.10). The freedom of movement on the Great Plains was made possible by the base-rich soils. Bison on shortgrass prairies did not use licks, an indication to these authors of high levels of calcium, magnesium, sulphur, and sodium in the shortgrass diet. Likewise, as pointed out earlier, northern cervids on Alaskan acid soils readily chew on shed antlers, but this behavior is rare among cervids living on base-rich Texas soils (Krausman and Bissonette 1977).

There is an ongoing controversy as to whether bone-pica and mineral-lick use is due to magnesium deficiency (Jones and Hansen 1985) or sodium deficiency (Weeks and Kirkpatrick 1976). Most researchers in Alaska have concluded that sodium loss is mainly responsible for these behaviors (Tankersley 1981). The comparatively low incidence of bone and antler chewing by herbivores during the

Pleistocene (fig. 8.8) indicates sufficient plant mineral quality and lack of seasonal mineral deficiencies among herbivores. Recently Ostercamp (pers. comm.) has found low sodium content in Holocene soils and exceptionally high sodium content in frozen Pleistocene sediments.

Bright Days and Cold Clear Nights

The different albedo of glacial ice and the exposed outer continental shelf seems to have created stable high pressure areas over much of the northern landscape. In interior Alaska these produced low rainfall and low snowfall, mainly clear bright skies in the summer, and cold night skies in the winter. In addition to moisture loss by evaporation, sublimation would have increased. In most habitats, resulting aridity would have favored vegetation with incomplete ground cover, exposing raw soil.

Because the Mammoth Steppe was so far north, it was influenced, either directly or indirectly, by periglacial activity throughout much of its area. Continental ice lowered sea levels, changed upper atmospheric circulation, produced windier conditions, increased erosion, and created loess-forming situations, dramatically changing soil and vegetation. Of course the Mammoth Steppe vegetation occurred in areas quite distant from actual glaciers, but it was always indirectly affected by glaciers.

This high seasonality, new soils, and inhospitable conditions for woody plants favored grasses and grasslike plants. Today there are many cold- and dry-adapted tundra plants that could and did flourish in such an environment. This low sward, predominantly of herbs, created a varied habitat, with both xeric facies and better-watered facies near streams and downslope from persistent snowdrifts. It would have looked similar to shortgrass regions, or steppes as they are often called in Asia, which today occur at a variety of latitudes (figs. 8.11 and 8.12). Arid late Pleistocene conditions in the north combined xeric grassland forms with xeric tundra species, probably in mixes both within and between local communities.

Likewise, the relative lack of snow and warm middays in autumn would have extended the other end of the growing season. Autumns would have been characterized by striking diurnal temperature fluctuations: bright days and cold clear nights. Clear summer days would have raised summer soil temperatures near the surface, but at deeper levels the ground would have remained cool because bare or almost bare winter soil would have given its heat to winter skies to a much greater extent than the snow- and vegetation-

Fig. 8.11. Asian steppes. The steppes in central Asia are the closest thing remaining to the Pleistocene Mammoth Steppe; and they may serve as analogue. Horses, hemiones, saiga antelope, ferrets, and lions, which became extinct in Alaska, continued to live and thrive in the Asian steppes until the latter part of the Holocene. (Photo by David Murray)

Fig. 8.12. Steppe grasses in central Asia. The grasses are *stipa,* a common grass of the steppes. (Photo by David Murray)

insulated soils of today. But if summer thaws were fast and deep in these conditions, the opposite was true of winter. Heat was pulled from great depths, creating and maintaining permanently frozen ground below reach of the summer thaw. Winters must have been cold, as they are today, creating permafrost conditions that we now see in fossil form relatively far south of present-day permafrost activity. Winter winds would have disturbed the temperature inversions that create our most extreme winter temperatures today, so actual temperature readings might not have been so low, but the chill factor calculated by including winds would have made for very cold conditions.

We have to consider the possibility that the Mammoth Steppe had the unusual combination of relatively deep summer thaws and an accumulation of deeply frozen ground (fig. 8.13). Today that combination seems odd because permafrost now exists where summer soil temperatures are low—north-facing slopes, high altitudes, and in the Arctic lowlands—making it easy to assume that the widespread creation of permafrost farther south had to involve colder summers. But permafrost is produced and maintained by a series of phenomena not distilled in mean summer temperatures. I propose that the Pleistocene pattern was a reversal of what we find today throughout the subarctic or even most of the Arctic. Today summer thaw is shallow, heat is no longer being withdrawn from deep ground, and there has been a Holocene reduction in the extent of permafrost (fig. 8.13).

Although there are some significant temperature (insolation) differences between those of today and those that prevailed during full glacial (isotope stage 2), I argue that, in the far north, insulation is also a significant force. Today's soils are well insulated all year. Snow limits the amount of heat extracted from the ground in winter, and during summer the vegetation mat decreases the amount of heat gained. These insulators buffer soil temperatures.

If, as we have proposed, there was exposed soil within the open vegetation cover during the glacials, more heat would have been absorbed by the soil in summer. Less rain in summer meant clear summer skies, which meant drier surfaces, which altogether meant greater heat accumulation in the soil during summer—deeper thaws. (This is in the face of a reduced sun angle 18,000 years ago and hence less potential maximum insolation.) It is possible that summer soil temperatures were warmer during the full glacial and that the air temperatures were cooler than those of today. Clear summer skies would have allowed soil temperatures to exceed those of

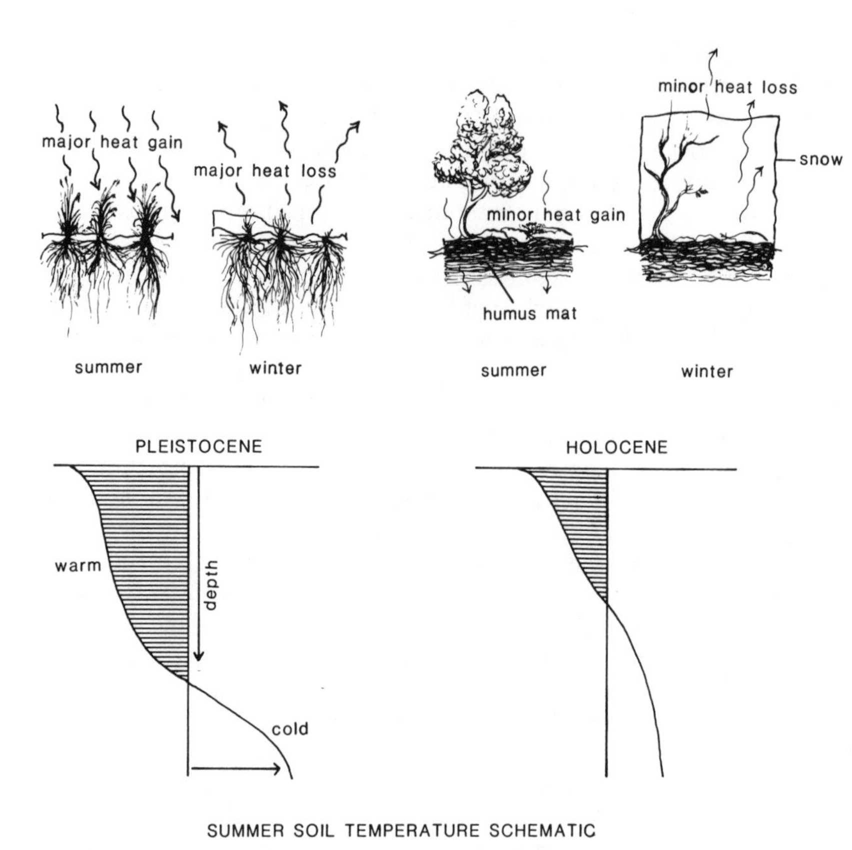

Fig. 8.13. Pleistocene and Holocene soil temperatures. Meager snow cover and soil litter during the Pleistocene promoted the removal of large amounts of heat creating permanently frozen ground (permafrost). Less litter also allowed deeper summer thaw. The Holocene situation is reversed.

today, while termperatures a meter above ground may have been cooler. Dry soil surfaces would also have lost less heat from evaporation.

On subarctic roads, where snow cover is removed, there is deep penetration of frost, and often considerable frost cracking as these cold-soaked areas contract. Contraction cracks appear down the road, like ice wedges and polygonal ground in the making; however, the dark pavement evidently absorbs so much heat in spring, summer, and autumn that it pays back this winter heat debt and no permanently frozen ground accumulates.

Permafrost specialists are familiar with this phenomenon of reduced insulative cover producing permafrost (Washburn 1980), but it has not been emphasized in reconstructions of Pleistocene soil heat dynamics.

Snow considerably retards spring in the north today. At the vernal equinox, when we have the same length of day as the rest of the world, Fairbanks is still in the throes of winter. Snow takes weeks to sublimate and melt. Breakup occurs during seventeen hours of daylight in late April, a full month after equinox (fig. 8.14). Variations of this phenomenon occur all across the north. Most northern plants are very hardy to night frost and would begin to grow in the warmth of the spring sun much earlier if no snow were present. This delayed uncovering of plants, a dramatic boom of plant growth almost immediately after snowmelt, creates a sense that there is no spring in the far north. Indeed, it is less than two weeks between breakup and

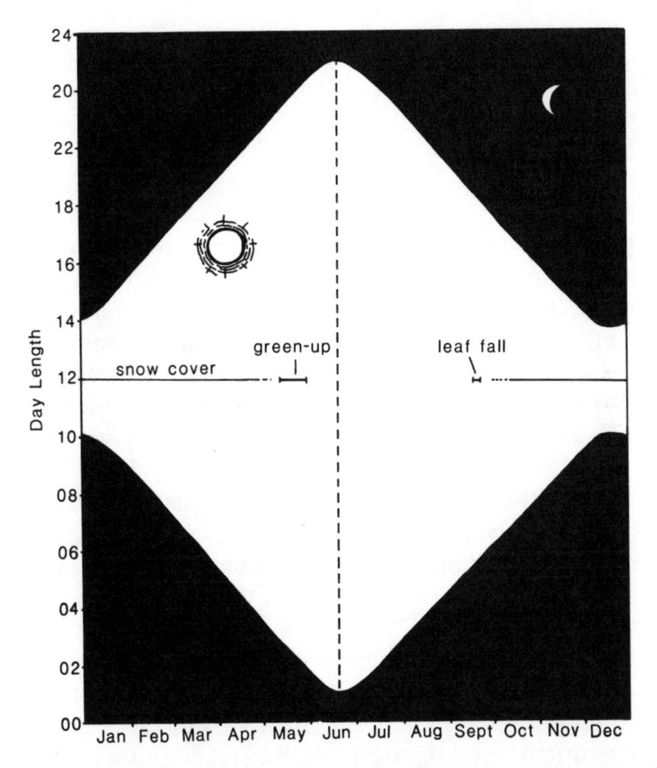

Fig. 8.14. Hours of sunshine received at 65° north (Fairbanks, Alaska). This diamond shows how little sunlight is available during the winter. The asymmetric relation of the summer solstice and the four-month green season is also clearly apparent. Long summer days provide considerable photosynthetic potential for plants that are adapted to use the extra hours of light.

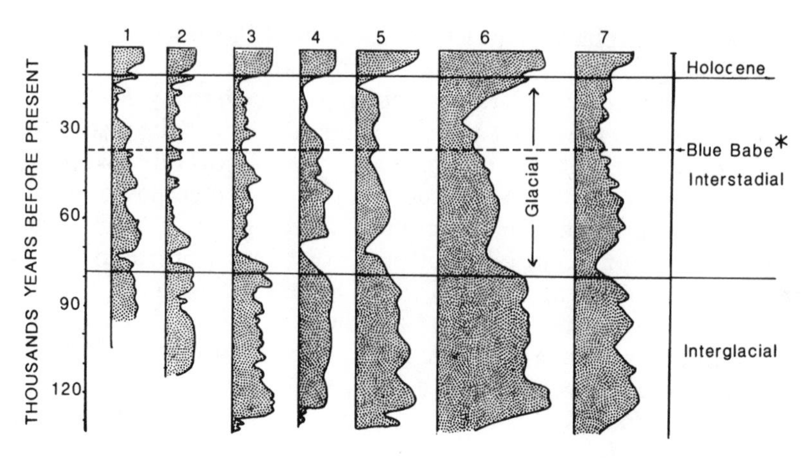

Fig. 8.15. Proxy climatic curves summarized from Wijmstra and Van der Hammen 1974: (1) pollen (Fuquene); (2) pollen (Macedonia); (3–5) deep-sea cores of % fauna (V23-82, V23-83, and 180-73); (6 and 7) oxygen isotope ratios (V28-283, 280) from deep-sea cores. Blue Babe is shown in relation to these cold-warm episodes as falling between the two glacial peaks of the last major glacial event.

green-up. This is one of the advantages "Holocene" plants such as spruce and Labrador tea have over Pleistocene grasses. Spruce and Labrador tea are evergreens which can begin to photosynthesize early, losing little by the postponed spring. In Fairbanks, trees are green by the end of May, while the grass sward still wears its winter brown. Grasses have their biomass hidden below the frozen surface and must wait until the soil is thawed; hence they get off to a late start.

Although the points emphasized in this and the following chapter deal mainly with glacial episodes, they pertain almost as much to the interstadials, which differ from full glacials only by degree (fig. 8.15). Interglacial episodes on the other hand, are marked by more qualitative change.

Snow would have been minimal and redistributed by wind when it did occur, leaving much land exposed, increasing heat loss, and destroying woody plants, but making the standing dead sward available to ungulates. Moreover, exposed soils would have advanced spring thaw. Deep snows now consume the sun's energy a month to two months past the spring equinox, but exposed ground absorbs the sun's heat much faster than snow. Earlier thaw would have lengthened the time nutritious plants were available, thus increasing the growth season for large herbivores.

Pleistocene eolian activity made a dusty landscape: dust in the air, dirty snow, and hazy skies. Sunrises and sunsets must have been spectacular, as the northern sun rolled along the horizon. The dust would have diffused the light and given the north a richer tone.

9
ARGUMENTS AND CONTROVERSIES ABOUT THE MAMMOTH STEPPE

For a number of years (Guthrie 1966b, 1968, 1980, 1982, 1984a, 1984b) I have argued that data from Pleistocene large mammals indicate an arid steppe existed across northern Eurasia and Alaska during the glacials. Mammalian fossil evidence suggests that during the Pleistocene, ice-free northern areas had both greater carrying capacity and diversity of large mammals than we see today. This was a vast area, with complex communities and wide geographic variations, yet it seemed, nevertheless, to have an integrity that justified my use of a general name: the Mammoth Steppe. Old World faunal evidence and faunal data from Alaska and the Yukon Territory have supported that portrayal (e.g., Matthews 1979; Harington 1978). Sketchy evidence provided by relics within modern vegetation has, at least, not been in conflict with the Mammoth Steppe concept (e.g., Yurtsev 1974; Young 1982).

Several palynologists (pollen specialists), however, have taken issue with the idea of a Mammoth Steppe, arguing on the basis of pollen data that a grass-dominated vegetation did not exist during glacial episodes, nor were large mammalian grazers associated with it (Cwynar and Ritchie 1980; Cwynar 1980; Ritchie and Cwynar 1982; Ritchie 1984; Colinvaux 1980, 1986). At first, I (and other paleoecologists) accepted their data, if not their interpretations, as presented (Guthrie 1982; Matthews 1982) and tried, rather unsuccessfully, to account for the apparent conflict between pollen data and other fossil evidence. During my reconstruction work with Blue Babe, yet another attack on the Mammoth Steppe concept (Colinvaux and West 1984) prompted me, for the first time, to take a critical look at the palynological data itself. I found, certainly to my surprise, that some palynologists have misread their own data. Rather than being at odds with the Mammoth Steppe concept, the pollen record strongly supports the idea, or so I propose in this chapter.

First, however, let me explain why this question and our shared efforts to understand Pleistocene environments in the far north are important. The area in question is immense, truly circumpolar, but this debate about late Pleistocene events is not simply a question of how to paint a large panel of stage scenery. Our reconstructions of this past critically shape our present ecological understanding. The pulls and tugs of ideas in this debate involve fundamental questions in ecological theory, for example, how species and communities react to different climatic changes, as well as more specific issues, such as the ecology of Blue Babe. Finally, the character of the late Pleistocene north is closely linked to the question of when early peoples colonized the New World.

Palynologists have argued against the term and concept of a Mammoth Steppe primarily on the basis of the meager pollen influx (the comparative numbers of pollen in each cubic centimeter of sediment) during peak glacials. They concluded from these small amounts of pollen that Alaska was a barren landscape, or polar desert, uninhabitable or nearly so by the "mammoth fauna" we find as fossils. Rather, they propose that these animals must have been in Alaska mainly during the wetter and warmer interstades and interglacials.

Ritchie, Cwynar, and Colinvaux state or imply several propositions which I review in detail. I think sufficient data exist to test them. The propositions are as follows:

A. The amount of pollen produced (the pollen influx in lake-bottom cores) was so low during the glacials that it can only be matched by a present-day high Arctic polar desert or barren tundra.

B. Thus, the glacial environment, as reconstructed from pollen data, was so harsh it could not possibly have supported the "mammoth fauna"; this is said to be borne out by the lack of dated large mammal fossils from full glacial times.

C. The taxa of glacial (Duvany Yar) pollen show that species from southern steppes were not present; glacial taxa were restricted to tundra species found throughout the north today. Thus, glacial vegetation was not similar to a steppe but was barren tundra like that now existing in the very far north.

After examining these propositions I review arguments supporting the existence of an arid grassy steppe during both the glacials and their interstades (Blue Babe's time). My argument for this steppe takes the following outline:

1. Contrary to some palynologists' contentions, pollen cores that meet minimum acceptable standards for influx analysis do

show herb pollen production peaking during the glacial maximum (Duvany Yar) and declining dramatically from the glacial maximum to late Holocene. Herb pollen influx during the peak Duvany Yar is not definitive of barren tundra but is in fact comparable to that of some arid grasslands today.

2. There are sufficient numbers of dated fossils to show that the dominant members of the "mammoth fauna" (bison, horse, and mammoth) were present continuously in the far north during the late Pleistocene (at least to radiocarbon-dating limits), a span of time covering part of the last interstadial (Boutillier Interval), the full glacial peak (Duvany Yar), and the late glacial (Birch Period). Additionally, most of the less common Beringian species in the fossil record have radiocarbon dates associating them with the full glacial.

3. There are, in the Pleistocene fossil record, flora and fauna that we associate with more xeric environments farther south. These taxa are now generally absent in the area once occupied by the Mammoth Steppe. Those that do occur in the far north are now found in rare, special habitats.

4. The paleoenvironments of Alaska and Beringia should not be analyzed totally separately from those of northern Eurasia. Pleistocene large mammal communities in Alaska and the Yukon Territory were the easternmost part of mammalian communities, extending across Eurasia to the shores of the Atlantic. Evidence from that faunal complex suggests a generally similar cold and dry environment where graminoids predominated, although with locally diverse facies.

5. The mammals (particularly large mammals) taken individually and as a community were grazing specialists.

6. Teeth and stomach contents of frozen mummies of these same species show they were indeed mainly grazers.

7. Details of foot morphology indicate they were adapted to a firm substrate, quite unlike the moist and yielding soils over much of the region today, apart from mountainous terrain.

8. Their body size indicates that they were not on marginal tundra or polar desert summer range but were able to acquire large quantities of quality forage during the summer growth season.

9. There is a continuity and coherence to the Mammoth Steppe fauna. The fossils of most members of this large-mammal grazing community are ubiquitous, distributed throughout the unglaciated north from high areas adjacent to alpine glaciers to the broad lowlands, quite distant from hill country.

10. The inability of most of these grazing large mammals to tolerate present snow depths means that, due to low snowfall, wind

redistribution of snow, or both, they had widespread access to adequate winter range. Snow depths are now limiting to all northern large mammals; a low snow cover would have enabled a larger diversity and density (standing biomass) of large mammals than exists today.

I shall include the first three theses above with my discussion of the palynological propositions A, B, and C because they are related. The remaining points are then discussed under their appropriate headings.

Proposition A
The amount of pollen produced (the pollen influx in lake bottom cores) is so low during the glacials that it can only be matched by present-day high Arctic polar desert.

Ecological inferences from mammals points unwaveringly toward a Pleistocene grassland environment throughout the north and yet the pollen evidence is in apparent disaccord. The problem lies not with the fauna, but with certain interpretations of pollen evidence.

One fundamental difficulty is that pollen can be well identified among arboreal species, but identification is very poor for monocots. In fact, graminoids can only be separated into grasses and sedges with no greater taxonomic refinement. This has produced an understandable preoccupation among palynologists with arboreal species; I would argue, however, that this orientation has led to a misinterpretation of Beringian pollen, exacerbated by inappropriate notions about biome integrity through time.

Palynologists have asked how one can propose the presence of productive grassland during the full glacials when today's unproductive northern landscape produces much more pollen. But this question, which fails to discriminate between arboreal and nonarboreal pollen, reflects a misreading of their own data. The more accurate question is why several times more herb pollen was produced during the full glacials than today.

Simply, I propose that pollen data do not support certain conclusions regarding pollen influx during the glacials. Pollen influx is rather abundant in the Holocene sections of pollen cores (in the thousands per cm^3) and sparse (in the hundreds) during the peak of the last glacial, 17,000–22,000 years ago. The relative scarcity of pollen during the last peak glacial (Duvany Yar) is dramatic. But if we compare Holocene and Duvany Yar pollen influx with two necessary corrections—(1) excluding from our analysis two pollen cores that exhibit dramatically unequal sedimentation rates, and (2) looking at pollen influx of herbs only, particularly graminoids and sage, *Artemesia*—then the pollen record offers surprisingly strong support for the concept of a Mammoth Steppe.

Two pollen profiles constitute the main evidence on which Ritchie, Cywnar, and Colinvaux base their objections to a Pleistocene Beringian vegetation that could support a complex community of large mammals during the glacials. One is from Hanging Lake, in northern Yukon Territory (Cwynar 1982) and the other is from St. Paul Island, one of the Pribilofs in the Bering Sea (Colinvaux 1981). Profiles of pollen influx are most credible when the sediments are similar and sedimentation rates are comparable throughout the core. Both cores get poor marks in each category. The Hanging Lake core has a long Holocene section, but the earlier 13,000–30,000 years are crammed into a short lower segment. The sharp break in deposition rate occurs at about 270 cm, dated at 12,800 + − 320 (GSC 2,846). Below that point there is a little more than half a meter of core; immediately beneath that point Cwynar begins to identify significant amounts of Tertiary pollen, presumably weathered from bedrock. There are also bedded aquatic peat zones beneath that break. Organic content is virtually nonexistent, and it is almost sterile of pollen. These characters suggest that during the Pleistocene the pond was dry and thus experienced mostly aerobic conditions or, at most, held intermittent annual or seasonal standing water during glacial times. Whatever its degree of sedimentary continuity, the Hanging Lake site does not seem an optimum core to use for a general model of pre-Holocene vegetation reconstruction in Beringia. No other core in the north shows such a dramatic discrepancy in herb influx numbers during Holocene and peak glacial, with one exception—the core taken from St. Paul Island (Colinvaux 1964).

The deep core from St. Paul Island (Cagaloq Lake) also has a well-dated Holocene section down to about 10,000 years ago. But farther down, the core is mainly sand, suggesting, as Colinvaux proposes, that a permanent pond was only established at the beginning of the Holocene, with the onset of the oceanic climate of the flooded land-bridge. Like the Hanging Lake core, most of the St. Paul core is Holocene; the condensed earlier portion of sand has virtually no pollen, as one would predict from aerobically exposed, wind distributed sand. Although there are intermittent layers of silt in the sand it is not possible to determine influx in these bands because the sedimentary rates during these times are unknown. But we do not have to depend on these two cores for paleoreconstruction. There are pollen profiles from other Alaskan lakes which better meet sedimentary constraints for pollen influx studies.

Before turning to these other pollen profiles, let us discuss the lumping of arboreal and nonarboreal pollen to produce total pollen concentrations or influx. One characteristic of northern trees is their ability to produce large volumes of pollen. Hiking through an alder thicket in the early spring, one can literally turn golden with pollen dust. I would argue that to compare total pollen influx of Holocene forests with glacial-aged (Duvany Yar) herb communities is not very meaningful. Ostensibly, influx studies allow one to separate chronological patterns of different taxa; this cannot be done in percentage diagrams. Thus, I use these influx studies where arboreal and nonarboreal pollen influx are listed separately, and set aside the arboreal pollen.

Most herbs in the north are not big pollen producers. Many herb families that produce significant quantities of pollen at temperate latitudes are rare or minor elements in northern vegetation. These include ragweeds, *Ambrosieae*, goosefoots, *Chenopodiaceae*, and composites (except for *Artemesia*), *Compositae*. Furthermore, the ability of graminoids to shift from sexual to vegetative reproduction, in response to grazing pressure, means grasses can be silent in the pollen record.

Despite these problems, influx profiles, within taxa, allow us to compare herb pollen from glacial times with that of the late Holocene. We know there is a large herb biomass in many parts of Alaska today. In fact, herbs probably constitute as much of the ground cover as trees, if we count understory cover, muskegs, lowland tundra, alpine tundra, and so forth. Thus we must ask if there was more or less herb biomass during the glacials. Comparing pollen accumulation in that way we could formulate a more meaningful influx comparison of herb biomass between today's herb vegetation and the unknown herb community of the peak glacial.

In figures 9.1–9.4 I have compared the four published cores extending back beyond 24,000 B.P. and the beginning of the Duvanny Yar Interval, the last stage of the last glacial (isotope stage 2), in which influx data were calculated. Other published cores with influx data either do not go beyond the Birch period (14,000) or go only sightly beyond it at the very bottom of the core. Judging from these cores one might conclude that there were few permanent lakes during the full glacials.

Following pollen influx down the core, the amount of herb pol-

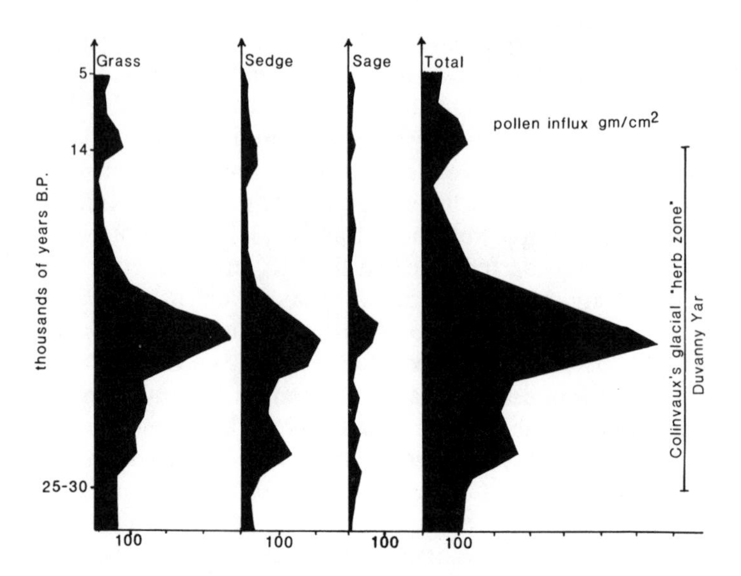

Fig. 9.1. Imuruk Lake herbaceous pollen. Colinvaux's (1964) deep core from Imuruk Lake was the first core to reach into the interstade (Boutellier Interval). I have used only the portion of the core that is within the radiocarbon range and have calculated influx from his pollen concentration figures. The greatest influx of herb pollen is during the Duvanny Yar Interval, or the last glacial maximum. This contradicts the idea that pollen influx was almost nonexistent during full glacial as it is in a polar desert.

len increases by a factor of three once past the Holocene and into the glacial. If the implicit assumption that pollen influx correlates directly and positively with volume of herb vegetation were correct, then one might conclude that full glacial herbaceous vegetation was much greater in volume than that of today.

Dating in these three cores is relatively good, although there are a few inverted radiocarbon dates. Fortunately, we know enough about patterns of postglacial arboreal recolonization to use that as a cross-check on the carbon dates. All three cores show a taxonomic profile consistent with the well-dated chronology of other Holocene and late glacial-aged cores. Squirrel River (Anderson 1982, 1985) dates back beyond 23,000, and the base of the Kaiyak Lake core is beyond the limits of radiocarbon range (Anderson 1982, 1985); Joe Lake is at 28,000 (Anderson 1988), while Colinvaux's (1964) Imuruk Lake core extends back at least to the last interglacial (but I have only used the portion pertaining to the late glacial and Holocene). Dates in this core are unclear. I have assumed, as did Colinvaux, that the penultimate mode of arboreal pollen represents the last interstadial or Boutillier (oxygen isotope stage 3). Colinvaux gave percentage and total pollen concentration, so I combined his figures to calculate the influx of individual taxa and herbs (120 cm per 15,000 years for the full glacial; the top of the core represents wave-disturbed sediments). In all cases, I used the author's chronological assessments. All three cores were taken from the central part of Beringia, with Anderson's cores closer to the Brooks Range. We can assume that they represent a general case for much of Beringia.

The picture presented by these three cores is fairly consistent. They show a pattern of relatively high herb influx during peak glacial, as compared to late Holocene segments. Grass pollen reaches its peak during full glacial, contrary to the conclusions of Ritchie and Cwynar (1982). *Artemisia,* or sage, also peaks during the full glacial; the same is true for the nongraminoid herbs. Influx of sedge pollen is less consistent. Sedge pollen is slightly more abundant during the wetter late glacial and Holocene segments of Anderson's two pollen cores (sedge forms a common vegetation on the margins of lakes in Alaska today, although a number of species are found only in dry habitats). Herb pollen is least in all three cores during the late

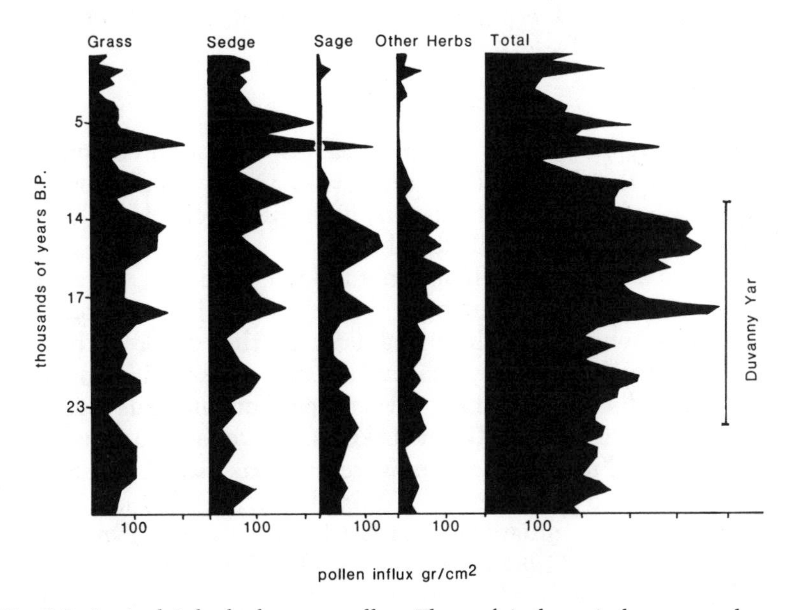

Fig. 9.2. Squirrel Lake herbaceous pollen. Three of Anderson's deep cores show a pattern similar to Colinvaux's core at Imuruk Lake. Squirrel Lake (Anderson 1982) shows almost twice the herb pollen influx during the full glacial (Duvanny Yar) as in the late Holocene.

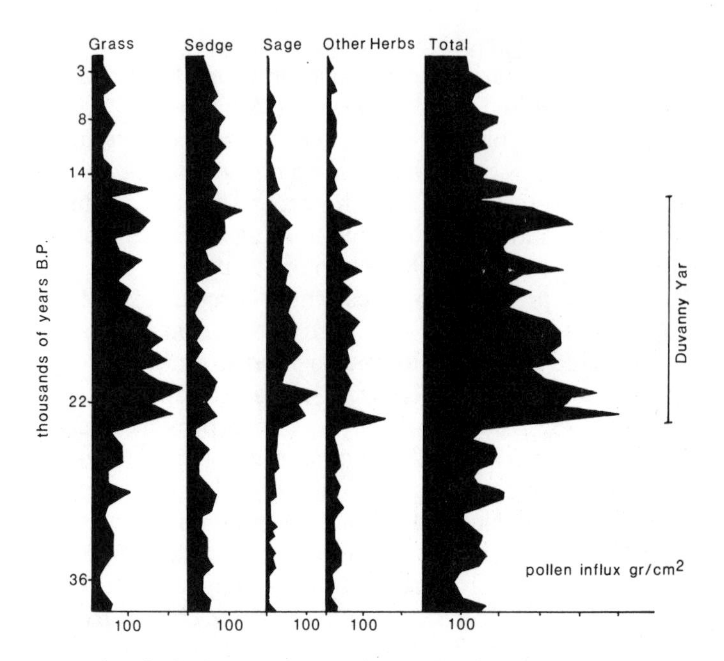

Fig. 9.3. Kaiyak Lake herbaceous pollen. The second of Anderson's (1982) cores, at Kaiyak Lake, exhibits a herb pollen pattern in the Duvanny Yar which is similar to the core from Squirrel Lake.

Holocene, despite the fact that herbaceous vegetation comprises a large part of the flora in those areas today. One can only conclude that today, and in the late Holocene, herbs produced relatively little pollen, compared to the full glacial.

Pollen assemblage still does not tell which species of grass and sedge were present, but one can propose (on the basis of southern steppe grass phytoliths, discussed later in this chapter) that increased aridity during the glacials would have caused different composition and distribution of herb vegetation than one finds today.

The Squirrel River, Kaiyak Lake, Joe Lake, and Imuruk Lake cores, which all reach past the most recent peak glacial, exhibit relatively large herb pollen influx during the early Holocene. In this, the deep cores are similar to more time-shallow cores (Cwynar 1982; Ager 1975; Ritchie 1984; Brubaker, Garfinkel, and Edwards 1983), but the latter do not reach beyond the sawtooth swings of herb pollen that precede the higher influx during the full glacial.

One thing is clear from these cores: herb pollen influx during full glacial times, as compared to the Holocene, was not reduced as

Colinvaux, Ritchie, and Cwynar have proposed; rather, several times more herb pollen was produced during peak glacials than in the late Holocene sections of the cores. This allows us to reject the first part of the palynologists' Proposition A.

The comparatively low herb pollen influx during the late Holocene and the last interstadial may be due to the increase of sites with humic soils. As I discussed in the last chapter, the acidity of the humic mat decreases available nutrients and its mulch insulates the soil, decreasing summer thaw and, in turn, reducing access to soil nutrients. Reduced nutrient and mineral access would produce more conservative reproductive effort (Grime 1979). Contrary to one's intuitive sense, the warm and wet conditions of the interstades-interglacials may have increased standing biomass but may not have greatly increased productivity, especially that useful to large mammals.

Now let us look at the second part of Proposition A: the argument that low annual pollen influx rates during the peak glacial indicate barren tundra, or polar desert, when compared with modern

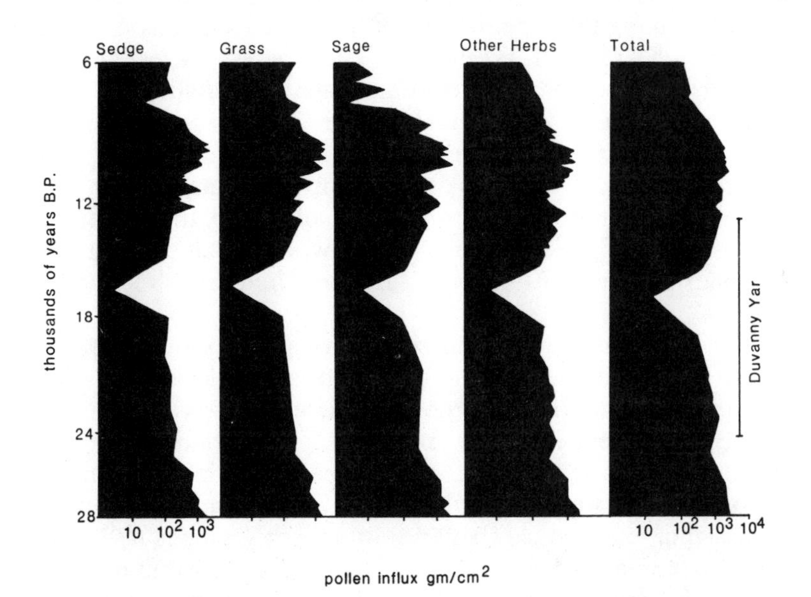

Fig. 9.4. Joe Lake herbaceous pollen. Unlike Squirrel Lake and Kaiyak Lake, the deep core at Joe Lake (Anderson 1988) has disparate sedimentation rates. Maximum sedimentation occurred in the Boutellier Interval and the Holocene. One of the six or so samples from the short Duvanny Yar core segment had little pollen, giving the graph a large bite at 16,000 yr B.P. However, the general pattern from the other five samples shows no dramatic pollen reduction during the Duvanny Yar.

vegetation. From diagrams in figures 9.1–9.4, one can see that annual influx rates are about 400 grains per cm^2 per year during full glacial from central Beringia. Ritchie and Cwynar (1982) compare these low productions with a grassland farther south. They chose a previously published study of five different sites from southern Manitoba. Unfortunately, southern Manitoba is not the best choice for a comparison because it is not thinly vegetated, shortgrass country as one might find near, say, Medicine Hat, Alberta. Southern Manitoba is characterized by tall and mixed grasslands extending northward from the midwest United States. Biogeographic maps even show big bluestem (*Andropogon gerardii*) to be one of the native dominants in parts of southern Manitoba. Big bluestem is a signature species of tallgrass prairie which once occupied the Iowa-Illinois corn belt.

Ritchie shows that annual pollen influx in the Manitoba sites ranges around 2,000 grains per cm^2 per year. Although his comparison with such a grassland is not quite appropriate, we can, for lack of other data, dissect this total as we did his Holocene northern totals and eliminate trees as well as, in this case, domestic cereals and groups comprising introduced European weeds (certainly one cannot justify including these pollen producers in the comparison). We can then focus on the influx values of wild grasses, sedges, and sage, which are most relevant for our comparisons. Surprisingly, the Manitoba sites are roughly comparable to the Alaskan fossil sites. When one looks only at these three dominant taxa from grasslands, total production at each location is not far from 400 grains per year. These are plotted in figure 9.5. I am not saying, of course, that Pleistocene Alaska was a tallgrass prairie or that it was as productive as a tallgrass prairie, only that influx data are very blunt and often misleading tools.

To determine whether these modern samples are really representative of influx volumes of preagriculture Holocene grasslands we must look at another core, north of these modern collecting sites. Ritchie's (1969) pollen profile from Riding Mountain, Manitoba, is informative about the meaning of pollen influx in grasslands. He interprets the changes from glacial spruce-dominated boreal forest, to grassland, to aspen parkland (for Ritchie this zone is rather enigmatic), to pine-birch boreal forests of today. Going bottom to top, the pollen influx in grains per year for all graminoids averages about 450, 450, 250, and 400, respectively, for these four different environments (fig. 9.6). These graminoid influx amounts are about the same as those from glacial portions of Alaskan pollen cores summarized in the same figure. Remember that pollen being sampled from the grasslands of Manitoba are east of (and hence downwind in the

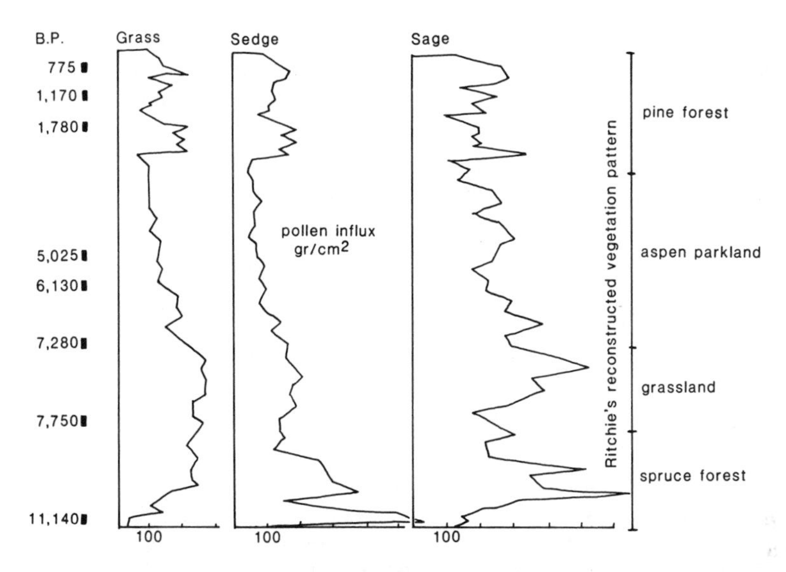

Fig. 9.5. Riding Mountain Manitoba, herbaceous pollen. To illustrate the limited usefulness of pollen influx data to our understanding of Mammoth Steppe vegetation, I used data from Ritchie's (1969) Riding Mountain core in Manitoba to plot influx of grass, sedge, and sage pollen. This site is downwind from the largest wild grassland in North America, and yet influx values for these wild herb taxa are modest. Ritchie's reconstruction of the dominant physiognomy (*right*) on the basis of the arboreal pollen is not readily apparent from influx of herb taxa. I conclude from this that influx data is a poor paleoecological tool in the far north.

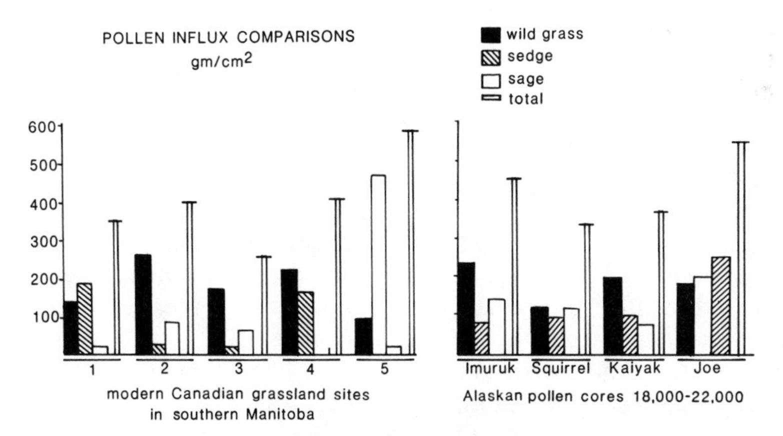

Fig. 9.6. The meaning of influx data from the Mammoth Steppe. This comparison of herb pollen influx at five Holocene grassland sites in Manitoba (Ritchie 1969) with Alaskan full glacial sites (Duvanny Yar Interval) shows that the patterns of pollen influx are not very different. This does not mean herbaceous vegetation was similar; it suggests, however, that we should use influx data with great caution.

Pacific Air Mass) one of the greatest grasslands on earth. If pollen from the general region is being broadly sampled, one would expect sage and grass pollen to be overrepresented in these cores rather than underrepresented.

Borrowing Ritchie's assumption that pollen influx is an indication of plant biomass, one could say that in Alaska plant biomass for graminoids and *Artemisia* per year during peak glacial was about the same as it now is on the mid- and tallgrass prairies of Manitoba. Although my purpose in these exercises is to show that one cannot take the details of influx values too seriously in reconstructing vegetation patterns, they do allow us to reject the remainder of Proposition A.

Actually, even a cursory review of the variations and idiosyncrasies of pollen influx figures is enough to make one cautious about paleoecological reconstruction on the basis of influx alone. Variations of influx values from within and between different sites, even when looking at the same taxa, range from the hundreds to the thousands (Waddington 1969; Davis 1983), with probably little difference in biomass or percentage of ground cover over large segments of the profile. Even on a long-term scale, taxa do not seem to have fixed reproductive efforts. But whatever factors enter the equation, we can say that influx data from Beringia do not require a polar desert. On the contrary, herb influx figures and mammalian fossils are quite consistent with the concept of a Mammoth Steppe, a rangeland suitable for large-mammal grazers.

As a vertebrate biologist, I do not purport to use these pollen influx data as the sole basis to argue for a grassland environment. The vertebrate fossils do that well enough. Rather I wish to show that existing pollen data for Alaska cannot credibly be used against such a concept. These herb influx totals for Pleistocene Alaska in figure 9.6 are around 400 per year. And we should remember that many forbs are insect pollinated and remain invisible in the fossil pollen record. The closed sward of alpine and low-land tundras today does not produce more pollen than found in Alaskan Pleistocene lake sediments, yet the standing biomass of these areas exceeds most arid steppes. If today we were to change the vegetation of Alaskan tundra to more xeric dominants, which winter cure with more nutrients, make them more widespread, and leave them exposed without much snow cover, they would be a productive rangeland for northern wild mammals of the mammoth fauna. I do not have to say "would," because mammalian fossils illustrate that Mammoth Steppe grazers flourished here.

Proposition B

The glacial environment, as reconstructed from pollen data, was so harsh that it could not possibly have supported the "mammoth fauna"; this is borne out by the lack of large-mammal fossils dating from full glacial (Duvanny Yar) times.

For nearly a century miners have been removing the silt overburden from hundreds of placer deposits in upland interior Alaska. Seldom have these efforts failed to produce large-mammal fossils. Some mines unearth a few bones; at many the fossils are measured in tons. Most of these fossils do not end up in museums, but are kept in Alaskan homes and sheds. Others leave the north and are scattered across the country, taken home as souvenirs. Many are sold on the curio market. No areas of unglaciated Alaska have not produced fossils of Pleistocene large mammals. The concentration of their distribution is almost completely correlated with mining activity. I gather from talking to Soviet paleontologists that this is true for Siberia as well. There are literally hundreds of Pleistocene large-mammal fossil localities in Beringia, and these invariably produce the same grazers, particularly bison, horses, and mammoths.

This is not the entire story, however. Not only are these fossil grassland mammals found in a "coarse grain" scattered across such a large area, but normally the dominant species occur together locally. The common Pleistocene large-mammal species are found in virtually all of the terrestrial Pleistocene paleontological sites. This is true whether one is in the high hill country (Guthrie 1968) or the lowlands (Bonnichsen 1979; Harington 1978). Species composition does not always retain the same proportions, for example, bison and sheep tend to increase in the uplands (Guthrie 1968). In some areas, a species drops out (such as the woolly rhino, which has not been found in Alaska) or changes in a geoclinal gradient so that a niche is occupied by another related species (red deer, *Cervus elaphus,* in the west and wapiti, *Cervus canadensis,* in the east), but for the most part the mammalian community retains its character. This is especially true for mammoths, bison, and horses. Their coexistence in individual sites suggests that these key species of the Mammoth Steppe were not mutually exclusive in spatial distributions, or chronologically exclusive, but were indeed adapted to coexist on a very local scale.

The impressive diversity of this large-mammal grazing community suggests that the Mammoth Steppe was (1) a more productive rangeland for large mammals than the natural vegetation in those areas today, (2) a vegetation pattern dominated by grasses, and (3) an environment without an appropriate modern analogue in the north (Matthews 1979; Guthrie 1982). However, Schweger and Habgood (1976), Ritchie and Cwynar (1982), and Colinvaux and West (1984) have argued that most of these mammals did not exist together, that the apparent high diversity is artificial, produced from a compilation of species from deposits that cover a long time span. Colinvaux and West, for example, argue that musk-oxen and bison probably did not coexist at all. Instead, musk-oxen took the place of bison during the interstades and interglacials. In European and Asian portions of the Mammoth Steppe there are many localities, both paleontological and archaeological, with these two species together, and in fact a wide assortment of these species is present in a given horizon in each site. However, the nature of deposition in Alaska and the Yukon Territory means that most bones do not occur in high concentrations in a typical "fauna," as one finds elsewhere in cave deposits. There are some exceptions, like Blue Fish Cave (Cinq-Mars 1979), with concentrations of bones from around 13,000 to 15,000 years ago, Loess Den (Guthrie 1988) at around 24,000 years ago, and Dixon's (1984) Porcupine Cave at around 21,000 years ago. These concentrations, like late Pleistocene deposits in Europe and Asia, do contain the characteristic species of the mammoth fauna (see synthesis in Hopkins et al. 1982).

Aside from the concentrations of bones with specific dates mentioned above, most of the thousands of Pleistocene bones now in collections occurred singly and have been dated in a haphazard manner for specific occasions and purposes. However, a number of dates have accumulated (Matthews 1982; Morlan and Cinq-Mars 1982). These dates, and others not published in those reviews, clearly show that the main members of the mammoth fauna were in Pleistocene eastern Beringia during the entire period covered by radiocarbon-dating and beyond. This is contrary to Ritchie's (1984) criticism of my original statement (Guthrie 1968) that virtually all of these fossil mammals date from late Wisconsinan. About 80% to 90% of these dates fall in the finite radiocarbon range of less than 40,000 years (although some dates in the high 30,000's and greater may actually have been brought into the finite range by contamination).

In Siberia and in Old Crow, Yukon Territory, the scatter of dates is most dense during the interstadial (around the 30,000's) and late glacial (14,000–12,000). This has prompted Ritchie and Cwynar (1982), Colinvaux and West (1984), and others to argue that the full glacial was unoccupied by large mammals or that the few here were "itinerants" from farther south. But the Siberian and Old Crow dates cluster during the two warm episodes because of major dating biases. In the Yukon Territory the dates are mainly from the once well-funded Old Crow early-man archaeological surveys. Full glacial dates from fossil mammals in the Old Crow lowlands cannot exist, because sediments of that age are all lacustrine. Proglacial lakes flooded the area during the full glacials (28,000–12,000), and hence large-mammal fossils do not occur in these sediments. In Siberia most radiocarbon dates are taken in association with archaeological sites, and there are few, if any, archaeological sites in the far north during the peak glacial. Furthermore, Soviets prefer to date archaeological sites (and other bone accumulations) with wood, and there was little or no wood in the Asian far north during the peak glacial times. Little wonder that there are so few dates from full glacial age on mammals from Siberia and the Old Crow.

There are, however, other means to test the proposition that large mammals were not in the far north during the Duvanny Yar full glacial. Since we cannot ever determine even approximate densities, except by some relative comparison or, in extreme cases, by absence, in a fairly sampled chronological scatter of dates the real question is whether there are fossils dating from the full glacial. Since there are a number of large mammal fossils of full glacial age, we can reject the latter part of Proposition B outright. A more useful question would be whether there are dates of this mammoth fauna during the full glacial, and whether they scatter without bimodial cluster during warm wet periods. We could fairly test chronological distribution of fossil mammals by examining dates from Alaska and the Yukon Territory, away from the proglacial lakes of Old Crow Flats. This would test Proposition B and the Mammoth Steppe concept.

We can use bison for our first example. I collected the dates in table 9.1 from a brief survey of Alaskan Pleistocene bison.

The existence of glacial-aged (14,000–25,000 B.P.) bison fossils allows us to reject Proposition B immediately. Furthermore, we can respond to the question of nonrandom clustering of dates outside of the peak glacial. The probability of such clustering is beyond the normal acceptable level at $p = .028$. There is a continuous scatter through the full glacial, contrary to the proposition of Ritchie and Cwynar (1982), Ritchie (1984), and Colinvaux and West (1984) that large mammals were confined to, or even concentrated in, the interstades and interglacials. It is apparent from these data that there are no striking absences of bison from Alaska during the last full glacial (Duvanny Yar Interval), from about 25,000 to 14,000 years ago.

Horses (*Equus* sp.) in Alaska and the Yukon Territory (minus Old Crow) show much the same pattern as bison (see table 9.2).

Equid samples used for radiocarbon dating were submitted precisely because most had no stratigraphic context; thus the dates can be considered as random samples from late Pleistocene sediments. The equid dates, like those of bison, do not show a marked bimodal curve on either side of the full glacial. The probability that these dates cluster outside of the 14,000–25,000 B.P. full glacial span is unacceptably low at $p = .018$.

Table 9.1 Chronological Distribution of Alaskan Pleistocene Bison

Date	Confidence Limit	Lab No.	Locality
>40,000		SI-291	L. Eldorado Creek
>39,000		SI-840	Cripple Creek
>37,000		I-9273	Tanana Bluffs
>35,000		SI-844	L. Eldorado Creek
>35,000		RL-402	Ester Creek
36,425	+2575/−1974	QC-891	Pearl Creek (Blue Babe)
32,700	980	ST-1632	Dome Creek
32,340	+1070/−1250	DIC-2123	Ikpikpuk River
31,980	4490	SI-843	Fairbanks area
31,400	2040	ST-1721	Dome Creek
29,295	2400	SI-842	Cripple Creek
26,900	2400/3400	AU-90	Baldwin Peninsula
26,760	300	SI-355	Lost Chicken Creek
25,750	+910/−1040	DIC-2125	Richardson Highway
25,090	1070	SI-850	Upper Cleary Creek
24,140	2200	SI-445	Fairbanks Creek
22,540	900	SI-292	Fairbanks area
21,780	310	DIC-1334	Porcupine Cave
21,300	1300	L-601	Fairbanks area
20,445	885	SI-837	Fairbanks Creek
18,000	200	SI-841	Manley Hot Springs
17,695	445	SI-851	Dome Creek
17,210	500	SI-454	Fairbanks Creek
17,170	840	SI-838	Fairbanks Creek
16,400	2000	M-38	Fairbanks area
15,380	300	SI-453	Fairbanks Creek
14,410	315	K-1210	Trail Creek Cave 9
12,460	320	SI-290	Upper Cleary Creek
11,980	135	ST-1633	Fairbanks Creek
11,735	130	ST-1631	Cleary Creek
10,715	225	SI-1561	Dry Creek
5,340	110	SI-845	Goldstream Creek
3,000 approx.		GX-6750/GS/6752	Delta Junction
470	90	SI-852	Chester Creek

Table 9.2 Chronological Distribution of the Horse

Date	Confidence limits	Lab No.	Locality
>40,000		I-9320	Ikpikpuk River
>40,000		I-9319	Ikpikpuk River
37,320	+1780/−2300	DIC-3095	Lost Chicken
35,620	+1530/−1900	DIC-3100	Fairbanks Creek
32,270	1500	I-9275	Ikpikpuk River
30,250	+890/−990	DIC-3092	Cleary
28,600	+690/−760	DIC-3099	Goldstream
26,830	+1230/−1450	DIC-3097	Lost Chicken
25,750	+910/−1040	DIC-2125	Richardson Highway
24,070	680	I-9320	Birch Creek
24,070	+380/−410	DIC-3098	Lost Chicken
23,920	620	I-9321	Fairbanks area
23,910	470	I-9318	Ikpikpuk River
23,340	388	DIC-3094	Lillian Creek
22,370	+300/−310	DIC-3093	Lillian Creek
21,050	+320/−340	DIC-1333	Porcupine Cave
20,810	410	I-9274	Ikpikpuk River
19,250	360	I-9371	Ikpikpuk River
18,640	205	DIC-3096	Lost Chicken
18,450	+200/−210	DIC-3091	Cleary
17,190	240	DIC-2418	Titaluk River
16,270	230	I-9271	Dominion Creek
15,750	350	K-1210	Trail Creek Cave
14,990	220	I-9316	Dominion Creek
13,640	410	I-9422	Fairbanks Area
12,900	100	GSC-2881	Bluefish Cave

Although thousands of mammoth bones have been found in Alaska, few have been dated. Dated mammoth bones show no tendency for a thinner spread during the full glacial (Duvanny Yar Interval). In fact, when we look at all the mammoth dates from Alaska, including those from the uplands of the Northwest Territory and Yukon Territory, the dates still do not show any modality away from the full glacial (see table 9.3).

As with the dates from bison and horse bones, one could conclude that mammoths were present in eastern Beringia throughout the Pleistocene covered by radiocarbon range. Sample size of dated mammoth bones is not sufficient to do a valid probability test, but one can see from the scatter that there is no hint of clustering outside the full glacial. The dates show that mammoths existed in eastern Beringia during the full glacial, again allowing us to reject Proposition B. And as I concluded from a comparison of several thousands of bones found in different late Pleistocene sediments

Table 9.3 Chronological Distribution of the Mammoth

Date	Confidence Limits	Lab No.	Locality
>35,000		I-9342	Ikpikpuk River
32,700	980	ST-1632	Fairbanks Creek
32,340	+1070/−1250	DIC-2123	Ikpikpuk River
31,000	2,040	ST-1721	Dome Creek
30,300	2000	I-3576	Whitestone River
26,770	+490/−520	DIC-2124	Richardson Highway
22,850	250	V-48-152	Colorado Creek
21,900	320	GSC-1760	Melville Island
21,050	310	DIC-1333	Porcupine Cave
19,440	290	I-8578	Tununuk
15,550	130	GSC-3053	Bluefish Cave
15,090	170	B-5691	Colorado Creek
13,340	115	DIC-2130	Teklanika River

(Guthrie 1968), these three large grazers account for the largest proportion of mammalian fossils.

Proposition B proposes that the region must have been barren tundra or polar desert, and as such large grazers could not have been present during glacials or were at least not present contemporaneously. This perception is voiced by Colinvaux and West (1984): "in early interglacial times . . . game herds harried by large felids and canids may have existed in a complex landscape . . . but the Beringia of land bridge (full glacial) times was an unproductive place" (p. 12); Ritchie and Cwynar (1982) state: "The large and diverse ungulate populations probably were present during Pleistocene interstadials more than 30,000 years ago . . . rather than during the time of the herb zone, 30,000 to 14,000 years ago" (p. 113). The radiocarbon dates above clearly do not support this perception.

The logic of Colinvaux and West's (1984) comment that the scarcity of radiocarbon dates on large mammal bones (of one date per thousand years) show that few animals were around is peculiar; this is comparable to saying that the few pollen cores studied from Alaska mean that there are few lakes. It is as obviously false to conclude from the small number of dated fossils that few bones of Pleistocene mammals have been found in Alaska. In fact, the radiocarbon-dated bones are a tiny sample of the tens of thousands of mammal bones removed from natural cut banks and mining exposures. Interior Alaska is extraordinarily rich in late Pleistocene vertebrate fossils; bones from the Fairbanks area are scattered in museums around the world and even predominate in some museums. Several floors of Pleistocene mammal fossils in the Frick Wing of the American Museum of Natural History are dominated by Alaskan material. It is true that hundreds of dates for each species would give us more information on a variety of paleoecological topics. Paleontologists working in Beringia can certainly be criticized for not having enough dates on fossil mammals, but the dates listed above, while few, are probably more numerous than for individual large-mammal species and for individual localities from other regions.

There are fewer dates on other species of the mammoth fauna. In some instances this is because the species are comparatively rare and dating would involve sacrificing the whole specimen. Museums are reluctant, for example, to donate Pleistocene lion or saiga specimens for dating. Species that did not become extinct are also poorly dated, as the chronology of their disappearance has not been an issue. Finally, fossils in most other places are found in concentrated sinkhole or cave deposits, where one radiocarbon date can be assigned to a number of fossils in the assemblage. Such concentrated faunal assemblages are uncommon in the north; usually each fossil must be individually dated.

There are some dates on species other than bison, horse, and mammoth. Colinvaux and West (1984) and Ritchie (1984) have argued that these other members of the mammoth fauna must have existed during interglacials or interstadials, that they could not have lived together during glacials. Yet published dates exist showing that in addition to horse, bison, and mammoth, many other species were living in Beringia during the full glacial (14,000–25,000): lion (*Felis*), musk-oxen (*Ovibos*), bonnet-horned musk-oxen (*Symbos*), camel (*Camelops*), sheep (*Ovis*), woolly rhinoceros (*Coelodonta*), and caribou (*Rangifer*) (Matthews 1982; Dixon 1984; Vereshchagin and Baryshnikov 1984). Proposition B ignores this evidence.

One cannot dismiss these radiocarbon dates as unreliable. Virtually all are from bone collagen or mummified soft tissue from frozen sediments. Alaskan Pleistocene bones generally retain over 90% of original bone collagen, and many still have some marrow. (Bone carbon from the mineral, or apatite, fraction is much less reliable for dating.) Likewise one cannot argue that there was a taphonomic bias to the preservation of bones of glacial age. Taphonomic bias, if it exists, probably runs against preservation during full glacial because this was a time of less moisture, and hence of less slope-wash redeposition of silt to cover bones.

Proposition C

The taxa of glacial (Duvanny Yar) pollen show that species from southern steppes were not present; glacial taxa were restricted to tundra species found throughout the north today. Thus, glacial vegetation was not similar to a steppe but was barren tundra, like that which now exists in the very far north.

Ritchie and Cwynar (1982) argue that the Mammoth Steppe did not exist and that the term *steppe* or *grassland* is inappropriate because "so far no taxon has been recorded in any late Pleistocene steppe element . . . or to the tundra steppes mapped . . . in western Greenland and in the Kolyma and Indigirka River watersheds of northern Chukotka" (p. 121). My argument that there was a grassland in the north during much of the Pleistocene does not imply that the same grassland community we now see to the south was then located in the far north. There is evidence of steppe species in interior Alaska during the full glacial (Bombin 1984), for example, *Stipa,* needlegrass, the main indicator species of extant steppe communities in North America and Eurasia. Needlegrass has retreated southward since the Duvanny Yar and no longer grows in Alaska. It was found in a ground squirrel nest from the Fairbanks area, dated at 18,230 +/− 410 B.P. (QC-668). On that important identification alone we can reject Proposition C, but there are additional data.

Indeed, there are numerous cases of northern Pleistocene sites containing faunal and flora elements that are now found only in grasslands farther to the south. This has provoked some interesting interpretations to avoid the conclusion that such species were once native to the far north. When Tikhomirov (1958) found graminoids that now grow very far to the south associated with Siberian mammoth mummies, he could only explain this by having the mammoth migrate far into the barren north with a stomach full of southern grasses. But this is, of course, impossible as proboscidians have a very short gut transit-time, on the order of twelve hours (Laws, Parker, and Johnstone 1975).

Failing to see how the far north could sustain mammoths and resorting instead to long-range migrations sounds familiar; indeed that is what Churcher (1980) and Colinvaux and West (1984) recently have proposed. However, the several-thousand-mile winter migrations of mammoth are not very probable. Mammoths had heavy graviportal, distally muscled legs, and they required much more energy to walk than other mammals. Robert White (pers. comm.) has calculated that, at lean weight, it takes a proboscidian, such as a mammoth, twice the energy to walk as to stand. In a more gracile ungulate, such as a caribou, however, walking requires only a 20% increase in energy over standing. It would have been energetically more efficient for mammoths to avoid long travel over a short period. Thick subcutaneous fat deposits on mammoth mummies suggest that, like other Alaskan ungulates, they did not undertake thousands-of-mile migrations to warmer climates farther south for the winter. Nomadic ranging is not the same as a mass exodus in a long migration. Besides, during glacial maxima, mammoths would have to go all the way from Alaska to Montana over several thousand miles of glaciers. Churcher theorized (1980) that Great Plains mammoths migrated from Alberta and Montana south to the warm Gulf Coast.

In any case, the "migrating mammoth" theories are internally contradictory, for if the north's summer environment were as barren as proposed, mammoths probably would not have undergone an expensive seasonal migration to get there. Again we do not have to argue solely from theory; numerous Alaskan fossil mammoth teeth (bison and horse as well) show an annulus constriction at the end of the root, indicating a winter death (Guthrie, pers. obs.).

Among modern northern ungulates the more sedentary species generally put on more fat than do the long-distance migrators. The former store fat; the latter adopt a strategy of traveling light to reach a range where winter resources are available. Northern winter ranges are well below nutrient and caloric maintenance requirements for extant ungulates (this was probably true in the Pleistocene as well), so a common strategy is to decrease activity, where possible. Most northern ungulates actually enter a dormant state in the winter. By not trying to grow and by expending as little energy as possible they can survive on below-maintenance rations for a long time. As winter proceeds they lose body fat and their condition slowly declines; most enter spring with deficits, but alive, and quickly recover on rich early summer greenery.

Even for the more mesic interstadial (Boutellier Interval) there are extralimital steppe forms in the far north. Plant macrofossils (e.g., Ukraintseva 1981) and insects (Kislev and Nazarov 1985), which now live only on dry steppes far to the south, are commonly identified at northern Siberian sites. The same is true for Alaska and the Yukon Territory (e.g., Matthews 1979; Nelson 1982). Plant species and beetles most characteristic of the Mammoth Steppe seem to have remained in the area or somewhat south in "relict" habitats, if we define that term liberally to include habitats more similar to dry Pleistocene conditions and do not necessarily imply that particular plant species have remained on a site as continuous occupants.

Fig. 9.7. Two steppe mammals now extinct in Alaska. American badgers (*Taxidae taxus*) and black-footed ferrets (*Mustela eversmanni* ~ *M. nigripes*) were present in Alaska and the Yukon Territory during the late Pleistocene. Their presence is strong corroborating evidence of a steppe environment. The steppe ferret, *M. eversmanni*, is the Asiatic remnant of this Alaskan relative, as is *M. nigripes* on the American steppes. Their disappearance in the far north supports the theory that ecological change rather than human overkill led to the large mammal extinctions in Beringia. Humans would not likely have hunted these two small carnivores to extinction.

Kislev and Nazarov (1985) point to the steppic character of many insects in deposits from the last glacial on the Kolyma Lowland in northwest Asia. The genera *Conioclenonus* and *Stephanochelonus* are very frequent in those deposits, whereas today they occur primarily in southern steppes with a few isolated populations in relict habitats in the north. One insect group more abundant in the Pleistocene than at present was the dung beetle, indicating that greater quantities of large-mammal dung were present than today (Matthews 1982).

Large- and medium-sized mammals provide the most striking examples of species found in Alaskan late Pleistocene deposits but which now live much farther south. In addition to the bison, horses, hemionids, and saigas already mentioned are American badgers (*Taxidae taxus*) and Eurasian steppe ferrets (*Mustela eversmanni*) (fig. 9.7), whose counterparts are today restricted to more arid mid-latitude grasslands (fig. 9.8). We do not have direct dates on these two species as their bones are too small for conventional radiocarbon dating, but we do know that their main prey, ground squirrels (*Spermophilus*), lived in the same localities near Fairbanks only during the full glacial (marine isotope stage 2). Ground squirrel nests offer abundant and suitable material to date. Dates we have for ground squirrel nests are listed in table 9.4.

Like the fossil dates discussed earlier, the ground squirrel materials were collected in an unbiased manner from late Pleistocene sections. It is unlikely that badgers and ferrets were present without ground squirrels, which is not to say that where there are ground squirrels there are badgers and ferrets. Ground squirrels now live in alpine and tundra areas of Alaska, which do not meet other requirements of badgers and ferrets.

Colinvaux and West (1984) argue that the failure of small mammals from southern grasslands to colonize the north shows that there was no northern grassland and that large mammal remains have given us a false picture. Small mammal fossils, indeed, do not present the same picture as those of large mammals, but environmental limits for small mammals are not the same as limits for large mammals.

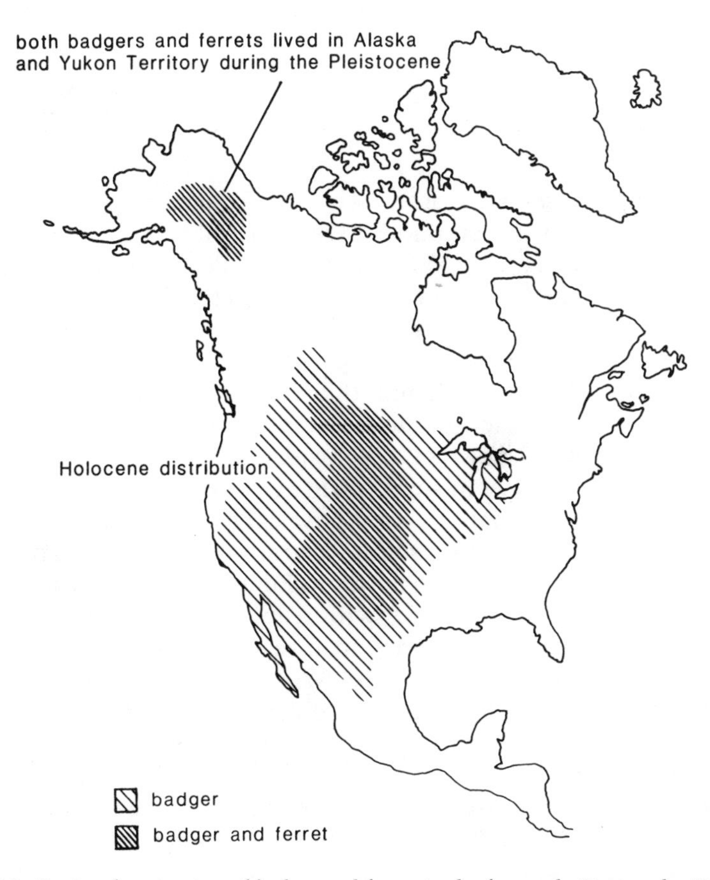

Fig. 9.8. Regional extinction of badger and ferret in the far north. During the Holocene, badger (*Taxidae taxus*) and black-footed ferret (*Mustela nigripes*) have been confined to the midcontinent grasslands. But in late Pleistocene times both species lived as far north as Alaska and the Yukon Territory.

Table 9.4 Chronological Distribution of the Ground Squirrel

Date	Confidence limits	Lab No.	Locality
26,950	+6030/−3405	QC-672	Cripple Creek
24,525	+1680/−1390	QC-675	Esther
24,000	+1500/−1200	QC-666	Goldstream Creek
23,380	+450/−2680	QC-667	Cripple Creek
23,130	+1765/−1445	QC-670	Goldstream Creek
22,280	+870/−770	QC-669	Esther
21,750	+1400/−1200	QC-660	Cripple Creek
19,660	+/−30	QC-673	L. Eldorado Creek
18,230	+/−410	QC-668	Engineer Creek
17,980	+/−575	QC-661	Cripple Creek
14,860	+/−860	GX-0251	Chatanika River
13,350	+/−265	QC-664	Lower Sulphur Creek, Yukon Territory
11,170	−(+/−205)	QC-662	Glacier Creek, Yukon Territory

There are several reasons why small mammals are not good indicator taxa for a Mammoth Steppe. Small mammals are more limited by cold (fig. 9.9). Most large mammals have a broad thermoneutral zone and do not have to increase metabolic rate until many degrees below zero. Large mammals are more sensitive than small mammals to the availability of appropriate foods and snow depth. Thin snow cover on the Mammoth Steppe aided large mammals but had the reverse impact on mouse-sized small mammals which require considerable snow cover as insulation. Such small mammals do not now overwinter in areas blown free of snow.

During full glacials, rigorous physical barriers separated Alaska from environments farther south, barring small-mammal colonization. Glacial connections to northern Eurasia were open for colonization, and, despite difficulties with reduced snow insulation, characteristic steppe species of small mammals moved into the Eurasian portion of the Mammoth Steppe. Vangengeim (1975) pointed out that the steppe vole, *Lagurus*, was much farther north than it is today. Likewise, the Asian ground squirrel, *Spermophilus major*, steppe hamster, *Cricetus cricetus*, and steppe pika, *Ochotona pusilla*, moved north and far west of their present range, reaching England. The jerboa, *Allactaga jaculus*, an occupant of the Asian steppes, also moved into central Europe during the last glacial (Kahlke 1975).

Instead of southern small mammals moving into Alaska during the glacials, small mammals with northern cold and snow adaptations moved south. The brown lemming, *Lemmus*, and the collared lemming, *Dicrostonyx*, moved into the central part of North America, outcompeting local microtines (fig. 9.10). In addition to the barriers formed by continental glaciers, the combination of cold and thin snow cover probably made it difficult for southern small mammals to colonize the far north, despite the fact that some of those species might have been well adapted to northern vegetation. The same happened in the Old World; *Lemmus* and *Dicrostonyx* are common Pleistocene fossils in Europe all the way to the Mediterranean. Today these arctic lemmings dominate small-mammal communities only in the far north.

Colinvaux and West (1984) are correct in observing that the Alaskan Pleistocene small-mammal community had many of the same species as today, but the conclusion they draw from this, that the vegetation was much the same, does not necessarily follow. Alaskan small mammals (*Dicrostonyx*, *Lemmus*, and northern *Microtus* species) are also found in the caves of southern France among most of the same Mammoth Steppe fauna (bison, horse, mammoth, reindeer, lion, and even saiga) that occur in Beringia, and certainly no one would want to label the Perigord, the type site of the Magdalenians, as polar desert on the basis of small mammals. As in the Würm

Fig. 9.9. The effect of snow depth on large and small mammals. Deep snow is generally detrimental to large mammals—it restricts their mobility and ability to obtain food, especially for species that feed on plants under the snow. Small mammals are usually more sensitive to winter temperatures. Deep snow insulates small mammals from extreme cold, creating a special environment called the subnivian space at ground level.

IV of the Perigord, fossil small mammals from glacial (Duvanny Yar) Alaska are not as good an indicator of climate and vegetation as are large mammals.

At this point I turn from testing propositions of the palynologists to my list of arguments for the existence of a Mammoth Steppe. I have been discussing the Mammoth Steppe climate at its most extreme (during full glacials), but we know aridity and grasslands could

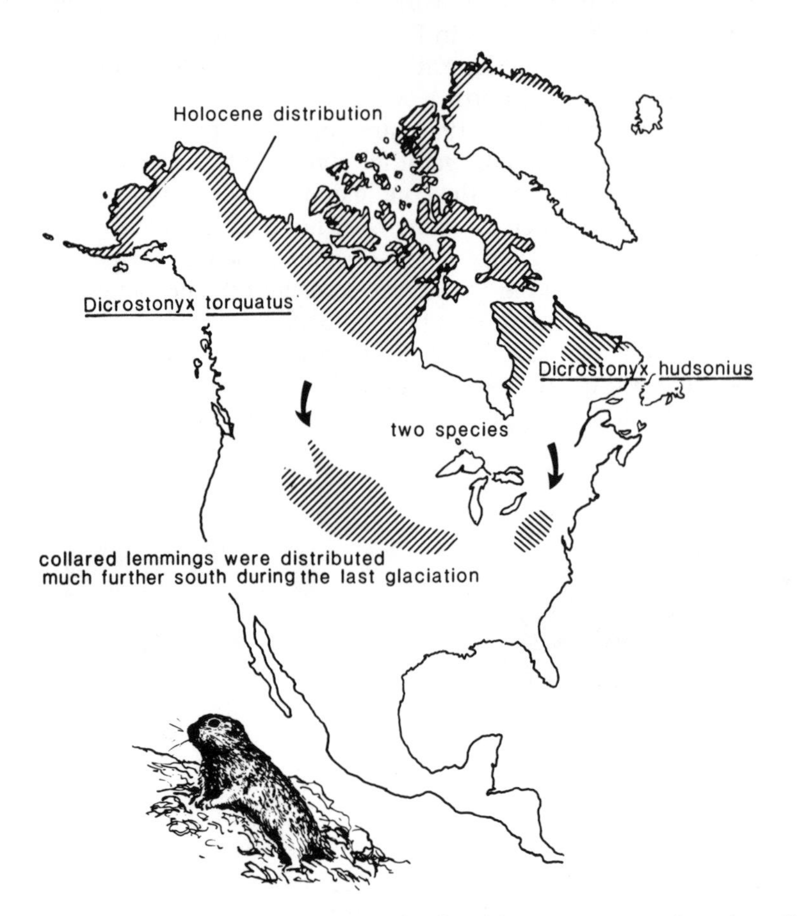

Fig. 9.10. Pleistocene range extension of collared lemmings. A number of small mammal species now found in the far north extended their range southward during the last glacial. Collared lemmings, *Dicrostonyx*, are divided into two species in North America (here shown in hatch marks of different directions). Both of these species are found as late Pleistocene fossils far south of their present ranges. *Dicrostonyx* has a number of cold adaptations which allow it to tolerate thin snow cover and severe cold. Because of low snow cover, species of small mammals adapted to southern steppes did not move northward during this cold, dry episode.

Fig. 9.11. Grassland—boreal forest ecotone. This southern margin of boreal forest seems more directly limited by aridity than temperature. Here aspen trees (*Populus tremuloides*) are able to grow on the more moist, north face of these grassy hills in central Alberta. Aridity may also have been a critical factor eliminating trees from interior Alaska during the late Pleistocene.

interfinger with trees as they did during the interstadial when Blue Babe was alive, because bison, horses, mammoths, and even saigas, main elements of the mammoth fauna, continued right through this period (Matthews 1982). In fact grasslands mingle with boreal forest today in both North America and northern Asia, but this mixing of grassland and taiga is more apparent along the southern border of the boreal forest than along its northern border (fig 9.11). This suggests that the aridity controlling the southern limit of the boreal forest was a more important force limiting trees during the Pleistocene than was cold, which today sets northern tree limit.

Pleistocene Paleoecology of Beringia as the Northeast Wing of the Eurasian Mammoth Steppe

I have argued (Guthrie 1980, 1982, 1984a, 1984b) that the character of Pleistocene Beringia was a special case of a larger, more inclusive biotic zone, the Mammoth Steppe. The concept of Beringia tends to confine our vision of the larger unit, as Beringia was only the eastern wing of the Mammoth Steppe (fig. 9.12).

The ubiquity of the mammoth fauna over the northern part of three continents suggests that their habitats—available grass, shallow snow cover, dry summers, and windy climate—were also wide-

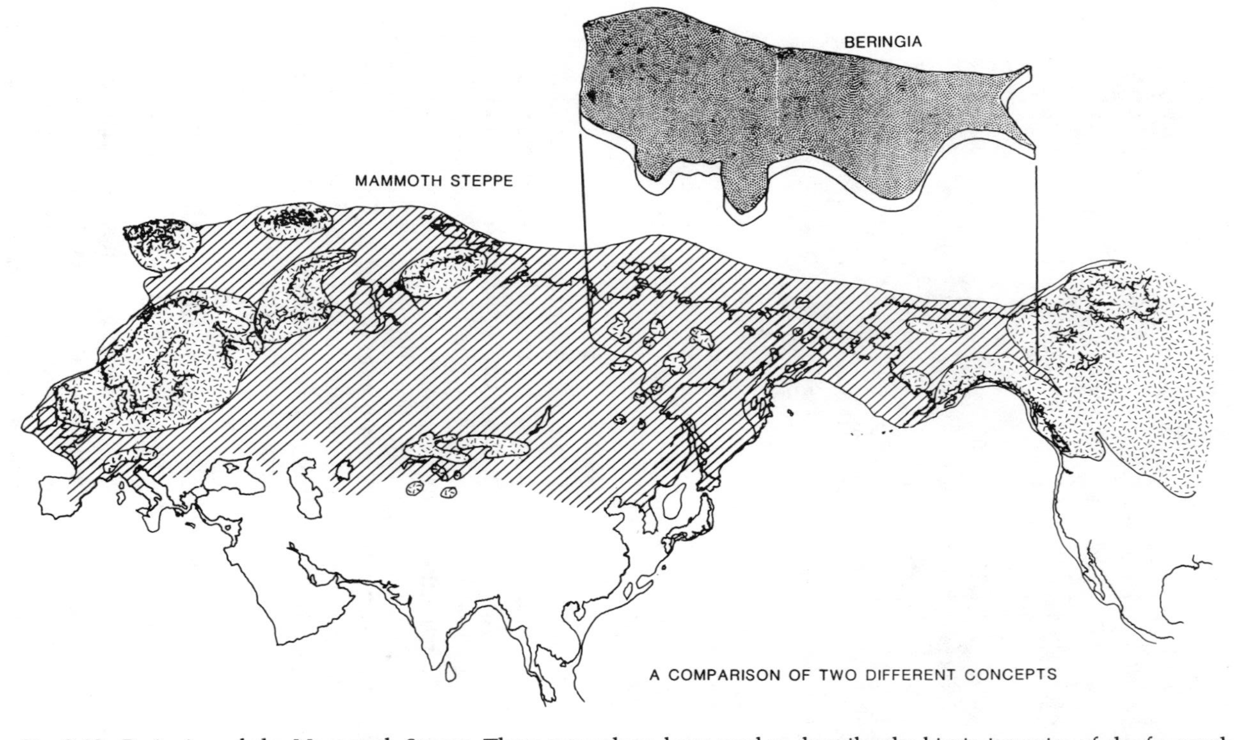

Fig. 9.12. Beringia and the Mammoth Steppe. These terms have been used to describe the biotic integrity of the far north during the late Pleistocene. The earlier concept of Beringia is especially useful in describing Siberian-Alaskan connections. Beringia extends from the Lena River in the east to the McKenzie River in the west. However, most major plant and animal taxa in Alaska and Siberia during the late Pleistocene also had affiliations with the entirety of northern Eurasia and Alaska–Yukon Territory.

spread, with a diverse permutation due to local influences. Indeed, these vegetational and climatic patterns have been reconstructed elsewhere throughout the Mammoth Steppe (e.g., Kowalski 1967b; Sher 1974; Yurtsev 1974; Vereshchagin and Baryshnikov 1982). The general character of Pleistocene vegetation in Alaska and the Yukon Territory cannot be fully discussed in isolation from its Eurasian relationships.

In that regard, Ritchie and Cwynar (1982) have implied that my proposed extension of the Mammoth Steppe westward from Beringia to Europe is incorrect and propose that Europe was also a polar desert. They use Woillard's (1978) pollen evaluation of the famous Grand Pile core in northwestern France as typical, stating that, "she identifies nine stadials . . . interpreted as representing open, discontinuous tundra and polar desert" (p. 120). That is not how I read Woillard's interpretation, nor do I find her using the terms *tundra* or *polar desert;* instead, she uses the pollen profile to interpret glacials as, "periods . . . characterized by the marked development of a *Graminae-Artemisia* steppe in which grow other steppe elements . . . and also by the presence of subarctic-alpine species" (p. 16). As in Beringia, the dominant pollen of the full glacial Grand Pile is grass and sedge.

A Gut-Level Argument for the Mammoth Steppe

The proposition has been put forth by many that most of the "mammoth fauna" were grazing specialists. Colinvaux and West (1984) and Ritchie (1984) have argued that bison, horses, and mammoths were not grazers and thus could get by on barren tundra or polar desert vegetation comprised of cushion plants, lichen, and miniature herbs and shrubs. Two lines of faunal evidence that could help resolve this issue are (a) some comparisons of glacial fauna with their nearest modern analogues and (b) information about anatomical specializations. Large-mammal groups have long evolutionary histories of partitioning dietary resources, and, given large enough samples, they can be good indicators of paleoenvironments (e.g., Vrba 1976).

Although we must be extremely cautious with arguments of analogy involving diets and relating diets to extinct species, there is least likelihood of error with bison, horse, and mammoth. These three species also comprised most of the large-mammal numbers and biomass (Guthrie 1968).

As I discussed in an earlier chapter, bison are adapted to open grasslands. Since the colonizing of central North America, bison south of Alaska have been concentrated in two areas: along the rain shadow of the Rocky Mountains where aridity created open plains; in the eastern part of the Gulf Coast in Florida, which was also apparently a more open savanna or grassland in late Pleistocene times. Bison in the east were restricted to more open parks in deciduous woodlands and were infrequent on the northwestern side of the Rocky Mountains. Remnant populations of European wisent, *Bison bonasus*, are found in atypical areas. Like the last feral horses and cattle confined to the Carmargue salt swamps in southern France, wild European bison now live in the last remaining fragments of wilderness—the forests of eastern Europe. Other areas have all been taken by agriculture; what once were bison steppes is now farmland. Bison lived on grasslands of those areas within historic times, including the valley parklands of Lithuania, the Carpathian uplands, the entire Caucasus, Transcaucasia, and the highlands of northern Iran (Flerov 1977). Peden et al. (1974) found that bison, in comparison to cattle, are more efficient digesters of low-quality grasses. Cattle and bison seem to have partitioned habitat: woodland and woodland parks for cattle; highland meadows, woodland parks, and open grasslands for bison (Guthrie 1980).

The last wild horses in Europe, the tarpan (*Equus ferus*), lived on these same eastern European steppes, away from intense agricultural development of central Europe, until they too were eventually displaced by human use of the land within historic times. The wild horse in Mongolia (*Equus przewalskii*) is closely related to the domestic horse and interbreeds freely with it, despite a difference in one chromosome. The habitat of the Mongolian horse was high arid steppes, where it has not been seen for over a decade. Horses closely related to these ranged into the northernmost parts of Siberia and eastern Beringia in the late Pleistocene. Likewise, hemionids (*Equus hemionus*), adapted to the high dry steppes, from Iran to Mongolia, were also present in northernmost areas of Siberia and eastern Beringia. The bones of *Equus*, both horses and hemionids, are frequent fossils throughout all Beringian habitats where bones are preserved, both uplands and lowlands. This seems to be the case for all times during the radiocarbon range within the Pleistocene. All living members of the genus *Equus* are specialized grazers, all occur in plains or open environments, and their Pleistocene bones are found in these same environments throughout the Holarctic.

Unlike bison and horse, mammoths have no close living analogue, at least not within the same genus. We do know that the main food of the two living species of proboscidians is grass (Laws, Parker, and Johnstone 1975; Olivier 1982). Judging from mammoth morphology, the woolly mammoth seems to have been a more grass-dependent proboscidian than either species of living elephant (Maglio 1973), but we need not rely solely on analogy and morphological features for a dietary reconstruction of these three species. Mummies of the Beringian bison, horse, woolly rhino, and woolly mammoth do exist, and we can see what these individuals ate before they died.

Several caveats must be mentioned before attempting to reconstruct past diets from gut contents. These are cautionary notes that come from having performed and seen performed over two thousand mammalian necropsies, including gut analysis. First, when animals take large mouthfuls of small-sized items they invariably incorporate secondary material along with their target food. Whatever is adjacent is likely to be ingested. A grizzly bear inadvertently eats many leaves and twigs along with berries in the fall, and grazers ingest a variety of plants in one bite. We cannot assume that the presence of a plant species as a minority dietary item in stomach contents means it was actively selected.

Second, animals choose from a variety of items ranging up and down a scale of preference, depending on what is available. Under starvation conditions they resort to the most unpalatable, uncharacteristic part of their diet. At the bottom of the scale are foods so toxic, fiberous, or low in nutrients that they cannot be digested. Ironically, most large herbivores that die of starvation are found with bulging-full rumens. This puzzled biologists until they realized the animals did not die from simple lack of food, but were reduced to eating undigestible food which clogged their digestive tract.

Third, there are different rates of decomposition in the gut; woody plants with an undigestible lignatious skeleton remain in the gut longer than leaves and can thus be overrepresented in a sample of gut contents. Also, animals often shift from their staple, representative diet for a short season to take advantage of transient resources. Most northern large ungulates are adapted to very specific winter diets, but their summer resources are more variable, with a higher interspecific overlap. Season of death thus becomes important in appraising gut contents. These warnings do not mean we cannot study gut contents to understand diet, but only that we must approach with caution.

As discussed in the first chapter of this book, large mammal mummies from the Soviet Union have been found with identifiable gut contents. At least two frozen mammoth mummies have voluminous quantities of identifiable plants in their stomachs. Both deaths occurred during autumn, in an interstadial period (Boutellier Interval, the same as Blue Babe). There is also a horse mummy, the

Selerikan horse from Siberia, dating in the same time range. Vereshchagin and Baryshnikov (1982) and Hopkins et al. (1982) in their syntheses have explained the concentration of mummies in this interstade by the greater quantities of silt moving downslope. We have to assume that the Boutellier Interval interstade was wetter in summer (most but not all of the carcasses are autumn deaths) than the Duvanny Yar (glacial), but not as wet as today, presumably with a more incomplete ground cover and exposed silt. Greater quantities of silt are needed to cover a large carcass than to cover individual bones. Of course this is only a statistical bias, as some frozen mummies and bones with soft tissue attached come from other times, even from peak glacials.

The Beresovka mammoth stomach contained mostly grass by volume (Ukraintseva 1981; Tikhomirov 1958) (additionally, samples taken from the stomach contents contained mostly grass pollen). There were, in addition, small percentages of a wide variety of other plants, including a variety of forb macrofossils. The other almost complete frozen carcass, the Shandrin mammoth, from the U.S.S.R. (Ukraintseva 1981), had gut contents that consisted of 90% grasses and sedges. In addition, there were twig tips of some woody plants—willow (*Salix*), larch (*Larix*), birch (*Betula*), and alder (*Alnus*). With the above evidence and the distinguishing morphology of mammoths, Olivier (1982) followed every other zoologist who has studied mammoths and concluded that the animals were grazers. Megaherbivore species such as mammoths, using the coarsest end of the grazing spectrum, make use of a poor-quality resource, one relatively undefended by antiherbivory compounds but deficient in nutrients and minerals (Guthrie 1982, 1984a; Olivier 1982). Agenbroad (1984), studying the plant remains in dung of late Pleistocene mammoths from the Great Plains, found grass and a supplement of minor herb and arboreal components. Elephants, like horses and bison, use poorer quality fibrous resources as a dietary staple and add supplemental dicots to complete their diets. Woody species in the frozen mammoth stomachs are twig tips, the very growth stage according to Bryant and Kuropat (1980) highest in nutrients. These poorly defended twig tips are high in the nutrients grasses lack (Olivier 1982). However, to avoid poisoning, the animal must limit consumption of any one species of these dicots so as not to exceed its ability to detoxify the secondary plant compounds (Bryant et al. 1985). That is especially true for cecal digesters (like horses and mammoths) which do not have rumen to assist in detoxification (Guthrie 1984b).

We can see this same pattern, using grasses as a dietary staple with supplemental dicots, in the stomachs of frozen horse and bison mummies. A Pleistocene horse, found at Selerikan in the Indigirka River basin and dating about 37,000 years before present, had a stomach and intestines containing 90% herbaceous material, of which *Festuca* grasses predominated, along with the sedge, *Kobresia.* The latter was identifiable to species, *Kobresia capilliformis,* by seeds (Ukraintseva 1981). This xeric sedge species, Ukraintseva remarks, is not present in the Indigirka area today, but it is a typical plant in the high dry mountains of central Asia, the Middle East, and Mongolia. She concluded that it had been abundant in the area where the horse died. Macrofossils in the gastrointestinal tract included small quantities of willow (*Salix*), dwarf birch (*Betula nana*) twigs, and very small quantities of moss (Ukraintseva 1981). This diet is reasonable for a northern grazer, especially given the interstadial vegetation available.

Stomach contents of a woolly rhino, *Coelodonta,* mummy from Yukutia (Churapachi) also contained mostly grass (Vereshchagin and Baryshnikov 1984). Woolly rhinos have the most complex teeth of any rhino, even more so than the African white rhino, *Ceratotherium,* which is a strict grazer; the Siberian and Polish woolly rhino mummies have a wide front lip for grazing, like those of African white rhinos.

Only one Pleistocene bison, located on the Krestovka River in the Kolyma Basin, has been found in Siberia with stomach contents. No percentages have been described in detail regarding volume of different plant groups, other than to say that graminoids predominated. Among vascular plants, *Kobresia* macrofossils, *Poaceae, Cyperaceae,* and *Ericaceae* were identified from the stomach (Ukraintseva 1981). For this same bison Korobkov and Filin (1982) looked at the pollen and spore content of the gut contents. In their largest sample the *Graminae* predominated with 77.7%, *Artemisia* with 18.3%, and Cruciferae with 0.9%; in another sample *Graminae* was 84% and *Artemisia* 10.9%. In another sample *Artemisia* predominated, and in another *Chenopodiaceae* was the most abundant. This information strongly resembles a diet typical of extant bison on a mixed grassland habitat (Peden et al. 1974).

Colinvaux and West (1984) proposed that some animals on the Mammoth Steppe were "itinerant" and ventured above the tree line of the boreal forest only for brief episodes. But this model of a sparse Beringian tundra, lacking enough grass to support a grazing proboscidian, led Colinvaux and West to transform the woolly mammoth into a "browser," despite the fact that mammoth gut contents clearly show that woolly mammoths were grazing specialists.

Dentition of the bison, horse, and mammoth also provides

clear evidence that these species, particularly the Beringian representatives, were specialized grazers. In special situations, when these species are not eating siliceous grasses, it is quite apparent from their teeth. Paleontologists (e.g., Simpson 1951) have recognized there is a high correlation among ungulates between the development of hypsodont, high-crowned teeth with elaborate crown patterns and a grassy diet. Grasses, especially when used for winter range, are very low in energy and nutrients, high in fiber, and quite siliceous because opaline phytoliths are embedded in the leafy tissue. These characteristics mean that grass must be thoroughly chewed to reduce particle size and facilitate digestion of fibers. A grazer actually chews a greater volume of this low-quality food than would a comparable browser. Abrasion from this longer mastication time and larger volume of food causes the teeth of grazers to wear at a faster rate. Mammals invading the grazing niche have responded to increased tooth wear by an evolutionary increase in the height of the tooth and by elaborating the crown pattern in tooth enamel. These changes increase the chewing surface, making chewing more efficient and allowing the tooth to last longer (McNaughton et al., 1985.)

Woolly mammoths in Beringia had the most complex teeth of any proboscidian past or present (fig. 9.13) and have been judged to be the most extreme form of grazing specialist (Maglio 1973; Olivier 1982). In another study based on cementum root annuli (Guthrie, unpub.), I showed that mammoths lived about as long as extant elephants and that they wore out teeth at a similar succession rate, despite much more complex tooth crowns and more hypsodont molars. Likewise, among bovids, bison teeth are some of the most complex; among rhinos, woolly rhinos had by far the most complex dental patterns. Horses and hemionids have the most complex teeth of any fossil equid; this is especially true in Beringia where they have an unusually elongated protoloph.

Tooth wear changes when grazers use a less abrasive dietary staple. An example discussed earlier showed slower tooth wear among bison that eat more sedges than grasses, and the crown surface assumes a vaulted steeple character (Haynes 1984). As we have already seen, teeth of the Beringian bison wore at a rapid rate, indicative of an abrasive grass diet. Again, this is in contradiction to the bison diet portrayed by Ritchie (1984).

The evolution of complex-crowned hypsodont teeth among grazers is associated with the evolution of silica defenses in grasses (McNaughton et al., 1985). There are two divergent pathways in the

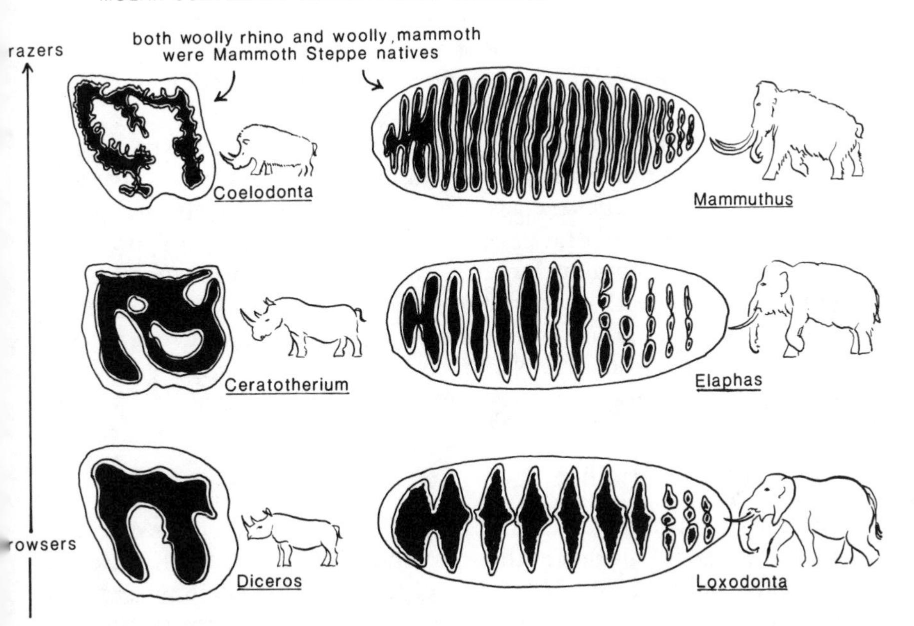

Fig. 9.13. Molar complexity. Large grazers of the Mammoth Steppe had very complex molars, adapted for chewing sparse winter forage into smaller, digestible fractions. Overall surface area of the crowns are similar; increased trituration is achieved by a more complex enamel pattern, similar to switching the grating surface of a food processor. (*Left*) *Coelodonta,* the woolly rhino, has the most complex enamel pattern among rhinos. The second most complex is the African white rhino, *Ceratotherium,* also a grazer. And on the bottom is the African black rhino, *Diceros,* a browser. (Right) the top molar is from a woolly mammoth (*Mammuthus primigenius*) of the Mammoth Steppe. The middle molar is from an Asiatic elephant, *Elaphas,* which is similar to the southern mammoth, *M. columbianus.* On the bottom is a molar from the African elephant, *Loxodonta,* a species that consumes much browse during the dry season.

evolution of plant defenses. Dicots have depended on elaborate secondary chemical defenses, and almost all pharmacologically active plants are part of that radiation. These dicotyledonary plants have their region of active growth on the terminal meristems, unlike graminoids which grow from the base and thus keep active growth tissue near the surface of the ground, protected from large herbivores. Grasses have a depauperate secondary chemistry: instead of producing poisons, they defend leafy photosynthetic tissue with a relatively cheap soil constituent—silicon—in the form of angular opaline crystals. This material is harder than tooth enamel and is the main determinant of tooth wear (Baker, Jones, and Wardrop 1959). Fiber content of different plants seems unrelated to tooth wear rates (Barnicot 1957). A woody-stem dicot eater such as a moose has very polished tooth crowns, but a rather slow rate of wear in its low-crowned teeth.

McNaughton et al. (1985) have stated, with regard to grass phytoliths, that no other plant trait putatively ascribed a protective role against herbivory due to coevolution has simultaneously such well-documented detrimental effects on herbivores, strong support from the fossil record, and evidence of natural selection in contemporary communities. Teeth of mammoths, horses, and bison in the Beringian Pleistocene allow us to state with as much confidence as one can ever obtain from the fossil record that a large component of the diet of these three species was graminoids, mainly grasses.

Small Feet, Firm Substrates, and Steppe Mammals

Another mammalian character available to help us evaluate paleoenvironmental conditions is the form and shape of the hoof. Animals respond evolutionarily to different conditions by developing larger or smaller feet, adjusting foot-loading characteristics to the softness of the substrate. Foot characteristics are one geographic difference between subspecies of extant ungulates, and there are chronological variations in foot shape among related fossils (Eisenmann 1984). Today the boggy lowland tundras of Beringia are not easily negotiable in the summer by small hoofed ungulates. Horses in a packtrain frequently get stuck (Guthrie, pers. obs.).

Plains mammals such as saigas, horses, and bison use speed to outdistance predators. But running over moist muskeg or tundra is simply not possible unless an ungulate has special foot and leg morphology similar to the caribou, musk-oxen, and moose, which is well adapted to negotiating wet landscapes. Neither horses, bison, saigas, nor mammoths have such morphology. Although it is difficult to compare leg morphology (such as the ability of a moose foot to be withdrawn vertically and the ability of the phalanges to fold along a narrow central axis), we can easily compare the loading on feet for a rough estimate of the requirements of summer substrate. Fortunately an easy comparison is possible because most of these foot-loading differences have been reviewed and published (Kuz'mina 1977; Telfer and Kelsall 1984).

Caribou are extremely well adapted to boggy substrate, with loadings of 140–80 (g/cm^2), compared to the heavy loadings of horses (625–830), saigas (600–800), elephants (510–660), and, very worst of all, bison (1,000–1,300). Medium-sized mammals found on lowland tundras today include wolves (89–114), wolverines (20–35), and arctic foxes (40–60). For an intuitive sense of what this means, I have a foot loading of about 200 g/cm^2 when barefoot. These are figures for walking animals; when these animals run, foot loading is increased considerably.

These figures do not tell the whole story, however, because other adaptations can affect ease of foot withdrawal and walking on boggy ground, especially for caribou and moose. For example, moose (420–560 g/cm^2) and musk-oxen (325–400) both have large dew-claw hooves and hooves with the ability to spread apart and form a broad surface. This is, of course, unavailable to horses and mammoths, which have rather fixed foot surfaces.

Pleistocene horses and bison in Beringia did not have larger hooves than their counterparts now living on firm substrates farther south. We know this from well-preserved mummies, as well as from comparisons of distal foot bones. Woolly mammoth feet, for example, are not larger than the feet of living elephants.

Ritchie (1984) and Colinvaux and West (1984) have argued that instead of being a firm-substrate grassland, Beringia was simply an expansion of dry tundra, more like a polar desert. Low Arctic tundra beyond the latitudinal margin of the trees, and even most upland tundra, is quite mesic, composed of either wet sedge meadows, tussock-producing sedges, mosses, or cushion plants. Farther north, in the high Arctic, bogs become fewer mainly because of a rocky substrate, a condition not characteristic of the loess-covered Beringia. Colinvaux (1984) argues that much of Beringia was unsuitable for bison, horse, and mammoth and that it was only along the rivers that one could find these grazing mammals. But in interior Alaska, more large mammal fossils are found in the uplands. A map of fossil localities of large mammals in Alaska or Siberia shows that bones are found everywhere conditions are suitable for fossil preservation, from valley bottoms to well above the present tree-line.

The Mammoth Steppe: A Rangeland That Produced Giants

The Beringian tundra as portrayed by Colinvaux and West (1984) and Ritchie (1984) is marginal habitat with almost no plant biomass to support a community of large mammals; further, they suggest that the few large mammals present were beyond their range of optimum adaptation. Animals on marginal summer range (growing season) can be quickly identified by their morphology; these animals are small bodied, and parts of their body not directly related to survival, such as horns and antlers, are poorly developed (Geist 1971b). Do the fossil remains of Beringian large mammals exhibit such signs of marginal summer habitat? The answer is mixed—no and yes. Ruminant and caecalid fossils show opposite trends.

Sheep, bison, caribou, and other ruminants on the Mammoth Steppe were giants (Guthrie 1984a). Ruminant species are adapted to take full advantage of high-quality and high-quantity resources in northern latitudes (fig. 9.14). Social paraphernalia such as horns and antlers of Mammoth Steppe ruminants were enormous, indicative of animals on superior summer range (Guthrie 1984b). Sheep, moose, caribou, bison, wapiti, and musk-oxen are now much smaller (fig. 9.15) than they were in the late Pleistocene, suggesting that range quality in the Holocene is not better, but worse.

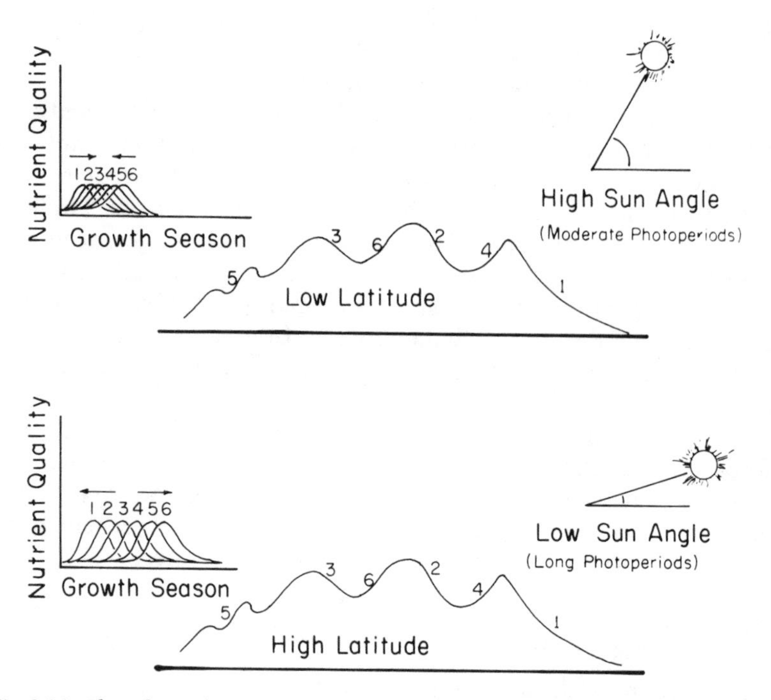

Fig. 9.14. Phenology of same plant species in different habitats. Many northern plants are unusually high in protein because of their rapid growth during the long summer days. Lower sun angles result in a longer season of early growth vegetation (the most nutritious). Farther south, the supernutritious flush of early growth peaks rather abruptly. Northern animals can move about the landscape, feeding on highest quality forage for much of the summer.

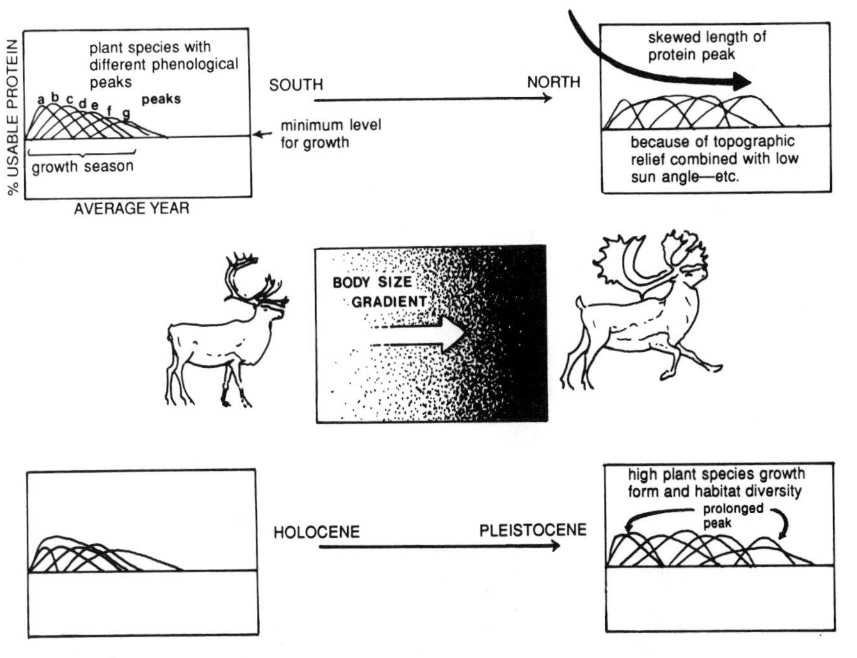

Fig. 9.15. Schematic illustration of body-size changes in time and space. Northern large mammals enjoy higher quality forage for a longer growth season than their southern counterparts (fig. 9.14). This phenomenon occurs on a historical axis; peak range quality was even more sustained during the Pleistocene. I propose this explains the body-size trends along both axes.

Unlike ruminants, late Pleistocene Beringian caecalids were comparatively small. Mammoths, rhinos, and horses do not have rumens, and these caecalids (fig. 9.16) operate with different developmental and reproductive strategies as well. Caecalids—animals that compost their food in a large diverticulum in the hind gut—are geared to managing on low to modest nutrient levels spread across a long growing season (Guthrie 1984b). In contrast, ruminants have more rapid growth rates and can take advantage of nutrient peaks. Lengthening these peak periods selects for large body size in ruminants. Caecalids have more conservative growth rates and simply cannot fully utilize nutrient peaks; instead they rely on a long trajectory of moderate nutrient levels.

On the Mammoth Steppe, nutrient levels were very low during the winter, but nutrients of high quality were abundant during the critical growing season. Ruminants were able to make the most of this steep-sided arch of nutrients, and they grew to large size. Caecalids simply needed a longer growing season to reach their maxi-

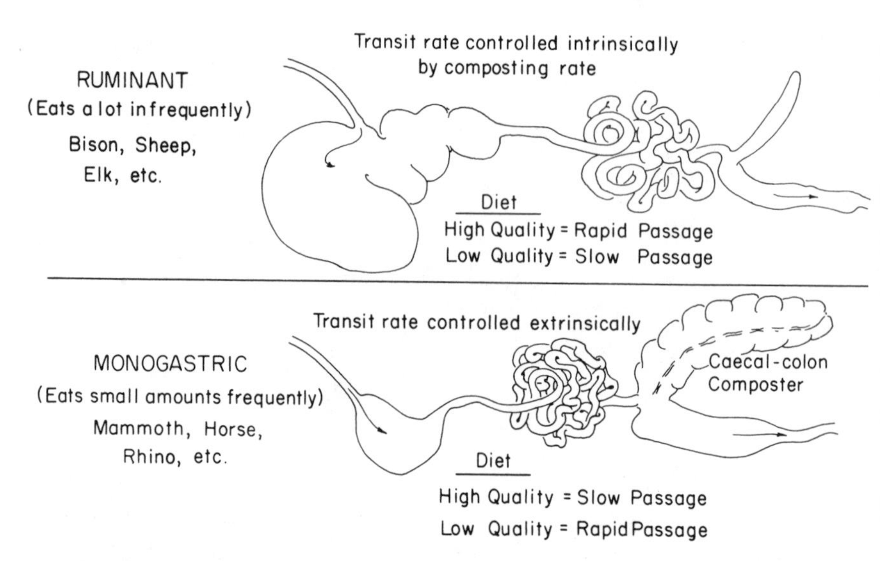

Fig. 9.16. Ruminants and caecalids. There are two main digestive strategies among large mammal groups. Most cloven-hoofed, artiodactyl species are ruminants (bison, sheep, moose, etc.), while the perissodatyls (horses and rhinos) and proboscidians (mammoths) use their hind gut to ferment cellulose. Generally, monogastric hind-gut fermenters (caecalids) are slower growing and reproduce more conservatively.

mum potential size. This, combined with a different response to antiherbivory compounds (fig. 9.17), is the reason late Pleistocene mammoths, rhinos, and horses all across the north were small compared to earlier counterparts (Guthrie 1984a, 1984b). Although the issue of a species' allocation of resources to reproduction, survival, and growth is complex, it ultimately relates to the quantity and quality of food available during the growth season. The limited northern growth season necessarily means that a ruminant, which is one of the largest representatives of its species, must be getting large quantities of high-quality forage. Growth-season pastures have to be optimal, not adequate. This logic must hold for Pleistocene Beringia as well (fig. 9.18).

Our comparisons of bison size based on interstade (Boutelliere Interval) Blue Babe showed that interstadial bison are smaller than those from the peak glacial (Duvanny Yar). Partial reforestation during the Boutelliere Interval probably selected against large bison just as it did later, in the Holocene. This may be true for other species of the mammoth fauna, but as yet we lack detailed comparisons.

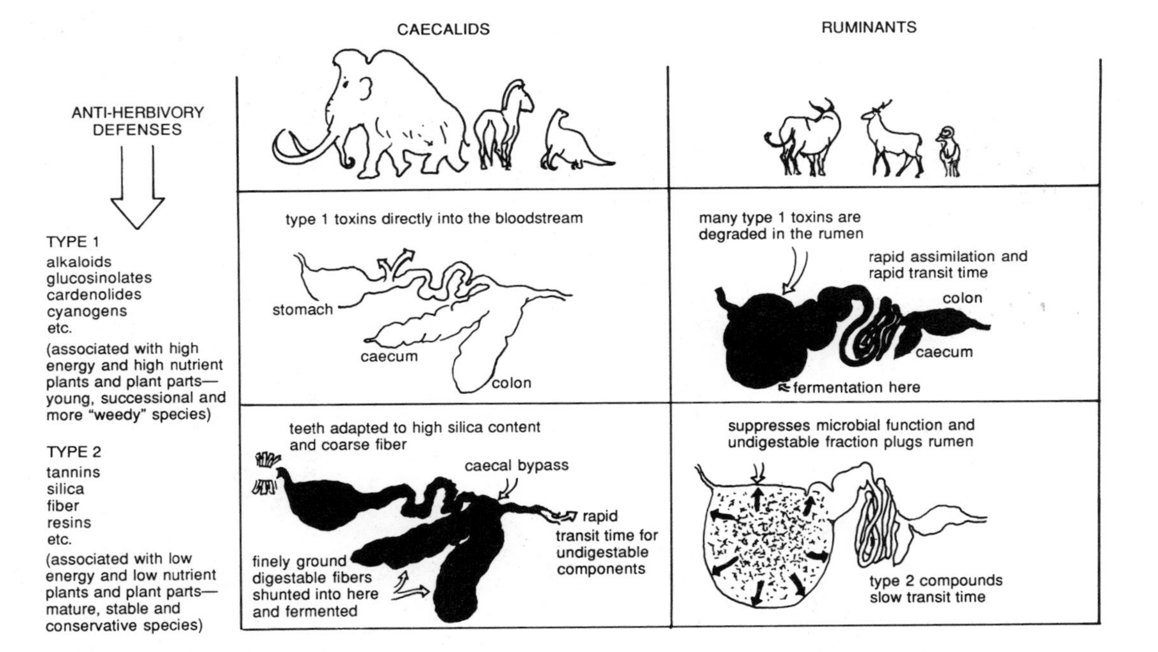

Fog. 9.17. Two main digestive strategies and antiherbivory defenses. The ruminants and caecalids portrayed in fig. 9.16 also differ in their response to plant defense compounds. In general, the rumen is a great detoxifier of chemical defenses, but has limitations in digesting high-fiber, poor-quality food. Caecalids can handle low-fiber food but have no rumen to detoxify compounds before they are absorbed. So caecalids and ruminants have their niches in which each excel, and though there is overlap, they are not quite the same.

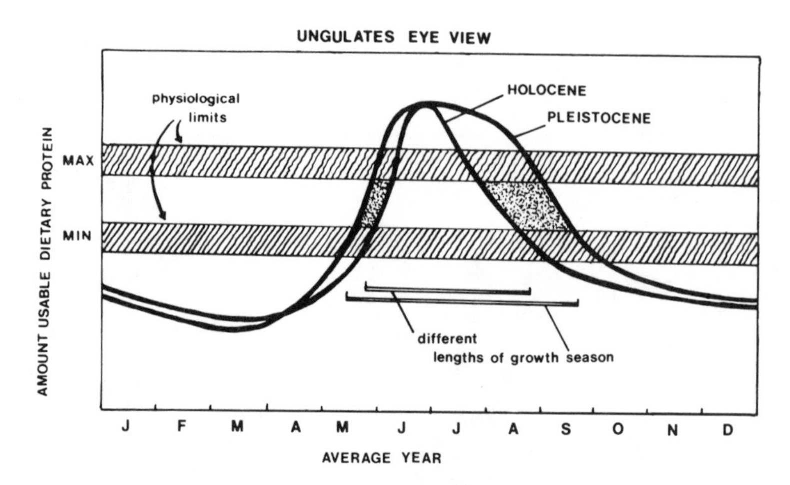

Fig. 9.18. Body size and available protein. Maximum body size among ungulates depends on (1) nutritional quality (primarily protein) sufficient for growth and (2) the seasonal duration of that protein level. There is a nutritional level below which an animal cannot grow and another which exceeds the animal's capacity. The annual nutrient curve greatly affects body size. Under a high and broad curve animals not only become developmentally large, but they are selected to increase their maximum growth potential—they evolve a larger body size.

One could argue that bison reintroduced are doing well because they are so large bodied. A number of bison from the Delta herd rank high in trophy record books. These seem to be on excellent range and at the maximum of their genetic potential, but we must remember that these bison use farmers' barley. Large mammals can be pushed to maximum genetic size by supplemental feeding during the growth season and thus can exceed the body and horn size of wild populations from which they were taken (Guthrie 1984a).

Faunal Coherence of the Fauna and Continuity of the Mammoth Steppe

The initial impetus for bringing the vast conglomeration of arid grass-dominated communities together under the name of Mammoth Steppe was the recurrent faunal melody over this area. The more common large herbivores (Chaline 1972; Stuart 1982) in the Pleistocene of western Europe during colder and drier episodes (and to some degree during warmer and wetter periods) are horses, steppe bison, woolly mammoths, reindeer, red deer, and woolly rhinos. These are almost exactly the species we find across eastern Europe (Kowalski 1967a), across Asia (Vereshchagin and Baryshnikov 1982) and northern china (Liu and Li 1984) in the south, to the polar areas on the north (Sher 1974). For the most part, these species also occur in Alaska and the Yukon Territory (Guthrie 1968, 1982; Harington 1978). This continuity is not fortuitous. It indicates an enormous expanse of similar communities, which can legitimately and usefully be characterized as a larger unit, despite obvious local and regional variants.

The present boreal forest is analogous in scale and biotic continuity to the Mammoth Steppe. No one would be purist enough to argue against the concept of a taiga, or boreal forest, because it can be shown to have different regional communities and facies of species composition due to aspect, drainage, altitude, latitude, history, and so forth, or major physiognomic forms in its different stages of succession. Locally and regionally it has very different subcommunities, with considerable geographic variation in species composition, but wherever one goes in the boreal forests, from Siberia to the Canadian northwoods, there is a commonality of character we understand in terms of "the boreal forest."

The Mammoth Steppe Concept

I think ecologists have difficulty imagining the Mammoth Steppe because it was a biotic zone that no longer exists. We meet the obverse difficulty in realizing that today's biotic zones are not quite the primeval structures we once believed them to be. We have only recently come to see biotic zones as more dynamic units, some of which had a quite different character during the Pleistocene. Today, deciduous and boreal forests are rather cleanly separated; this was not true during the last glacial. Likewise, steppes and tundra are now widely separated zones, yet they were not so distinct during the last glacial. Pleistocene mixtures may look strange to us, but we must not assume too much. Holocene biotic zones may be more exceptional than assemblages we see in the fossil record.

Not long ago, when most of today's senior scientists received their training, it was thought intuitively obvious that vegetational zones in North America and Europe were squeezed farther south during the glacials. Preliminary evidence, such as the boreal forests being pushed south, and the space between conifers and the ice sheet being occupied by tundra, seemed to corroborate that squashed sandwich image (Budel 1951). But more recent evidence shows, contrary to intuitive sense, that the sandwich theory does not hold.

Most glacial-aged biotic communities have no modern analogue (see Matthews 1979 and Guthrie 1984a as examples of literature reviews). In fact, most of Europe was a steppe during the last glaciation, even southern Europe (e.g., Kowalski 1967a; Beug 1968; Brunnacker 1974, 1980; Woillard 1978; Cassoli 1972). This glacial steppe (*Kaltsteppe* in German) was not, however, identical to the Holocene Eurasian steppe, although it shared many xeric steppe plant and animal species. Fossil data are in agreement that these northern steppes were strange mixtures of species living in the same communities—mixtures without a modern analogue. But this no-analogue steppe image is still a new way of thinking in Europe.

A parallel reconstruction of glacial steppe developed almost independently in eastern Europe and northern Asia. Beginning with an attempt to account for the widespread "mammoth fauna," early researchers proposed a more temperate Pleistocene climate. Serebrovskij (1935) argued against this and proposed taiga and tundra vegetation similar to that of today. Later, paleobotanists and paleontologists came to yet another conclusion, with which there is now general agreement—that most of unglaciated Eurasia was more like steppe than tundra, but again, not identical to extant steppes (e.g., Giterman and Globeva 1967; Vangengeim 1967; Frenzel 1968; Sher 1968, 1974; Yurtsev 1974; Vereshchagin and Baryshnikov 1982).

I view the development of ideas about a Pleistocene steppe in Alaska as part of that same debate, even though ideas about an Alaskan steppe developed rather independently, based primarily on large-mammal evidence. It is a new and, for me, truly exciting paradigm to see these debates interconnected.

Some critics of the Mammoth Steppe concept have not argued for a barren tundra or polar desert but instead have attempted to merge the two camps into a hybrid. They contend that there was neither steppe nor tundra but a mosaic of the two (Hopkins et al. 1982; Schweger 1982; Anderson 1982, 1985; Bombin 1984). This seems to me a spurious compromise. I think most parties agree that Beringia had no exact modern analogue; it was not exactly like any extant biotope, steppe or tundra (figs. 9.19 and 9.20), but their use of the "mosaic" category seems to misunderstand the concept of a physiognomic unit. Certainly all grasslands, tundras, boreal coniferous forests, deciduous forests, and so on are mosaics of different communities and subcommunities regulated by aspect, slope, drainage, altitude, fire history, herbivory pressure, and almost an infinite number of other variables. Species combine and mix individualisti-

Fig. 9.19. Lowland tundra habitat. Northern tundra is composed primarily of hydric and mesic plants that consist of relatively slow-growing, rather toxic species well defended against herbivores.

Fig. 9.20. The boreal forest. As it spreads across Eurasia and North America, the boreal forest has a variety of local communities, but these share a similar physiognomy. Plants in this biome are well defended against large herbivores. There are two exceptions: successional growth after a fire and disturbed, nutrient-rich communities along streams.

cally across the landscape to meet their specific needs and competitive edge; the result is a complex mix of communities and subcommunities. There is never a monoculture of sameness, nor was one implied in the concept of a Mammoth Steppe. The real issue is whether, taken as a whole, it was physiognomically more like the variety of things we lump under the heading of "tundra" or more like the variety of things we lump under the heading of "steppes." It does not seem to have been a zone of separate card-shuffled steppe and tundra communities, as the "mosaic camp" proposes, but more likely was a very complex amalgam of what we see today in quite separate communities, both on small and large grain scales.

Since the communities of the unglaciated north of Europe, Asia, and Beringia seem to have compromised an environment that had no adequate analogue, I did not want to confuse the issue by employing a label with a modern image, because it obviously was not floristically or faunistically like any existing grassland. Mosaic-mix terms such as *tundra-steppe* or *steppe-tundra* seem likewise inappropriate. The term *Mammoth Steppe* emphasizes that the concept includes both fauna and vegetation. What the Soviets call the "mammoth fauna" played a large part in our understanding of these northern glacial communities. The woolly mammoth prefix points to its northern character, approximating the distribution of woolly mammoths, while *steppe* suggests a predominance of low, thin sward. In Eurasia, and even in North America, we refer to arid grasslands of low-sward profile as steppes. Not that these must be homogenous to use that title; they are all made up of a mosaic of communities and subcommunities.

It is my conviction that the Mammoth Steppe environment was a dominant fact of life in the north in late Pleistocene times. Contrary to the views of Colinvaux, Cwynar, Ritchie, and West, the evidence supports this contention. The geographic vastness of this environmentally similar habitat commands a special appellation, and the double entendre of a Mammoth Steppe seems uniquely appropriate.

10
BISON HUNTING ON THE MAMMOTH STEPPE

Bison and Human Colonization of the Far North

Early peoples lived in Eurasia for hundreds of thousands of years (fig. 10.1), but they did not colonize the Asian far north until the waning phases of the last glacial. What kept them from utilizing the large-mammal resources on the Mammoth Steppe?

The evidence available, as I see it, argues for colonization of Beringia, north of the 60° parallel, at the close of the last glaciation (Duvany Yar Interval), and at no time prior to that. This means that human access to North America first occurred 12,000–13,000 years before the present. In fact, northward colonization of Eurasians during the late glacial seems to have coincided with a general ecological change that *reduced* diversity and abundance of large mammals, that is, people were extending their range northward *despite* a reduction in large-mammal resources.

I suspect there was one factor that kept eastern Asians from capitalizing on the herds of northern bison prior to that time, and it can be applied at two different times. In a word, it is *technology*—technology inadequate to cope with the severity of northern climates, cold and all its correlates.

The far north is an inhospitable place for a tropical hominid. We who live above 60° latitude do so with an advanced technology of well-insulated housing, tailored clothing, efficient heat sources, and a suite of tools that reduce danger, frustration, and expended energy. All peoples who have occupied the far north have had these in one form or another. One can survive in a lean-to shelter facing an open spruce fire, but more is required to live a productive life—to cope with newborn babies and illness, lay up secure stores, move to seasonally optimal habitats, and so on.

During most of the last glacial there were no plentiful supplies of wood (if any at all) north of 60° latitude. There were no ericad

Fig. 10.1. The cold wall to colonization. Early Paleolithic sites scatter across Eurasia in an irregular pattern, but with a definite boundary, indicated here by the thick black line. The lower Paleolithic, of course, covers a long period of time, and this northern boundary must have shifted with varying Pleistocene climate. This boundary reflected the limits of techniques developed for living in an extremely rigorous climate and landscape.

berries or mast crops from woody plants. Even with a highly sophisticated technology, life without wood is difficult, if not impossible; I would argue that such a technology (if it even exists) is hundreds of years old, not tens of thousands. No one lived in the far north in either Eurasia or North America during the height of the last glacial event (fig. 10.2).

However, during Blue Babe's time (throughout the last interstadial) there was wood in the far north. That warmer and wetter saddle within the last glacial stage lasted until around 28,000–30,000 years ago. Although there were anatomically modern people in Eurasia at

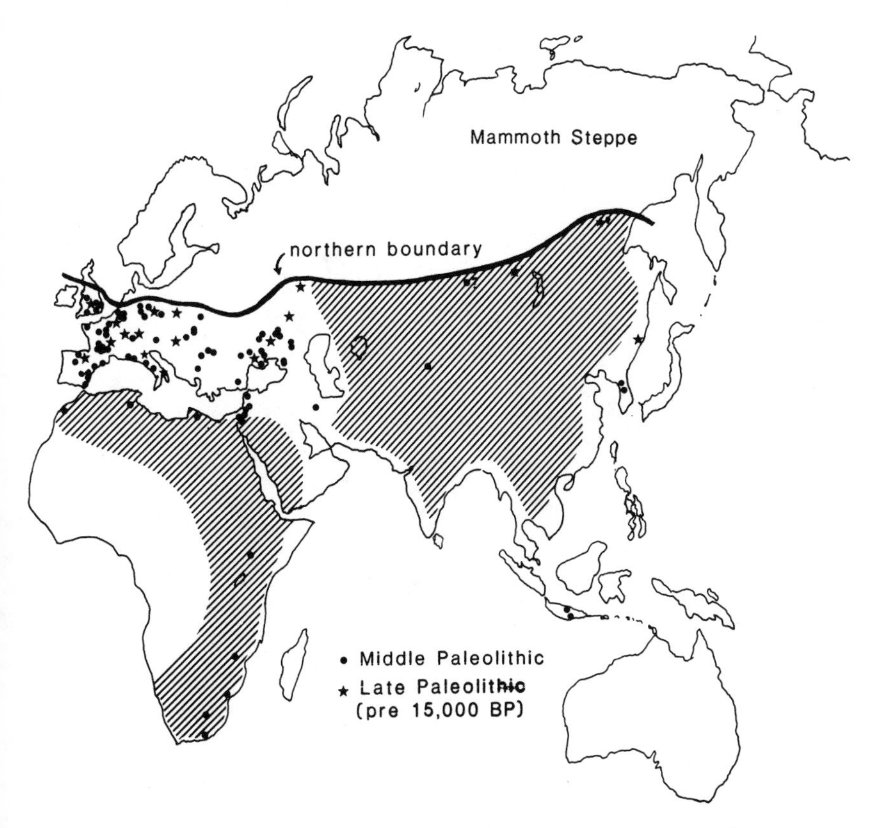

Fig. 10.2. Northern shift in the cold wall due to technological and climatic change. Middle and late Pleistocene sites scattered across Eurasia also show a northern boundary (black line). This boundary was not breached until around 15,000 years ago, and sites are not common north of that line until 2,000 or 3,000 years later. Some earlier sites have been described north of this line, but they are controversial.

that time, the sophistication of their northern-adapted technology was inadequate for the far north. For example, there is no evidence that 30,000 years ago people had sewing needles with which to make weatherproof seamed clothing. The lithic technology of most of the last interstade was crude. Lean and efficient stone projectile points came late during that time, and even the thrown spear itself, judging from projectile points, was probably fresh on the scene.

The presence of wood alone was not enough to permit access; interstade people 35,000 years ago still faced a limit at 60° north. Yet the warm and wet interstade probably allowed people to move far-

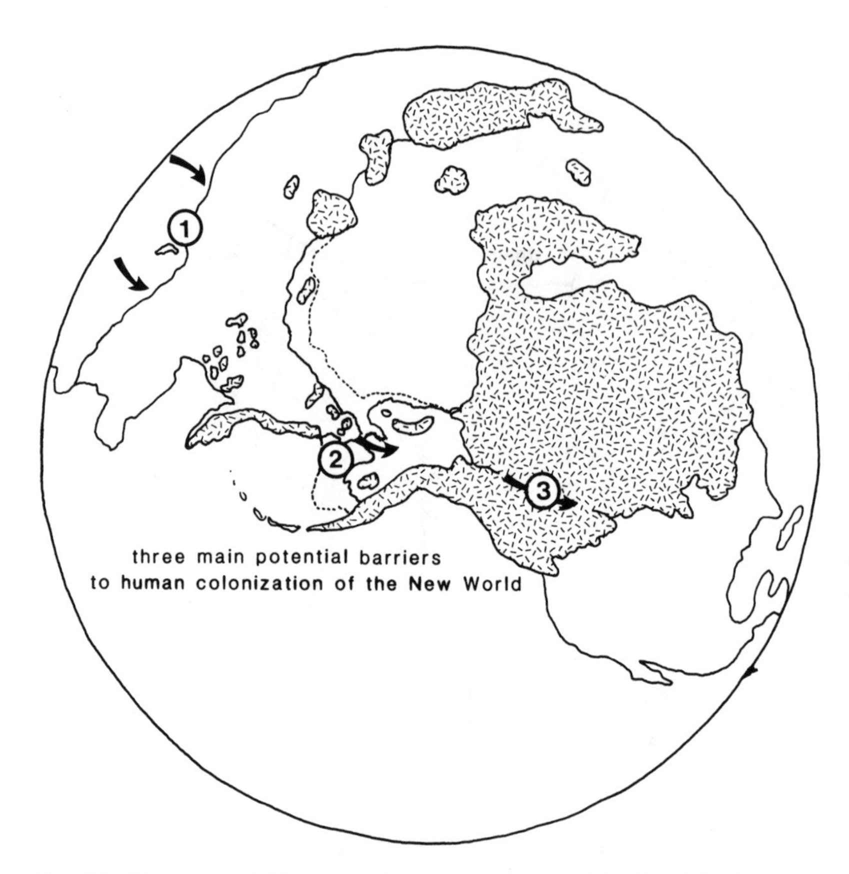

Fig. 10.3. Three potential barriers to human colonization of the New World.

ther north in Eurasia than they had ever been, an expansion that was slowly driven southward during the last glacial (Duvanny Yar Interval).

After the last glacial, trees again entered the north around 12,000–13,000 years ago, and people came north with the new trees; this time they were not stopped at 60° but pushed northeastward into the then exposed Bering-Chuckchi Platform and farther, into North America. We know from archaeological sites that these people were hunting the mammoth fauna, especially bison. Colinvaux and West (1984) unnecessarily assumed that the lack of archaeological evidence in the far north during the glacials was due to the lack of resident game to hunt. According to the fossil record, the game was there; something else, not large mammal resources, excluded humans.

Archaeologists have discussed three barriers (fig. 10.3) limiting human colonization of the New World: (1) The northern line of archaeological sites seems to conform to this limit of 60° N, mainly because there was insufficient technology to carry on life in far northern environments. (2) The Bering Strait has been considered a barrier, and some archaeologists have thought that the intermittent land bridge between Alaska and Siberia was the key to colonization. But there are places where one can cross over to Alaska on the winter ice, even without a land bridge. People adapted to life on one side could easily have crossed the few kilometers of ice to the other side. (3) Continental ice sheets did separate Alaska from ice-free regions to the south for thousand of years, but a corridor several hundred kilometers wide seems to have been open when Asian people first arrived in Alaska sometime between 12,000 and 13,000 years ago. Thus (1) seems to have been the obstacle of most significance. However, had these new colonists come 14,000 years ago or earlier, this corridor would have indeed been closed. We can see its pincer effects on the mammal groups it squeezed in two (fig. 10.4).

At its extreme development the Mammoth Steppe seems to have been a hostile environment for early people—winter temperatures and high winds, combined with few woody materials for shelter, fuel, or tools. The otherwise good hunting grounds were bad living grounds. Abundant wood is a necessary part of northern technology. The rivers flow northward in Beringia, however, so the running water may have aided dispersal of seeds (fig. 10.5). The northern barrier was breached, and people moved into land never seen by humans. Something new had happened. We can wager that these first people were bison hunters, as well as hunters of other mammals on the disintegrating Mammoth Steppe.

Not all archaeologists would agree with the timing or some of the interpretation of the above scenario. It is a view that many do accept, and if we proceed with this interpretation of archaeological and paleontological data, we can say that Blue Babe lived in a land untracked by human hunters. During his lifetime, thousands of kilometers to the west and south, however, people were dining on bison almost exactly like Blue Babe. Much later, the first colonists in the Fairbanks area would also be hunting bison, descendents of Blue Babe that were probably very similar to him in some ways.

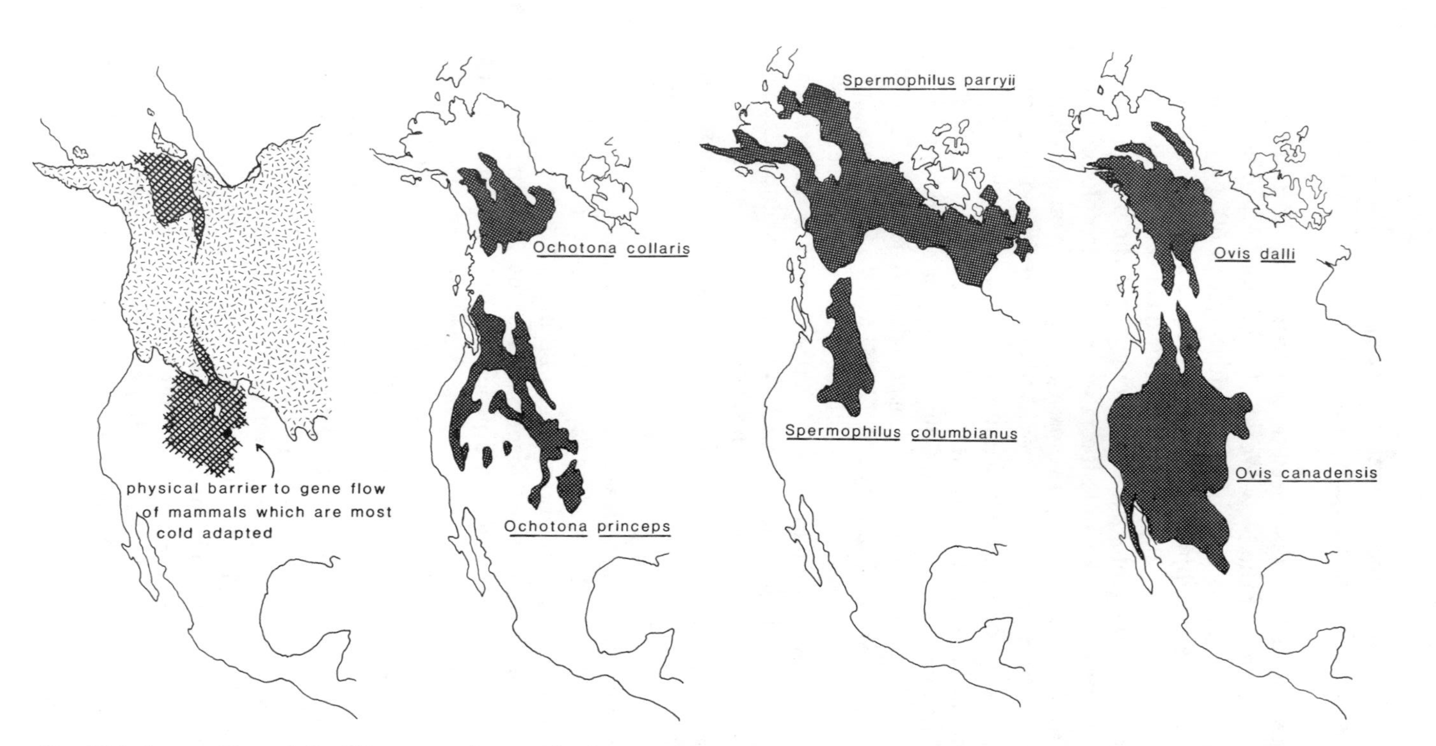

Fig. 10.4. Laurentide and Cordillerian ice sheets. This barrier separated some species into genetically different, northern and southern forms. Pikas (*Ochotona*), northern ground squirrels, (*Spermophilus*), and mountain sheep (*Ovis*) are three mammals that live and prosper in periglacial conditions. Their present distributions reveal a recent north-south separation. This suggests that the corridor between Alaska and the rest of the continent was closed to virtually all terrestrial Alaskan mammals, including humans if they had been present, during the last glaciation.

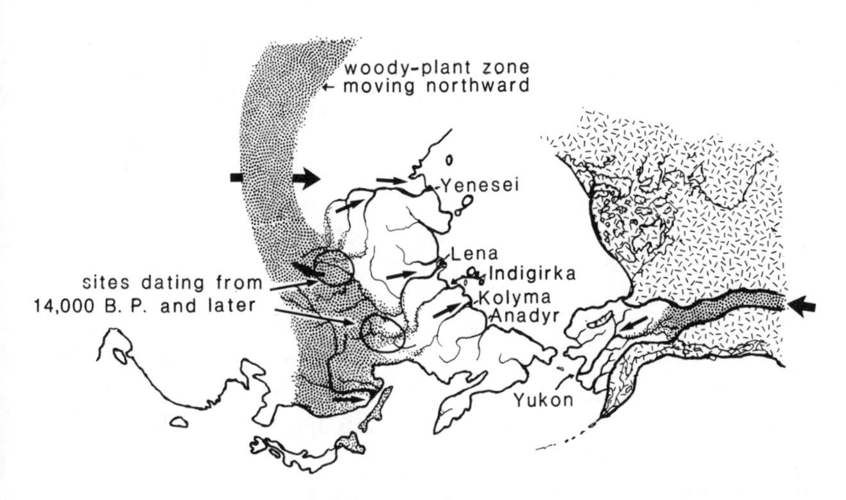

Fig. 10.5. Riparian model of woody-plant recolonization of the north. During glacials the northern part of the Mammoth Steppe was virtually devoid of trees. At the onset of the warm-wet interstadial and, again, with the beginning of the Holocene, trees recolonized the far north. The northward flow of these rivers would have assisted rapid colonization. Many northern trees make use of runoff water for seed dispersal.

Predation and Bison Defense Strategy: Pleistocene-Holocene Differences

A well-preserved Pleistocene bison with accompanying paleoecological data offers a occasion to reflect on the behavioral dynamics of human predation on bison in the Pleistocene. How extensive was bison hunting in the Pleistocene, and was it as easy as it later was during the Holocene bison drives on the American Great Plains? Were Pleistocene bison behaviorally similar to living bison, which lack notable ferocity toward humans? Insights about such hunting can be gained by comparing Pleistocene predator-prey strategies (based on information we have from Blue Babe) and the present.

When Blue Babe was found we carefully excavated the carcass, collecting surrounding sediments and wet screening them for associated artifacts, since there was the possibility that the mummy was part of an archaeological kill site. As it turned out it was not, but any late Pleistocene large-mammal carcass or special accumulation of bones should be excavated with the possibility in mind that it may be an archaeological site. In fact, most paleontological sites are now excavated by the same techniques and seek the same kinds of paleoenvironmental data as archaeological sites.

The entry of African hominids into Eurasia roughly corresponds to the rise of bison as a genus. A rather straight-horned creature called *Leptobos* was the early ancestor to most bovines: cattle, buffalo, bison, and so forth. *Leptabos* occurs in the earliest Pleistocene of Europe and Asia. Later in the Pleistocene (a little less than a million years ago), the first animals that we could call bison appeared. Like later bison, these were bovines with rugged skulls. Their horns emanated from the sides of the skull, and eye orbits telescoped out away from the skull.

The first *Homo* colonists arrived in Eurasia at a time of major fauna change, when the more primitive "Villifranchian" fauna of Eurasia was being replaced by the "Galerian" fauna (Turner 1982; Azzaroli 1983). Villifranchian large mammals, including the three-toed horse, *Hipparion,* and the primitive bovine mentioned above, *Leptobos,* were more Tertiary-like in their appearance, The Galerian fauna was more "modern" and persisted relatively unchanged in overall character until the end of the Pleistocene.

This change in faunas was major. New elements from the Americas, such as true horses *(Equus),* dispersed into Eurasia and Africa from Beringia. Also at this time African species, for example, lions (*Panthera leo*), dispersed throughout Eurasia. Hominids came along in this faunal movement up from Africa (Turner 1982). Timing of the Galerian Dispersal event is still not precisely clear. Azzaroli (1983) proposes that it happened before the last major pole reversal (before the beginning of the Bruhnes normal epoch), sometime before 700,000 years ago.

These first Eurasian hominids were not our own species. They have sometimes been referred to as *Homo erectus.* They had rather simple tools, and evidence suggests they had begun to use fire rather early in their occupation of Eurasia. They were ruggedly build, with a brain size smaller than that found today among any living peoples, but in many respects they were similar to modern peoples. These earliest Eurasians probably shared communities with bison, as some of the oldest bison records are rather far south (India). Bovine niches were soon divided latitudinally in Eurasia. Bison evolved as northern-adapted bovines, while the southern bovine niche was occupied by not-too-distant relatives: cattle (*Bos*) and buffalo (*Bubalus*) species.

Neanderthals were the first consistent hominid predators of bison. Neanderthals replaced early hominids in Eurasia (or were derived locally from them) during the last interglacial, about 120,000 years ago. Some authors would call Neanderthalers *Homo sapiens,* but Neanderthals were anatomically distinct from modern peoples, despite many similarities. Their brain size was as large if not larger, than living peoples, but like *Homo erectus,* they were very ruggedly built, even more so than *Homo erectus*. Neanderthal tool kits, at least the ones that were preserved, were also relatively simple, although more diverse than *Homo erectus*. The lithic industry associated with Neanderthals is called Mousterian; it is based on flakes as opposed to a core technology of *Homo erectus*.

Neanderthals were not bison-hunting specialists; judging by their bone refuse they hunted larger mammals, such as mammoths, rhinos, and horses, as well as bison. Bison bones in Neanderthal sites are usually thoroughly smashed, presumably to make complete use of the marrow. Neanderthals seldom made use of bone, ivory, and antler for tools, and despite hundreds of sites, there is no evidence Neanderthals ever used ornamentation. Neanderthals ranged rather far north during the interglacials and interstades, but during the glacial maxima 80,000 years ago (isotope stage 4) they were probably pushed far south. They occupied Europe and Asia widely during the interglacial of 100,000 years ago (isotope stage 5) and the interstadial centering around 40,000 years ago. Blue Babe was a contemporary of the last Neanderthals.

Between 35,000 and 40,000 years ago, Neanderthals were replaced in Europe by anatomically modern peoples; in Asia this change may have occurred a little earlier. These new peoples began to use long blades struck from stone cores instead of stone flakes. Ornaments of many kinds appear in the sites, especially the elaborate use of bone, ivory, and reindeer antler as a tool and artistic medium. The newcomers ranged farther north than had preceding peoples, well into the domain of bison. And bison occur in many of their sites. But with the exception of a few areas far to the south in the Caucasus (Vereshchagin 1959), there are few sites where bison predominate. Although there are numerous Eurasian Pleistocene archaeological sites where horses, reindeer, and mammoths make up almost the entire bone assemblage, bison seldom do. However, bison are present in low abundance in many sites of late Paleolithic age.

This same bison-use pattern continues during and after colonization of the New World. For example, there are no Clovis sites in the Americas where bison bones predominate. Clovis people seem to have hunted bison, yet not in great numbers. At the Dry Creek site near Healy, Alaska, which dated around 11,000 years ago, Roger Powers, John Hoffecker, and I found bison associated with stone microblades, sometimes only a little over a millimeter thick and less than an inch long (fig. 10.6). These miniature blades are struck from a small core (fig. 10.7). We know from sites in Siberia that such little blades could be inserted in the side of an antler or bone shaft to provide a cutting edge. Bone and antler points penetrate well, but

Fig. 10.6. The Dry Creek site. Bison were present in the Dry Creek site in central Alaska, which dated around 11,000 years ago. Work at other sites will help us understand the role of Mammoth Steppe fauna in the lives of these first North Americans.

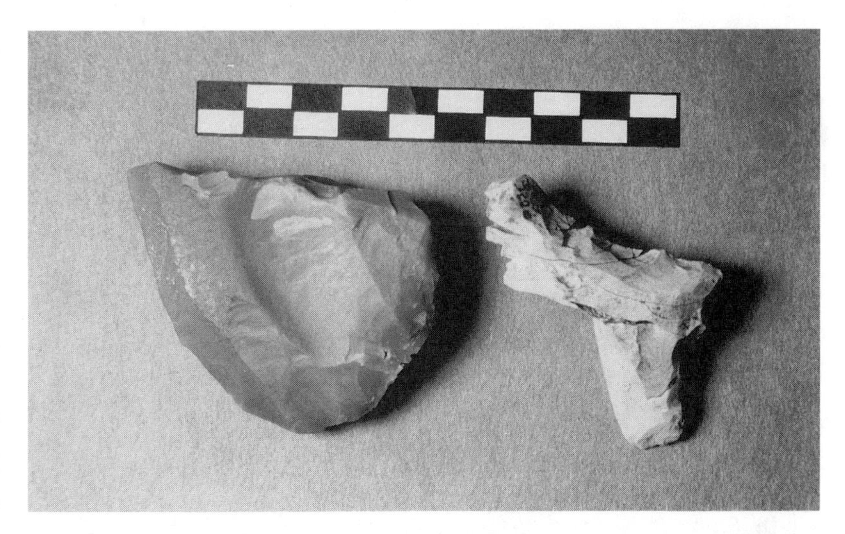

Fig. 10.7. Small stone cores used to strike microblades. This illustration shows (*left*) a blank core before microblades have been removed and (*right*) an exhausted core to produce more microblades. Near the exhausted core we also found the large flakes struck to produce the table from which the microblades were made. These flakes are glued back into place. Both specimens are from the Dry Creek site, which is about 11,000 years old.

without a stone cutting edge these do not cause enough bleeding to make an animal die quickly (Guthrie 1983). These microblades, their microcores, and/or small biface stone projectile points, and small burins apparently for working osseous projectile points are a characteristic signature of lithic assemblages across the north at the end of the Pleistocene. This style of stone technology, called Beringian tradition by West (1981), is found over an enormous area: Alaska, the Yukon Territory, eastern Siberia south into China, northern Japan, and across much of northern Asia. Similar technologies (including microblades) are found farther west to the Urals and on into Europe (fig. 10.8). But this does not necessarily mean that these areas were inhabited by the same peoples; undoubtedly the technique crossed biological and cultural lines. Bison are associated with many of these sites, but the microblade sites fit the general picture of the sites with a few bison bones; in no site are bison bones dominant.

In one Siberian site, Kokorevo, a composite projectile point of reindeer antler and microblades (fig. 10.9), was found sticking in the scapula of a bison (Abramova 1982)—solid evidence that people were hunting bison with this kind of weapon. Gönnersdorf, a site in

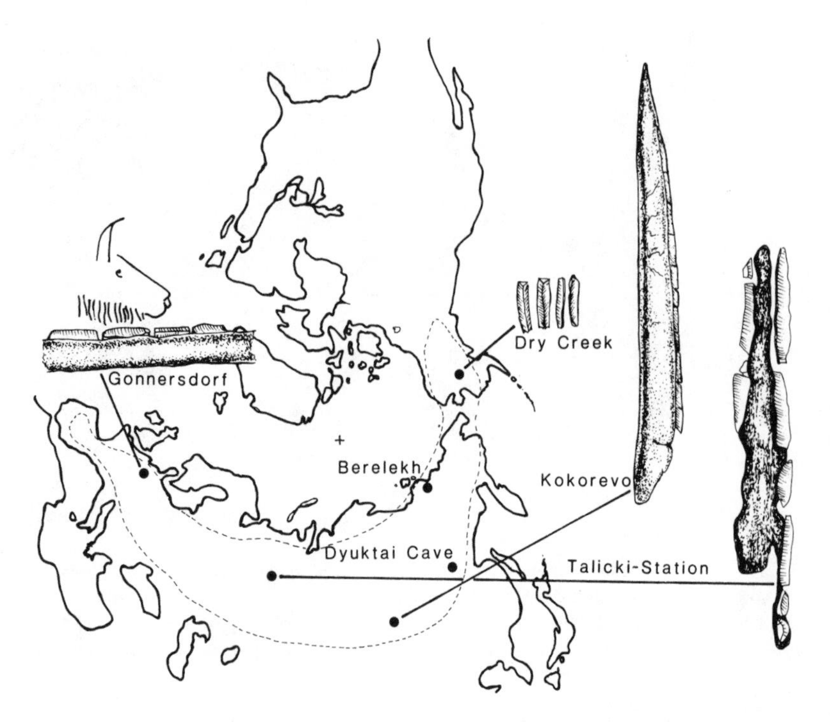

Fig. 10.8. Association of microblades and bison, 13,000–10,000 B.P. Within the range of *B. priscus* (dashed line), a hunting technology expanded across the waning Mammoth Steppe. The tools included burins, small stone projectile points, and rectangular microblades that apparently were inset into caribou-reindeer antler points. This technology obviously was part of several cultures and racial stocks. The sites shown here (from 11,000 to 13,000 years old) contain both bison remains and microblades.

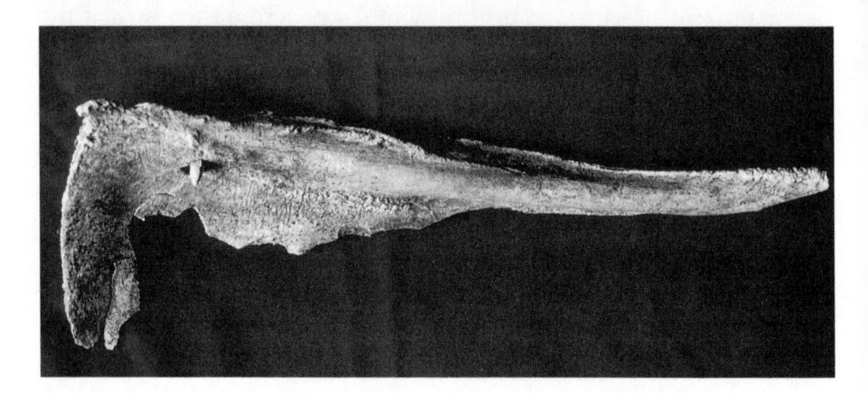

Fig. 10.9. Bison hunters and microblades. A reindeer antler projectile point with small grooves for microblades was found protruding from a bison scapula in the Kokorevo site in Siberia, illustrating that early Beringian people used such tools to hunt bison.

northern Germany (Bosinski 1981), contained bone evidence of at least two bison and a sketch of a bison face scratched in stone. Gönnersdorf also contained microblades not too unlike those in the Dry Creek site in Alaska. The Berelekh archaeological site on the Indigirka also contained microblades and bison bones. Most of these sites with microblades and bison bones date from 13,000 to 10,000 years ago, and perhaps a little earlier and later.

People seem to have been colonizing the far north at this time; it was a time of waning continental ice, increasing moisture, decreasing wind, and invasion of the north by shrubs and trees, producing a climate more habitable to humans. People using stone projectile points and microblade technology spread northward during the late glacial, hunting different species of the mammoth fauna. They persisted in that hunting mode after many of these species were completely extinct (e.g., woolly mammoth and woolly rhino) or regionally extinct (e.g., horse and saiga).

Several paleoecologists have argued that humans were instrumental in causing these extinctions (see Martin and Klein 1984 for a review of this subject), but the actual causes of extinctions in the far north, both regional and complete, are difficult to reconstruct because climate and vegetation were changing dramatically at the same time. Even if humans had not been present, most of these large-mammal species could not have survived the vegetational changes and increasing snow depth (Guthrie 1982).

Although I have argued that the large-mammal (horse, mammoth, etc.) extinctions in Beringia were a product of changing climate (Guthrie 1982, 1984b), a forceful argument has been made that these mammals were killed off by the new human immigrants from Asia. However, there are many problems with the "overkill" explanations of mammalian extinctions in Alaska; for example, it is unlikely that the demise of saigas, badgers, and black-footed ferrets can be explained directly by the incursion of human hunters.

Bison survived the extinction of many other large mammal species in the central North American continent. Bison numbers may have even increased (fig. 10.10) in the Holocene as a by-product of the reduction of species diversity and hence decreased competition. The immense bison herds on the Great Plains seem to be unique to the Holocene.

Bison seem to have also survived Pleistocene extinctions in Alaska and western Beringia (Vereshchagin and Baryshnikov 1984), while mammoths, horses, helmeted musk oxen, camels, and others died out in the far north. Bison survived because small patches of bison habitat remained in interior Alaska, as discussed earlier. Bison

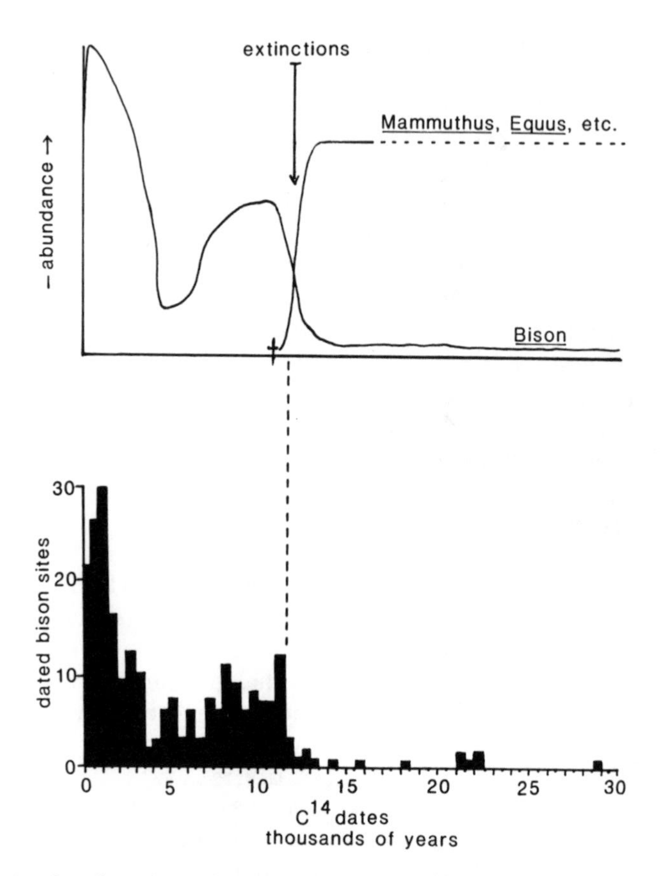

Fig. 10.10. The effect of megafaunal extinctions on bison numbers. In the lower graph the number of radiocarbon dates on bison from the continental United States is plotted against time. Prior to 12,000 years ago, there are few dated bison fossils. This number, however, rises sharply at 11,500 to about 7,000 years ago. I suggest this is a response to the disappearance of other grazers (*Mammuthus, Equus,* etc.) lost in late Pleistocene extinctions. The decrease in bison on the Great Plains between 4,000 and 7,000 years ago is probably due to the aridity of the Hypsithermal.

were probably hunted until the time of their demise, a few hundred years before the present, not long before Europeans and firearms entered the far north.

Thus bison experienced two quite different evolutionary contexts in the late Quaternary. Bison had originated in, and shared for almost a million years, a complex community of large predators and grazing competitors which changed abruptly about 11,000 years ago, at the Pleistocene-Holocene transition. These quite different evolutionary contexts produced two different kinds of bison, and I have argued that this watershed made for different anatomical optima (Guthrie 1980). And indirectly, Blue Babe's death contributes some insight into the response of bison to human hunters.

I discussed the comparative rarity of bison in Paleolithic and Clovis archaeological sites, yet bison remains dominate many Holocene sites. Bison are a common fossil and also a very common subject of Paleolithic art. Few caves with Paleolithic drawings lack bison pictures. Judging from these artworks and paleontological sites, bison seem to have been numerous, but hunted only lightly.

Neither American nor European bison are today particularly dangerous animals. They can be hunted without fear of being gored and killed. American plains Indians hunted bison and killed them by the hundreds with a variety of techniques. Bison of the Great Plains were not considered very dangerous. Their usual response to humans, even when mortally wounded, is to run. In that regard they are quite different from the African buffalo, *Syncerus,* which is one of the most dangerous of large mammals. When African buffalo are wounded, bulls frequently charge in violent attempts to kill their human hunters, and often the bulls succeed. African buffalo and bison species are within the same size range and same general ecological niche, yet their behavior in this respect is markedly different.

I propose that African buffalo are so fierce because lions are their major predators (Sinclair 1977). African buffalo cannot simply run away from lions; they must use strength and ferocity to ward off predation. Indeed buffalo occasionally kill lions (Schaller 1972). To take a buffalo, a lion must first throw the buffalo to the ground by grabbing hold and dragging it down, before the lion can actually kill the buffalo by strangulation. In a one-on-one situation a buffalo can turn toward a lion and use its sharp horns for defense. A fierce retaliation of horn goring forms an important deterrent to any but a very desperate lion. Lions must be judicious in their choice of prey; they can only attack in situations where they can make a kill and still have very little chance of self-injury (fig. 10.11).

For modern bison the situation is different. For more than 10,000 years bison have lived without any predator as large as a lion. Their only potential predators have been a large solitary cat *(Felis concolor)* and a social wolf *(Canis lupus).* Defense against wolves is quite different than defense against lions (fig. 10.12). It is unlikely that a bison could reasonably expect to catch and gore a wolf if it were to try to turn the attack, and even if caught, wolves are small enough that a hook with horns and head butt could not be expected to cause the same degree of damage. Unlike lions, wolves do not grab a bovine and try to throw it off of its feet; rather, wolves kill very

Fig. 10.11. A lone elephant scatters a pride of lions from their fresh kill near a water hole. Even a large pride of lions cannot kill healthy elephants and rhinos.

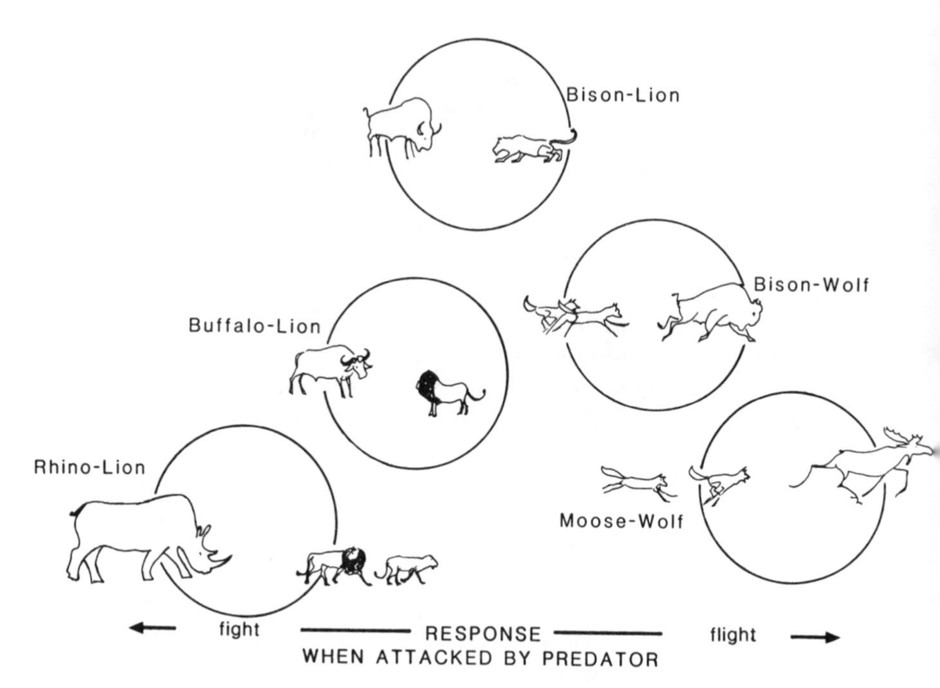

Fig. 10.12. Flight or fight. Flight or fight response depends on the nature of the predator (speed, strength, numbers) and prey (body size, speed, weaponry). At one extreme, rhinos and elephants will chase away the largest predators. On the other hand, extant bison and moose almost always run from wolves. African buffalo, which regularly experience predation by lions, are very pugnacious; they are one of Africa's most dangerous large mammals. With lions for predators, it is likely that the temperament of steppe bison would have been more similar to that of African buffalo.

large herbivores by a drawn-out series of quick attacks to the rear, resulting in the accumulation of many traumatic bites in the hind legs and hindquarters, until the animal is finally debilitated and can no longer defend itself from the front. Not surprisingly, the reaction of bison and moose to wolves is flight (Carbyn 1975). That is how they react to human predators as well, even in confrontational situations when wounded (there are some rare exceptions in the literature). Running moose and bison are a windmill of powerful hoofs, too formidable for an easy bite by wolves.

Moose *(Alces)* are also not formidable prey. Moose hunters have often voiced wonder than an animal as enormous as a bull moose, with many-tined antlers up to two meters across, never attacks a hunter. If moose were more ferocious they would be less likely to be killed by humans in many situations. This defense strategy of escape seems maladaptive and must be understood in context of wolf predators rather than human hunters with guns. Charging a wolf with the intent to kill would be a relatively unsuccessful strategy for moose. Most moose escape by running, but once winded, it is better to stand and let the wolves come within kicking range. For the kinds of predation that recent and Holocene moose and bison encounter, it seems strategically better to respond with flight and a static defense rather than with a charge.

One wonders then if Pleistocene bison, which experienced predation by lions, would also react aggressively to human hunters. I think the answer has to be yes. For an animal the size of a buffalo, a more ferocious defense would have paid off in confrontation with lions. These earlier bison would have been formidable creatures to hunt for supper, which is, I propose, the very reason bison were not a daily part of the menu of late Paleolithic hunters. Their frequent portrayal in Paleolithic art now takes on a different tone. These are not pictures of livestock for the table, but of very dangerous bovines that must have been greatly respected, sought after, but at the same time feared. The three pictures from Paleolithic art that clearly show humans being attacked by dangerous animals portray bison (Guthrie 1984c). The famous drawing from Lascaux of a bison bull attacking a downed man (Leroi-Gourhan and Allain 1979) shows the bison wounded, its intestines spewing out of the spear wound (fig. 10.13). Of the other two, one is from Le Roc-de-Sers and the other from Villars, both in France.

These Paleolithic pictures thus corroborate the theory proposed above—that bovines normally preyed on by lions would be selected to fight when wounded rather than flee. We have every reason to think that such behavior would also spill over in response to

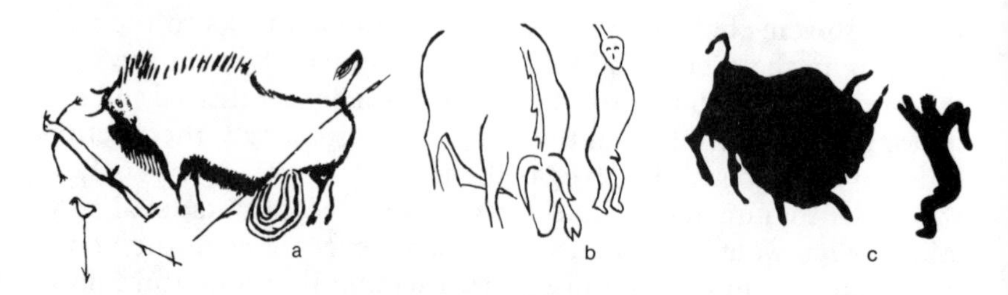

Fig. 10.13. Dangerous Pleistocene bison. Although living bison are not particularly dangerous, some living bovids such as African buffalo have different predators than living bison. Since predators of Pleistocene bison were similar to those of African buffalo, it is likely that those bison were aggressive and dangerous to hunt. Some Paleolithic drawings of bison support this idea: (*a*) Lascaux, (*b*) Le Roc-de-Sers, and (*c*) Villars.

human hunters. The aggressive attack of a wounded African buffalo was probably the reaction *Bison priscus* also had to human or lion hunters. It is unlikely Blue Babe went down on that cold early winter night without a fight.

11
PREPARATION AND EXHIBITION OF BLUE BABE

Blue Babe was found in the summer of 1979, just weeks before I was to begin a year's leave from the University of Alaska. Because European and Soviet paleontologists work with many of the same Pleistocene species and problems I do, I planned to travel to Europe and the Soviet Union to study museum collections and talk with colleagues. In addition, I wanted to see firsthand the Mammoth Steppe species portrayed in European Paleolithic art. In Alaska we have bones and mummies of these animals, but Germany, the Soviet Union, France, and Spain have drawings and sculptures of them. As soon as the bison was excavated, I deposited the entire carcass and many samples of silt from the site in a large freezer at the university. Locking Blue Babe away in a deep freeze for a year was frustrating—we had hardly taken a good look at him—but I knew it was important to establish a thorough program of analysis and preservation before thawing the carcass. I had never worked on such a well-preserved Pleistocene mummy, nor had anyone else in North America. I needed to correspond with and visit Soviet experts to learn from their work with Siberian frozen mummies. The sabbatical gave me the time I needed to think about the bison mummy and provided contacts with Soviet mammalogists and paleontologists so that I could do the best possible job with Blue Babe.

Scientific analysis of the carcass and sediments was more obvious to me than what to do with the mummy afterward. No large frozen Pleistocene mummy had ever been exhibited in North America in a mounted condition prepared by normal taxidermy methods, as are the Beresovka and Dima mammoths in the Zoological Museum in Leningrad. Furthermore our bison's skin was not complete, and much of the hair was missing. When I flew to Europe on sabbatical, Blue Babe was a gray-brown ball of frozen mud. Although various people were enthusiastic about mounting and exhibition, I suspected they envisioned a richly pelted creature just taken from clear ice, when in fact Blue Babe was a scavenged assortment of grime and mud. I was less than enthusiastic about a simple assignment to a conventional taxidermist.

American taxidermy was at its zenith around the turn of the century. Carl Akeley at the Chicago Natural History Museum and others at the American Museum of Natural History were pioneering methods to produce amazingly realistic forms. Akeley's new approach was to finely sculpt a clay mannequin of the original animal and then make a permanent cast of the mannequin in plaster and burlap. Tanned skin was stretched and sewn over this model of the animal, creating a lifelike mount. And although this technique spread throughout the world, the art of making original mannequins has all but been lost in the United States. "Stuffed animals" have gone out of vogue in large natural history museums and have been largely supplanted by ecology and Space exhibits. New exhibits that do contain mounted specimens are usually contracted to firms outside the museum. Few museums in America now have a full taxidermy staff; skills of mounting that once existed have been lost. The economics of production has forced most private taxidermists to send out skins to commercial tanneries and to use precast mannequins available in small, medium, large, and extra large. Labor is expensive, and a custom-made mannequin is a luxury few can afford. Thus, anatomical sculptural skills are unpracticed.

All of this left me in a quandary. I had done enough taxidermy to realize some of the problems this bison mummy might present. Fortunately for Blue Babe and myself, Bjorn Kurtén in the Department of Paleontology in Helsinki introduced me to Eirik Granqvist, who was then conservator at the Zoological Museum in Helsinki. The Helsinki museum operates on a low budget, but has one of the better large-mammal exhibits in Europe because of Eirik Granqvist's skilled hands. He is a taxidermist trained in the classic method, doing everything from beginning to end by himself: collecting the animals, skinning, tanning, model building, casting his mannequins, and all other processes of taxidermy. Granqvist also ran a school at the Helsinki museum, training apprentices for careers in other museums or private business. He was a good self-trained anatomist and biologist and was most enthusiastic about working with the bison. While he agreed to travel to Alaska and do the taxidermy work to mount Blue Babe, finding funds for the job proved difficult.

I would remind the University of Alaska museum director several times annually that we needed to mount the bison mummy soon, but the museum had no extra funds available. Funds were sought from the Alaska State Legislature to no avail. I kept writing

Granqvist that I hoped to find money soon. During the necropsy I had removed the skin where it was still attached to the head, legs, and lower part of the thorax (sternum and ribs) and, on Granqvist's recommendation, had put the skin into large vats of 80% ethanol. After literally years of delay, finally, in exasperation, I came to the unilateral conclusion that the skin would soon harden in the alcohol and that mounting had to be done in the next year or the specimen might be tragically ruined. At last some museum acquisition funds were reallocated for work on Blue Babe. We contacted Granqvist. A UNESCO-funded short course for African taxidermists in the Sahara and Sahel had fallen through when the U.S. had withdrawn from UNESCO, so Granqvist had an opening in his schedule for the spring of 1984. The Institute of Arctic Biology was willing to provide space and support for the taxidermy work as no space was available in the museum. The institute also contributed housing for Granqvist while he worked on the mummy.

During the time we waited for museum funding, I sculpted a three-dimensional scale model of Blue Babe using anatomical data from the carcass, pictures from European Paleolithic art, and a lot of trial and error and fussing. I sealed the plasticine clay model and made a mold of PVC rubber, into which a wax positive was poured. This wax bison was then cast in bronze by the lost-wax process.

When Granqvist arrived we took the bronze bison in hand and had a long session on how to mount Blue Babe. We concluded that a simple pile of head and skin, like the Dome Creek bison mummy on exhibit in the Smithsonian, was not sufficient, nor did we want a full standing mount. We decided that the position of the bison when it died, prone with legs gathered underneath, might be a good compromise.

With the scale model, Granqvist quickly used his artistic talents in laying out a plywood silhouette onto which he wired remaining limb bones and plywood cutouts of missing ones. He shaped chicken wire in the approximate contours of the body, and over this he sculpted a clay form of Blue Babe as he would have looked at death (fig. 11.1).

Granqvist made a rough plaster cast of the skull to attach to the life-sized clay form (see fig. 11.2, which diagrams the following description). I helped cast epoxy horns for the exhibit using PVC rubber molds made from Blue Babe's real horns because I wanted to keep the actual head and horns frozen, available for future research.

During this time Granqvist also thoroughly cleaned and split the skin so that the thinned outer portion could be stretched over the plastic form he was about to produce. It was during this splitting

Fig. 11.1. Preparing the museum exhibit. Eirik Granqvist is shown here modeling Blue Babe's form in clay, using a small bronze study by the author for reference. (Photo by Don Borchardt)

that Granqvist found the carnassial tooth fragment of a lion lodged inside the thick fibrous dermis of the skin. Alcohol was washed from the skin and the hide was placed in an acid bath tanning solution.

Once Granqvist was satisfied that his life-size sculpture was accurate, he made a plaster piece-mold about 1.5 inches (35 mm) thick over the entire body, without undercuts (fig. 11.3). When fully set, this was removed and the model dismembered. Granqvist then reassembled the plaster mold and commenced to line it with a plaster coat backed with strips of burlap dipped in wet plaster. These added strength to the plaster, producing a strong, thin-walled mannequin over which to stretch the split and tanned skin.

After being tanned, the skin was treated with a commercial preservative that chemically locks open protein bonds, making it difficult for insect or microbial decomposers to attack the hide. The skin was then treated with a relaxant to maximize its stretchability

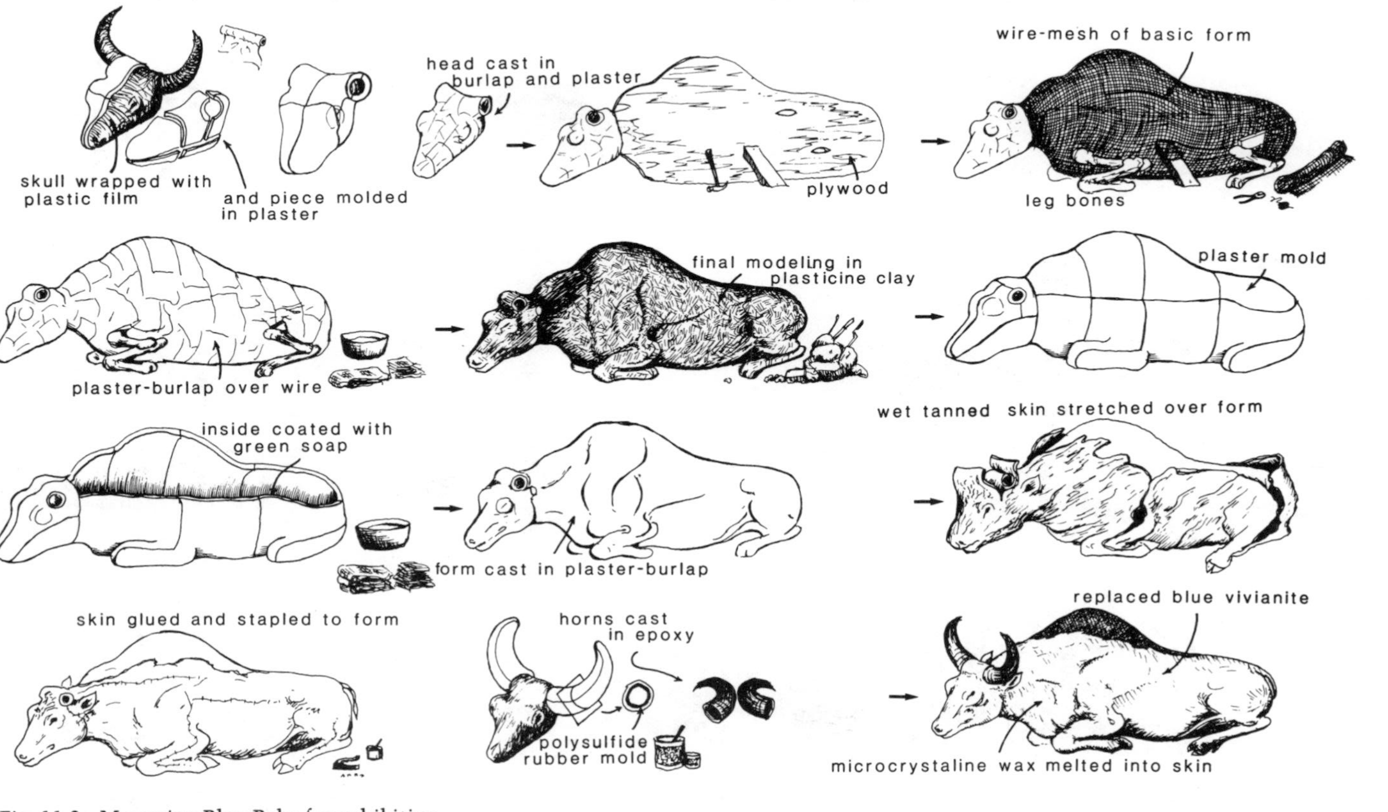

Fig. 11.2. Mounting Blue Babe for exhibition.

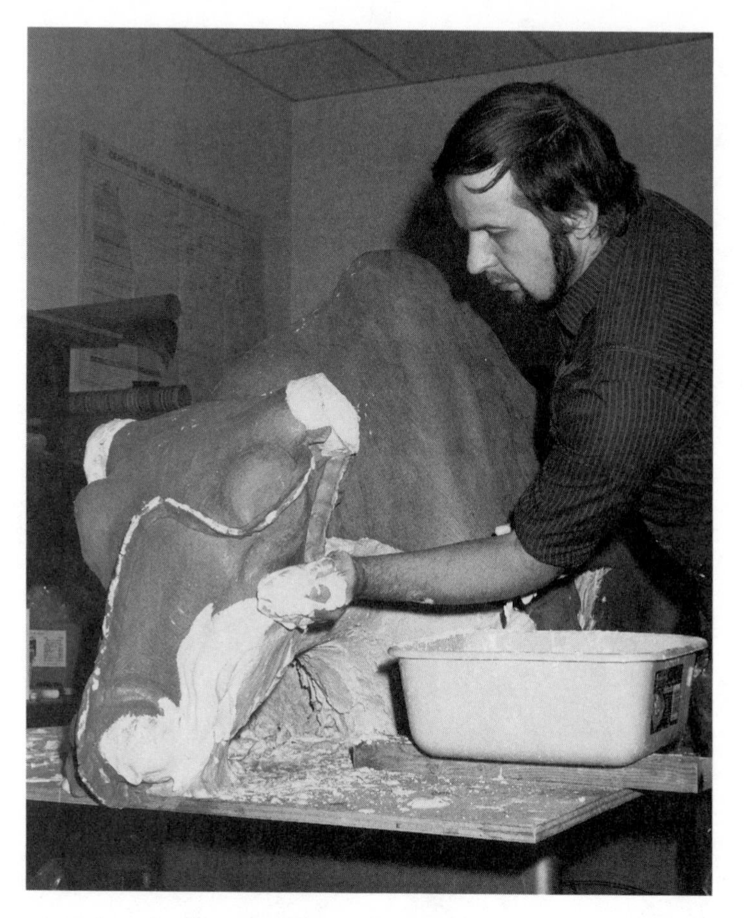

Fig. 11.3. Making the plaster mold. Once the clay model was finished, a plaster mold was constructed in which to cast the final plaster and burlap form. Blue Babe's tanned skin was then mounted on the completed form. (Photo by Don Borchardt)

and was finally pulled over the bison mannequin. It fit, almost. Some skin pieces were missing at the dorsal surface, where carnivores had first opened the carcass. But after the skin was relaxed and stretched it was obvious that there was more skin than I had first reconstructed, so I had to change my drawings. Thus, mounting directly helped my reconstruction. Granqvist tacked the wet skin in place with galvanized staples and allowed it to dry partially (fig. 11.4). A melted wax mixture was painted onto the skin and heated with a hair-dryer, causing the mixture to penetrate the dry skin. The

Fig. 11.4. Blue Babe mount nearing completion. The wet, tanned skin was stretched over the form, tacked in place, and allowed to dry. (Photo by Don Borchardt)

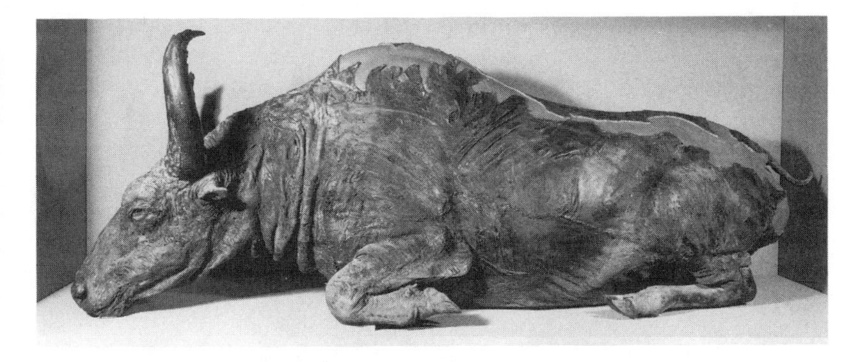

Fig. 11.5. Blue Babe on permanent display at the University of Alaska Museum, Fairbanks. (University of Alaska Museum photo)

wax filled spaces in the skin fibers and made the skin less likely to shrink or crack later. It also gave the skin a fresher appearance. Then we installed the cast horns and sealed their seam with wax. Seams in the skin were filled with colored wax, and the bison was ready to be recoated with vivianite, which had been removed when the skin was cleaned of its mud. Once again blue, the bison was carried across the street from the Institute of Arctic Biology to its case in the museum (fig. 11.5). Granqvist's work was done.

To climax and celebrate Eirik Granqvist's work with Blue Babe, we had a bison stew dinner for him and for Bjorn Kurtén, who was giving a guest lecture at the University of Alaska that week. A small part of the mummy's neck was diced and simmered in a pot of stock and vegetables. We had Blue Babe for dinner. The meat was well aged but still a little tough, and it gave the stew a strong Pleistocene aroma, but nobody there would have dared miss it.

APPENDIX A

JOHN V. MATTHEWS, JR.
(Geological Survey of Canada, Ottawa)

Fossil Arthropod Report No.: 81-17

Sample No.: Guthrie's Bison mummy (vial 2)

Fossils

Insecta Coleoptera ("beetles")
- Carabidae ("ground beetles")
  - *Notiophilus* sp.
  - *Diacheila polita* Fald.
  - *Pterostichus parasimilis* Ball
  - *Pterostichus (Cryobius)* sp.
  - *Pterostichus (Cryobius) tareumiut* Ball
  - *Pterostichus (Cryobius) parasimilis*Ball
  - *Pterostichus (Cryobius) ventricosus* Eschz.
  - *Pterostichus (Cryobius) brevicornis* Kirby
  - *Pterostichus (Cryobius) nivalis* Sahlb.
  - *Pterostichus agonus* Horn
  - *Amara sp.*
- Staphylinidae ("rove beetles")
  - *Olophrum* sp.
  - *Lathrobium* sp.
  - *Tachinus apterus* grp.
- Byrrhidae ("pill beetles")
  - *Simplocaria* sp.
- Chrysomelidae ("leaf beetles")
  - *Chrysolina* sp.
- Curculionidae ("weevils")
  - *Lepidophorus lineaticollis* Kirby
  - *Hypera?* sp.
  - *Lepyrus gemellus* Kirby

DIPTERA ("flies")
- Calliphoridae?

Comments

The assemblage has several paleoenvironmental implications. First, all of the taxa are either facultative or obligate tundra inhabitants, thus a treeless or nearly treeless environment existed at the time of deposition. . . . The dominance of the assemblage by members of the *Cryobius* group suggests mesic tundra, but note that fragments of the weevil *Lepidophorus lineaticollis,* a species of dry substrates, are also relatively abundant.

The assemblage resembles others from the Fairbanks mucks. For example, fragments of *Cryobius* are abundant, *Morychus* (= *Chrysobyrrhulus* of Siberian literature) is present but rare, *Tachinus* is rare, and the weevil *Vitavitus* is missing entirely. But it differs from those, as well as from the insects associated with the Dawson Cut mammoth, by a rarity of *Amara alpina* fossils—another indication of the mesic tundra implications of the Bison mummy assemblage.

Another striking difference from the Dawson Cut assemblage is the lack of fossils of carrion insects. The only possible exception is a single puparial fragment of a blow fly (*Calliphoridae*). The Dawson Cut assemblage contained an abundance of Calliphorid puparia plus fossils of two carrion feeding species of the beetle genus *Silpha.*

Silpha lapponica and several other carrion feeders (e.g., *Nitidula, Creophilus maxillosus* L., *Cleridae* and blow-fly puparia) were collected by Guthrie at the site of a modern Bison carcass. Some of these species do not ordinarily occur in a tundra environment, one possible explanation of their absence from the Bison mummy, but the absence of all carrion forms in the Bison mummy assemblage is remarkable. It must mean that the carcass was exposed for only a short time, probably during the winter of a single year, before being buried and frozen.

A rarity of carrion species is what one might expect of a well preserved carcass. Excellent preservation of soft tissues cannot occur if they are exposed for long to carrion feeders.

The insects associated with the Berelekh mammoth concentration also lack carrion species, but the samples are so small as to make such negative evidence very suspect. The Bison mummy sample, though also small, is large enough to make the absence of carrion species significant.

Appendix B

DR. JAMES H. ANDERSON (Institute of Arctic Biology, Fairbanks)

Table A.1 Pollen Sample from Silt Associated with Blue Babe

	Number	Percentage
Picea	9	13
Betula	24	35
Alnus	4	6
Salix	6	9
Juniperus	1	1
Populus	3	4
Graminae	15	22
Ericad	2	3
Artemisia	1	1
Sphagnum	11	16
Bryales	19	28
Equisetum	1	1
Filicales	1	1
Botrychium	1	1
Fungus spores	8	12
Caryophyllaceae	1	1

Notes: This sample represents the pollen from one slide—107 grains. Percentages exclude *Sphagnum*, Bryales, and Fungus spores.

REFERENCES

Abramova, Z. A. 1982. Zur Jagd im Jungpalaolithikum: Nach Beispielen des jungpalaolithischen Fundplatzes Kokorevo I in Sibiren. *Archäologisches Korrespondenblatt* 12:1–9.

Agenbroad, L. D. 1984. Chronology and diet of *Mammuthus* from Eurasia and temperate North America. Abstract of the International Geological Congress, Moscow.

Ager, T. A. 1975. Late Quaternary environmental history of the Tanana Valley, Alaska. Institute of Polar Studies Report No. 4. Ohio State University, Columbus.

———. 1982. Vegetational history of western Alaska during the Wisconsin Glacial Interval and the Holocene. In D. M. Hopkins et al., eds., *Paleoecology of Beringia*, 75–94. New York: Academic Press.

Alexander, R. McN. 1984. Elastic energy stores in running vertebrates. *Amer. Zool.* 24:85–94.

Alexander, R. McN., N. J. Dimery, and R. K. Ker. 1985. Elastic structures in the back and their role in galloping in some mammals. *J. Zool.* 207:467–82.

Anderson, E. 1973. Ferret from the Pleistocene of central Alaska. *J. Mammal.* 54:777–78.

———. 1977. Pleistocene Mustelidae (Mammalia, Carnivora) from Fairbanks, Alaska. *Bull. Mus. Comp. Zool.* 148:1–21.

Anderson, P. M. 1982. Reconstructing the past: A synthesis of archaeological and palynological data, northern Alaska and northwestern Canada. Ph.D. diss., Brown University.

———. 1985. Late Quaternary vegetational change in the Kotzebue Sound area, Northwestern Alaska. *Quart. Res.* 24:307–21.

———. 1988. Late Quaternary pollen records from the Kobuk and Noatak River drainages, northwest Alaska. *Quart. Res.* 29:1–14.

Azzaroli, A. 1983. Quaternary mammals and the "end-Villifranchian" dispersal event: A turning point in the history of Eurasia. *Paleogeography, Paleoclimatology, and Paleoecology* 44:117–39.

Baker, F., L. H. P. Jones, and I. D. Wardrop. 1959. Cause of wear in sheep's teeth. *Nature* 184:1583–84.

Barnicot, C. R. 1957. Wear in sheep's teeth. *New Zealand Journal of Science and Technology* 38:583–632.
Baskirov, I. 1939. Kavkazskij zubr. Glav. Upr. Po Zapov. Zooparks I Zoosadam. SNK. *RSFSR* (Moscow), 1–72.
Batzli, G. O., and H. G. J. Juna. 1980. Nutritional geology of microtine rodents: Resource utilization near Atkasook, Alaska. *Arctic and Alpine Research* 12:483–99.
Bertram, B. C. R. 1973. Lion population regulation. *E. Afr. Wildl. J.* 11:215–25.
———. 1975. Social factors influencing reproduction in wild lions. *J. Zool. Lond.* 177:463–82.
Beug, H.-J. 1968. Problems der Vegetationsgeschichte in Sudeuropa. *Deutche Bot. Ges. Ber.* 80:682–89.
Bohlken, H. 1967. Beitnog zur Systematik der rezenten Formen der Gattung *Bison. Z. f. zool. Syst. u. Evolut.-forsch.* 5:54–110.
Bombin, M. 1984. On information evolutionary theory, phytoliths, and the Late Quaternary ecology of Beringia. Ph.D. diss., Department of Anthropology, University of Alberta, Edmonton.
Bonnichsen, R. 1979. Pleistocene bone technology in the Beringian refugium. Canadian Museum of Man, Mercury Series (Archaeological Survey of Canada) No. 89. Ottawa.
Boriskovskij, P. I., and N. D. Praslov. 1964. *Paleolithic of the upper Dnepr basin and the Praizove.* Svodarkheloquicheskikh istochnikov A 1-5. Moscow and Leningrad: Academy of Sciences.
Borowski, S., Z. Kra, and L. Milkowski. 1967. Food and the role of the European bison in the forest ecosystems. *Acta Theriol.* 12:367–76.
Bosinski, G. 1981. *Gönnersdorf: Eiszeitjager am Mittelrhein.* Landesmuseum Koblenz: Rheinlandpfalz.
Bowen, D. Q. 1981. Quaternary geology: A stratigraphic framework for multidisciplinary work. New York: Pergamon Press.
Braack, L. E. O. 1986. Arthropods associated with carcasses in northern Kruger National Park. *S. Afr. Tydskr. Natuurnav.* 16:91–98.
Brandt, J. F. 1866. Zur Levensgeschichte des Mammuth. *Bull. Acad. Sci.* (Academy of Science, St. Petersburg) 10:111–18.
Brantas, G. C. 1968. On the dominance order in Friesian-Dutch dairy cows. *Z. Tierzucht. Zuchtbiol.* 84:127–51.
Brubaker, L. B., H. L. Garfinkel, and M. E. Edwards. 1983. A late Wisconsin and Holocene vegetation history from the Central Brooks Range: Implications for Alaskan Paleoecology. *Quart. Res.* 20:194–214.
Brunnacker, K. 1974. Results of Quaternary stratigraphy on the middle and lower courses of the Rhine. In V. Sibrava, *IUGS-UNESCO international geological correlation programme.* Geological Survey of Prague.
———. 1980. Young Pleistocene loess as an indicator for the climate in the Mediterranean. *Palaeoecol. Afr.* 12:99–113.
Bryant, J. and P. J. Kuropat. 1980. Selection of winter forage by subarctic browsing vertebrates. *Ann. Rev. Ecol. Syst.* 11:261–68.

Bryant, S. P., F. S. Chapin III, P. Reichardt, and T. Clausen. 1985. Adaptation to resource availability as a determinant of chemical defense strategies in woody plants. In G. A. Cooper-Driver, T. Swain, and E. E. Conn, eds., *Chemically mediated interactions between plants and other organisms.* New York: Plenum Press.

Budel, J. 1951. Die Klimazonen des Eiszeitalters. *Eiszeitalter u. Gegenw.* 1:16–26.

Campbell, B. H, and M. Hinkes. 1983. Winter diets and habitat use of Alaska bison after wildfire. *Wildl. Soc. Bull.* 11:16–21.

Carbyn, L. N. 1975. Wolf predation and behavioral interactions with elk and other ungulates in an area of high prey diversity. Ph.D. diss., University of Toronto.

Cassoli, P. F. 1972. Le Pteroclide (Aves, Pteroclidae) fossile nei livelli del paleolitico superiore e medionel Pleistocene dell'Italia meridionale. *Quaternaria* 16:225–45.

Chaline, J. 1972. *Le Quaternaire.* Paris: Doin.

Chaney, R. W., and H. L. Mason. 1936. The Pleistocene flora of Fairbanks, Alaska. *Amer. Mus. Novitates* 887:1–17.

Chapin, F. S., III. 1980. The mineral nutrition of wild plants. *Ann. Rev. Ecol. Syst.* 11:233–60.

Churcher, C. S. 1980. Did the North American mammoths migrate? *Can. J. Anthropol.* 1:103–6.

Cinq-Mars, J. 1979. Bluefish Cave I: A late Pleistocene eastern Beringian cave deposit in the northern Yukon. *Can. J. Archaeol.* 3:1–32.

Clutton-Brock, T. H., F. E. Guinness, and S. D. Albon. 1982. *Red deer.* Chicago: University of Chicago Press.

Clutton-Brock, T. H., and P. H. Harvey. 1976. Evolutionary rules and primate societies. In P. P. G. Bateson and R. A. Hinde, eds., *Growing points in ethology,* 195–238. Cambridge: Cambridge University Press.

Colinvaux, P. A. 1964. The environment of the Bering Land Bridge. *Ecol. Monog.* 34:297–329.

———. 1980. Vegetation of the Bering Land Bridge revisited. *Quart. Rev. Archaeol.* 1:2–15.

———. 1981. Historical ecology of Beringia: The south land bridge coast of St. Paul Island. *Quart. Res.* 16:18–36.

———. 1984. The Beringian ecosystem, a book review of *Paleoecology of Beringia* by D. M. Hopkins et al. (eds). *Quart. Rev. Archaeol.* 5:10–16.

———. 1986. Plain thinking on Bering Land Bridge vegetation and mammoth populations. *Quart. Rev. Archaeol.* 7:8–9.

Colinvaux, P. A., and F. H. West. 1984. The Beringian ecosystem. *Quart. Rev. Archaeol.* 5:10–16.

Cuvier, G. 1825. *Recherches sur les ossements fossiles, ou l'on retablit les charactères des plusieurs animaux dont les revolutions du globe ont détruit les espèces.* 3d ed. Paris.

Cwynar, L. C. 1980. A Late-Quaternary vegetation history from Hanging Lake, northern Yukon. Ph.D. diss., University of Toronto.

———. 1982. A Late-Quaternary vegetation history from Hanging Lake, northern Yukon. *Ecol. Monog.* 52:1–24.

Cwynar, L. C., and J. C. Ritchie. 1980. Arctic steppe-tundra: A Yukon perspective. *Science* 208:1375–78.

Davis, B. M. 1983. Holocene vegetational history of the eastern United States. In H. E. Wright, Jr., ed., *Late-Quaternary environments of the United States,* vol. 2, *The Holocene,* 166–81. Minneapolis: University of Minnesota Press.

Degerbol, M., and J. Iversen. 1945. The bison of Denmark. *Danmarks Geoloske Undesogelse.* 73:1–62.

Digby, B. 1926. The mammoth—and mammoth hunting in north-east Siberia. New York: D. Appleton.

Dikov, N. N. 1979. *Drevmiye kul'tury severo-vostochnoy Azzi* (The early cultures of northeastern Asia). Moscow: Nauka.

Dixon, E. J. 1984. Context and environment in taphonomic analysis: Examples from Alaska's Porcupine River caves. *Quart. Res.* 22:201–15.

Dolitsky, A. B. 1985. Siberian Paleolithic archaeology: Approaches and analytic methods. *Curr. Anthropol.* 26:361–78.

Duffield, L. F. 1973. Aging and sexing the post-cranial skeleton of bison. *Plains Anthropol.* 18:132–39.

Eisenmann, V. 1984. Sur quelques charactères adaptatifs du squelette d'Equus (Mammalia, Périssodaqctyla) et leurs implications paléoecologiques. *Bull. Mus. National Hist. Nat. Paris* 2:185–95.

Estes, R. F. 1974. Social organization of the African bovidae. IUCN Publication No. 24. Morges, Switzerland.

Ewer, R. F. 1973. *Carnivores.* Ithaca: Cornell University Press.

Farrand, W. R. 1961. Frozen mammoths and modern geology. *Science* 133:729–835.

Flerov, C. C. 1967. The origins of the mammalian fauna in Canada. In D. M. Hopkins, ed., *The Bering Land Bridge.* Stanford: Stanford University Press.

———. 1977. Bison of northeastern Asia (in Russian). *Trudy Akad. Nauk SSSR, Zoologicheskie Instituta* 73:39–56.

———. 1979. Bison systematics and evolution (in Russian). In V. E. Sokolov, ed., 9–127. Zubr. M. Izd. Nauka.

Frazer, A. F. 1968. *Reproductive behavior in ungulates.* New York: Academic Press.

Frenzel, B. 1968. The Pleistocene vegetation of northern Eurasia. *Science* 161:637–49.

Frison, G. C., and C. A. Reher. 1970. Age determination of buffalo by teeth eruption and wear. *Plains Anthropol.* 15, memoir 7.

Fuller, W. A. 1959. The horns and teeth as indicators of age in bison. *J. Wildl. Mgmt.* 23:342–44.

Garut, V. E, Ye. P. Metel'tseva, and B. A. Tikhomirov. 1970. New data on the food of the woolly rhinoceros in Siberia (in Russian). In *Severo ledovityy okean i ego poberez'e v Kainozoe.* Leningrad: Gidrometeoizdat.

Geikie, J. 1881. *Prehistoric Europe: A geologic sketch.* London: Stanford.

Geist, V. 1966. The evolution of horn-like organs. *Behavior* 27:175–214.

———. 1971a. *Mountain sheep: A study in behavior and evolution.* Chicago: University of Chicago Press.

———. 1971b. The relation of social evolution and dispersal in ungulates during the Pleistocene with emphasis on the Old World deer and the genus *Bison. Quart. Res.* 1:283–315.

———. 1974. On the relationship of social evolution and ecology in ungulates. *Amer. Zool.* 14:205–20.

———. 1977. A comparison of social adaptations in relation to ecology in gallinaceous birds and ungulate societies. *Ann. Rev. Ecol. Syst.* 8:193–208.

Geist, V., and P. Karsten. 1977. The wood bison in relation to hypothesis on the origin of the American bison *Z. Säugetierk.* 42:119–22.

Gentry, A. W. 1967. Pelorovis oldowayensis Reck, an extinct bovid from East Africa. *Bull. Brit. Mus. (Nat. Hist.), Geol.* 14:245–99.

Gilbert, M. B., and L. D. Martin. 1984. Late Pleistocene fossils in Natural Trap Cave, Wyoming, and the Climatic Model of Extinction. In P. S. Martin and R. G. Klein, eds., *Quaternary extinctions,* 138–47. Tucson: University of Arizona Press.

Gipson, P. S., and J. D. McKendrick. 1981. Bison depredation of grain fields in interior Alaska. *Fifth Great Plains Wildlife Damage Control Workshop* 5:116–21.

Giterman, R. E., and L. V. Globeva. 1967. Vegetation of eastern Siberia during the Anthropogene Period. In D. M. Hopkins, ed., *The Bering Land Bridge,* 232–44. Stanford: Stanford University Press.

Gonyea, W., and R. Ashworth. 1975. The form and function of retractable claws in the Felidae and other carnivores. *J. Morphol.* 145:229–38.

Gorlova, R. N. 1982. Macroremains of plants from the stomach of the Shandrin mammoth. In: N. K. Vereshchagin and A. I. Nikolaev, eds., *The mammoth fauna of the Asiatic part of the U.S.S.R.,* 34–35. Akademiya Nauk, Trudy Zoologischeskologo Institute, vol. 3.

Grime, J. P. 1979. *Plant strategies and vegetation processes.* New York: John Wiley and Sons.

Gromova, V. 1965. *Kratkiy obzor chetvertichnykh mlekopitaiushchikh Evropy (opyt sopostavleniia).* Moscow: Nauka.

Groves, C. P. 1980. Systematic relationships in the Bovini (Artiodactyla, Bovidae). *Z. zool. Syst. Evolut-forsch.* 19:264–78.

Guthrie, R. D. 1966a. Bison horn cores: Character choice and systematics. *J. Paleontol.* 40:328–340.

———. 1966b. Pelage of fossil bison: A new osteological index. *J. Mammal.* 47:735–37.

———. 1968. Paleoecology of a late Pleistocene small mammal community from interior Alaska. *Arctic* 22:213–24.

———. 1970. Bison evolution and zoogeography in North America during the Pleistocene. *Quart. Rev. Biol.* 45:1–15.

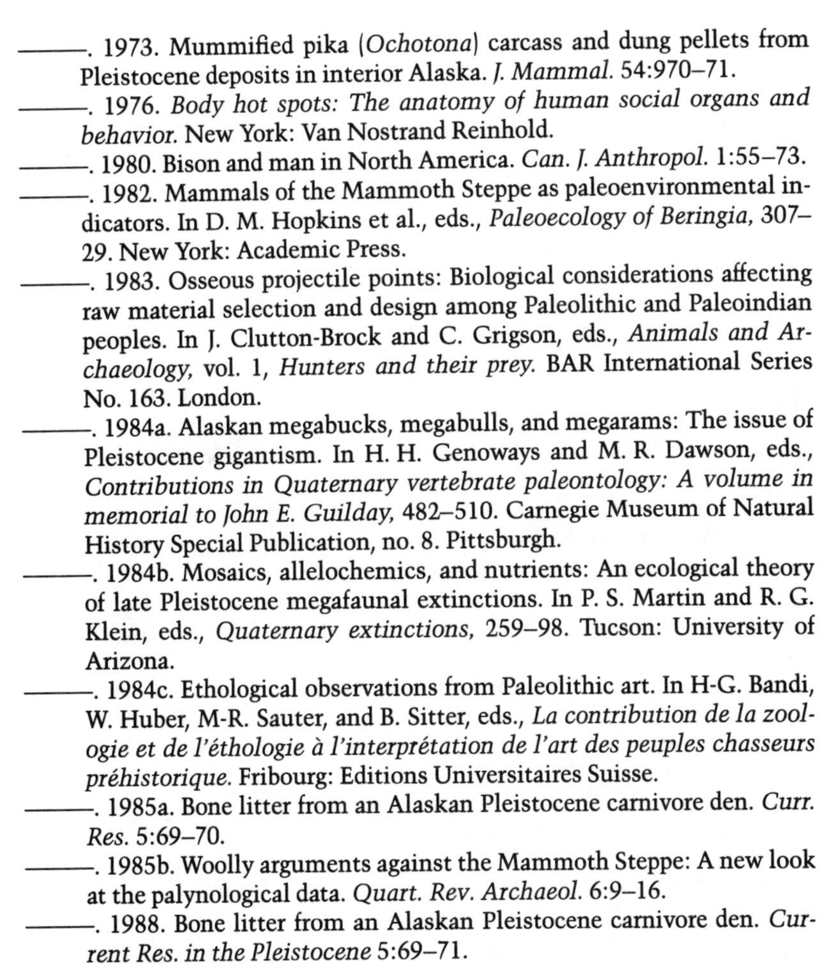

———. 1973. Mummified pika (*Ochotona*) carcass and dung pellets from Pleistocene deposits in interior Alaska. *J. Mammal.* 54:970–71.

———. 1976. *Body hot spots: The anatomy of human social organs and behavior.* New York: Van Nostrand Reinhold.

———. 1980. Bison and man in North America. *Can. J. Anthropol.* 1:55–73.

———. 1982. Mammals of the Mammoth Steppe as paleoenvironmental indicators. In D. M. Hopkins et al., eds., *Paleoecology of Beringia,* 307–29. New York: Academic Press.

———. 1983. Osseous projectile points: Biological considerations affecting raw material selection and design among Paleolithic and Paleoindian peoples. In J. Clutton-Brock and C. Grigson, eds., *Animals and Archaeology,* vol. 1, *Hunters and their prey.* BAR International Series No. 163. London.

———. 1984a. Alaskan megabucks, megabulls, and megarams: The issue of Pleistocene gigantism. In H. H. Genoways and M. R. Dawson, eds., *Contributions in Quaternary vertebrate paleontology: A volume in memorial to John E. Guilday,* 482–510. Carnegie Museum of Natural History Special Publication, no. 8. Pittsburgh.

———. 1984b. Mosaics, allelochemics, and nutrients: An ecological theory of late Pleistocene megafaunal extinctions. In P. S. Martin and R. G. Klein, eds., *Quaternary extinctions,* 259–98. Tucson: University of Arizona.

———. 1984c. Ethological observations from Paleolithic art. In H-G. Bandi, W. Huber, M-R. Sauter, and B. Sitter, eds., *La contribution de la zoologie et de l'éthologie à l'interprétation de l'art des peuples chasseurs préhistorique.* Fribourg: Editions Universitaires Suisse.

———. 1985a. Bone litter from an Alaskan Pleistocene carnivore den. *Curr. Res.* 5:69–70.

———. 1985b. Woolly arguments against the Mammoth Steppe: A new look at the palynological data. *Quart. Rev. Archaeol.* 6:9–16.

———. 1988. Bone litter from an Alaskan Pleistocene carnivore den. *Current Res. in the Pleistocene* 5:69–71.

———. In prep. The Holocene survival of bison and wapiti in Alaska.

Guthrie, R. D., and J. V. Matthews, Jr. 1971. The Cape Deceit fauna: Early Pleistocene mammalian assemblage from the Alaskan Arctic *Quart. Res.* 1:474–510.

Hafėz, E. S. E. 1975. *The behavior of domestic animals.* London: Baillière Tindall.

Hamilton, T. D. 1979. Quaternary stratigraphic sections with radiocarbon dates, Weisman quadrangle, Alaska. United States Geological Survey Open File Report 80–791.

Hamilton, T. D., J. L. Craig, and P. V. Sellmann. 1988. The Fox permafrost tunnel. *Geol. Soc. Amer. Bull.,* 100:948–69.

Hanks, J. 1979. *A struggle for survival.* London: Country Life Books.

Hansen, R. M. 1976. Foods of free-roaming horses in southern New Mexico. *J. Range Mgmt.* 29:347.

Hansen, R. M., R. C. Clark, and H. Lawhorn. 1977. Foods of wild horse, deer, and cattle in the Douglas Mountain area, Colorado. *J. Range Mgmt.* 30:116–18.

Harington, C. R. 1969. Pleistocene remains of the lion-like cat (*Pathera atrox*) from the Yukon Territory and northern Alaska. *Can. J. Earth Sci.* 6:1277–88.

———. 1978. *Quaternary vertebrate faunas of Canada and Alaska and their suggested chronological sequence.* Syllogeus No. 15. National Museum of Canada, Ottawa.

———. 1981. Pleistocene saiga antelopes of North America and their paleoecological implications. In W. C. Mahoney, ed., *Quaternary paleoclimate,* 193–225. Norwich, Conn.: Geoabstracts.

Haynes, G. 1982. Utilization and skeletal disturbances of North American prey carcasses. *Arctic* 35:266–81.

———. 1984. Tooth wear in northern bison. *J. Mammal.* 65:487–90.

Hedenstrom, M. 1830. *Otrivki o Sibiri.* Academy of Science, St. Petersburg.

Hofmann, R. R., and D. R. M. Stewart. 1972. Grazer or browser: A classification based on the stomach-structure and feeding habits of east African ruminants. *Mammalia* 36:226–40.

Hopkins, D. M. 1982. Aspects of paleogeography of Beringia during the late Pleistocene. In D. M. Hopkins et al., *Paleoecology of Beringia,* 3–28. New York: Academic Press.

Hopkins, D. M., J. V. Matthews, Jr., C. E. Schweger, and S. B. Young, eds. 1982. *Paleoecology of Beringia.* New York: Academic Press.

Hopkins, D. M., P. A. Smith, and J. V. Matthews, Jr. 1981. Dated wood from Alaska and the Yukon: Implications for forest refugia in Beringia. *Quart. Res.* 15:217–49.

Howorth, H. H. 1887. The mammoth in Siberia. *Geological Magazine* 7:408–561.

Hubbard, R. E., and R. M. Hansen. 1976. Diets of wild horses, cattle, and mule deer in the Piceance Basin, Colorado. *J. Range Mgmt.* 29:389–92.

Hultén, E. 1968. *Flora of Alaska and neighboring territories.* Stanford: Stanford University Press.

Irving, W. N. 1985. Context and chronology of Early Man in the Americas. *Ann. Rev. Anthropol.* 14:529–55.

Jaczewski, Z. 1958. Reproduction of the European bison *Bison bonasus* in reserves. *Acta Theriol.* 1:333–76.

Jones, R. L., and H. C. Hansen. 1985. *Mineral licks, geophagy, and biogeochemistry of North American ungulates.* Springfield, Ill.: Stackpole.

Kahlke, H. D. 1975. Der Saiga-Fundvon Bottrop/Wesphalen. *Quartar* 26:135–46.

Kislev, S. V., and V. I. Nazarov. 1985. Late Pleistocene insects. In A. A. Veichko, ed., *Late Quaternary environments of the Soviet Union,* 3:223–26. Minneapolis: University of Minnesota Press.

Koch, W. 1932. Über Wachstume und Alterveränderung am Skelett des Wisents. Abh. Bayer. *Acad. Wiss. Math.-naturwiss. Suppl.-Bd., Munchen* 15:555–678.

Korobkov, A. A., and V. R. Filin. 1982. An analysis of the plant remains from the digestive tract of the bison that was found in the deposits of the late Pleistocene of the Krestovka River (Kolyma Basin). *Bot. Zhur.* 67:1351–61.

Korockina, L. N. 1966. K voprosu o znacenii drevesnoj rostitelnosti v tanii zubrova Belovezskoj Pusci. *Vesci AN BSSR* 1:106–11.

Kowalski, K. 1967a. The evolution and fossil remains of European bison. *Acta Theriol.* 12:335–38.

———. 1967b. The Pleistocene extinction of mammals in Europe. In P. S. Martin and H. E. Wright, eds., *Pleistocene extinctions: The search for a cause.* New Haven: Yale University Press.

Krause, H. 1978. *The mammoth: In ice and snow.* Self-published.

———. 1983. *Elephants of the Arctic Sea.* Self-published.

Krausman, P. R., and J. A. Bissonette. 1977. Bone chewing behavior of desert mule deer. *Southwest Nat.* 22:149–50.

Kubiak, H. 1982. Morphological characters of the mammoth: An adaptation of the Arctic-Steppe environment. In D. M. Hopkins et al., eds., *Paleoecology of Beringia,* 281–90. New York: Academic Press.

Kurtén, B. 1985. The Pleistocene lions of Beringia. *Ann. Zool. Fennica* 22:117–21.

Kurtén, B., and E. Anderson. 1980. Pleistocene mammals of North America. New York: Columbia University Press.

Kuz'mina, Y. 1977. On the origin and history of the theriofauna of the Siberian Arctic. In O. A. Skarlato, ed., *The anthropogene fauna and flora of north-east Siberia.* Academy of Sciences of the USSR, Proceedings of the Zoological Institute, vol. 63.

Lachenbruch, A. H. 1962. Mechanics of thermal contraction cracks and ice wedge polygons in permafrost. *Geol. Soc. Amer.* Special Paper No. 70.

Lapparent, A. 1906. *Traité de géologie.* 5th ed. Paris.

Laws, R. M., E. S. C. Parker, and R. C. B. Johnstone. 1975. Elephants and their habitats: The ecology of elephants in north Bunyoro, Uganda. Oxford: Clarendon Press.

———. 1977a. Geological description of the place of burial of the carcass of the Selerikan horse. In O. A. Skarlato, ed., *The anthropogene fauna and flora of the north-east Siberia.* Academy of Sciences of the USSR, Proceedings of the Zoological Institute, vol. 63.

Lazarev, P. A. 1977b. History of the finding of the carcass of the Selerikan horse and its study. In O. A. Skarlato, ed., *The anthropogene fauna and flora of the north-east Siberia.* Academy of Sciences of the USSR, Proceedings of the Zoological Institute, vol. 63.

———. 1977c. New find of the skeleton of woolly rhinoceros in Yakutia. In O. A. Skarlato, ed., *The anthropogene fauna and flora of the north-east Siberia.* Academy of Sciences of the USSR, Proceedings of the Zoological Institute, vol. 63.

Leroi-Gourhan, A. 1982. *The dawn of Paleolithic art.* Cambridge: Cambridge University Press.

Leroi-Gourhan, A., and J. Allain. 1979. *Lascaux inconnu.* (20th suppl. to *Gallia prehistoire*). Paris: CNRS.

Liu Tung-sheng and Li Xing-guo. 1984. *Mammoths in China.* In P. S. Martin and R. G. Klein, eds. *Quaternary extinctions: A prehistoric revolution.* Tucson: University of Arizona Press.

Lott, D. F. 1974. Sexual and aggressive behavior of American bison *Bison bison.* In V. Geist and F. Walther, eds., *The behavior of ungulates and its relation to management.* IUCN Publication No. 24. Morges, Switzerland.

McDonald, J. N. 1981. *North American bison: Their classification and evolution.* Berkeley: University of California Press.

———. 1984. An extinct muskox mummy from near Fairbanks, Alaska: A progress report. *Biol. Pap. Univ. of Alaska,* Special Report No. 4:148–461.

McHugh, T. 1958. Social behavior of the American buffalo. *Zoologica* 43:1–40.

McNaughton, S. J., J. L. Tarrants, M. M. McNaughton, and R. H. Davis. 1985. Silica as a defense against herbivory and growth promotion in African grasses. *Ecology* 66:528–35.

Maglio, V. J. 1973. The origin and evolution of the Elephantidae. *Amer. Philosoph. Soc.* 62:1–149.

Martin, P. S., and R. G. Klein, eds. 1984. *Quaternary extinctions: A prehistoric revolution.* Tucson: University of Arizona Press.

Matthews, J. V., Jr. 1979. Tertiary and quaternary environments: Historical background for an analysis of the Canadian insect fauna. In H. D. Danks, ed., *Canada and its insect fauna.* Memoirs of the Entomological Society of Canada, no. 108. Ottawa.

———. 1982. East Beringia during late Wisconsin time: A review of the biotic evidence. In D. M. Hopkins et al., eds., *Paleoecology of Beringia,* 127–50. New York: Academic Press.

Meagher, M. 1971. Snow as a factor influencing bison distribution and numbers in Pelican Valley, Yellowstone National Park. In A. O. Haugen, ed., *Proceedings of the snow and ice in relation to wildlife recreation symposium.* Iowa Cooperative Wildlife Research Unit, Iowa State University, Iowa City.

Mech, L. D. 1966. *The wolves of Isle Royale.* Fauna of the National Parks, Fauna Series No. 7. U.S. Department of the Interior, Washington, D.C.

Middendorff, A. Th. 1848. Reise in den aussersten Norden und Osten Sibiriens Wahrend der Jahre 1843–1844.

Miquelle, D. 1985. Food habits and range conditions of bison and sympatric ungulates on the upper Chitna River, Wrangell-Saint Elias National Park and Preserve. National Park Service–Alaska Region Research/Resources Management Report AR-8.

Mochanov, Y. A. 1977. *Drevneyshie Etap Zaseleniya Chenlovekom Severo-*

Vostochnoy Azii. Akad. Nauk SSSR, Sibirsk. Otdel., Yakutsk Fil., Inst. Yazyka, Literatury, i istorii (Izd-vo) "Nauka," Sibir. Otdel.

Morlan, R. E., and J. Cinq-Mars. 1982. Ancient Beringians: Human occupation in the late Pleistocene of Alaska and the Yukon Territory. In Hopkins et al., eds., *Paleoecology of Beringia,* 353–82. New York: Academic Press.

Murie, O. J. 1954. *A field guide to animal tracks.* Reprint. Boston: Houghton Mifflin.

Navowkowski, N. S. 1965. Cemental deposition as an age criterion in bison and the relation of incisor wear, eye-lens weight, and dressed bison carcass weight to age. *Can. J. Zool.* 43:1–7.

Nelson, R. E. 1982. Late Quaternary environments of the western Arctic Slope, Alaska. Ph.D. diss., University of Washington, Seattle.

Olivier, R. C. D. 1982. Ecology and behavior of living elephants: Bases for adaptations concerning the extinct woolly mammoths. In D. M. Hopkins et al., eds., *Paleoecology of Beringia,* 281–90. New York: Academic Press.

Olsen, F. W., and R. M. Hansen. 1977. Food relations of wild free-roaming horses to livestock and big game, Red Desert, Wyoming. *J. Range Mgmt.* 30:17–20.

Peden, D. G., G. M. van Dyne, R. W. Rice, and R. M. Hansen. 1974. The trophic ecology of *Bison bison* on shortgrass plains. *J. Appl. Ecol.* 11:489–98.

Petersburg, S. J. 1973. Bull bison behavior at Wind Cave National Park. Ph.D. diss., Iowa State University, Ames.

Péwé, T. L. 1975a. Quaternary geology of Alaska. U.S. Geological Survey Professional Paper No. 835.

———. 1975b. Quaternary stratigraphy and nomenclature in unglaciated central Alaska. U.S. Geological Survey Professional Paper No. 862.

Pfitzenmayer, E. E. 1926. Mammulteichen und Urwaldmenschen in Nordost-Sibirien. Leipzig: F. A. Brockhaus.

Poplin, F. 1984. Sur le profil dorso-limbaire des bisons dans l'art Paléolithique. In H. G. Bandi, W. Huber, M.-R. Sauter, and B. Sitter, eds., *La contribution de la zoologie et de l'éthologie à l'interprétation de l'art des peuples chassuers préhistoriques,* 217–42. Fribourg: Editions Universitaires Fribourg Suisse.

Powers, W. R., R. D. Guthrie, and J. Hoffecker. In press. *The Dry Creek Site.* Fairbanks: Alaska Anthropological Association.

Quackenbush, L. S. 1909. Notes on the Alaskan mammoth expeditions of 1907 and 1908. *Bull. Amer. Mus. Nat. Hist.* 26:87–130.

Reher, C. A. 1974. Population study of the Casper Site bison. In G. C. Frison, ed., *The Casper Site,* 113–24. New York: Academic Press.

———. 1977. Adaptive processes on the shortgrass plains. In L. R. Binford, ed., *For theory building in archaeology,* 13–40. New York: Academic Press.

Ritchie. J. C. 1969. Absolute pollen frequencies and C14 age of a section of

Holocene lake sediments from Riding Mountain area of Manitoba. *Can. J. Bot.* 47:1345–49.
———. 1984. Past and present vegetation of far northwest of Canada. Toronto: University of Toronto Press.
Ritchie, J. C., and L. C. Cwynar. 1982. The late Quaternary vegetation of the northern Yukon. In D. M. Hopkins et al., eds., *Paleoecology of Beringia,* 113–26. New York: Academic Press.
Roe, F. G. 1951. *North American buffalo: A critical study of the species in its wild state.* Toronto: Toronto University Press.
Roskosz, T. 1962. Morphologie der Wirbelsäule des Wisents. *Acta Theriol.* 5:113–64.
Roskosz, T., and W. Empel. 1961. The size of the head and the height of the spinous processes in the region of the withers of the European bison. *Acta Theriol.* 6:63–69.
Rutter, N. W. 1978. Geology of the ice-free corridor. American Quaternary Association Abstracts, Fifth Annual Meeting, University of Alberta, Edmonton.
Salter, R. E., and R. J. Hudson. 1980. Range relationships of feral horses with wild ungulates and cattle in western Alberta. *J. Range Mgmt.* 33:266–71.
Sanderson, I. T. 1960. Riddle of the frozen giants. *Saturday Evening Post,* January 16.
Schaffer, W., and C. Reed. 1972. The co-evolution of social behavior and cranial morphology in sheep and goats (*Bovidae Caprini*). *Fieldiana Zoology* 61:1–88.
Schaller, G. B. 1967. *The deer and the tiger.* Chicago: University of Chicago Press.
———. 1972. *The Serengeti lion.* Chicago: University of Chicago Press.
Schloeth, R. 1961. Das Sozialleben des Camarrgue-Rindes: Qualitative und quantitative Untersuchungen über die sozialen Beziehungen-insbesondere die soziale Rangordnung des halbwilden französischen Kampfrindes. *Z. Tierpsychol.* 18:574–627.
Schweger, C. E. 1982. Late Pleistocene vegetation of eastern Beringia: Pollen analysis of dated alluvium. In D. M. Hopkins et al., eds., *Paleoecology of Beringia,* 95–112. New York: Academic Press.
Schweger, C. E., and T. Habgood. 1976. The late Pleistocene steppe-tundra in Beringia: A critique. *AMQUA Abstracts* 4:80–81.
Schweger, C. E., and J. A. P. Janssens. 1980. Paleoecology of the Boutellier non-glacial interval, St. Elias Mountains, Yukon Territory, Canada. *Arctic and Alpine Res.* 12:309–17.
Serebrovskij, P. V. 1935. *A history of the animal world in the USSR.* Moscow: Nauka.
Sher, A. V. 1968. Fossil saiga in northeastern Siberia and Alaska. *Inter. Geol. Rev.* 10:1247–60.
———. 1971. *Pleistocene mammals and stratigraphy of the far northeast USSR and North America* (in Russian). Geological Institute, Moscow

Academy of Sciences. Trans. by the American Geological Institute, in 1974, in *Inter. Geogr. Rev.* 16:1–284.
———. 1974. Pleistocene mammals and stratigraphy of the northeast USSR and North America. *Inter. Geol. Rev.* 16:1–284.
Shult, J. M. 1972. American bison behavior patterns at Wind Cave National Park. Ph.D. diss., Iowa State University, Ames.
Sikes, S. K. 1971. The natural history of the African elephant. American Elsevier Publishing.
Simpson, G. G. 1951. *Horses.* Natural History Library Ed. Garden City, N.Y.: Doubleday.
Sinclair, A. R. E. 1977. *The African buffalo.* Chicago: University of Chicago Press.
Skarlato, O. A., Ed. 1977. *The anthropogene flora and fauna of the north-east Siberia.* Academy of Sciences of the USSR, Proceedings of the Zoological Institute, vol. 63.
Skinner, M. F., and O. C. Kaisen. 1947. The fossil bison of Alaska and preliminary revision of the genus. *Bull. Amer. Mus. Nat. Hist.* 89:126–256.
Smuts, G. L. 1976. Population characteristics and recent history of lions in two parts of Kruger National Park. *Koedoe* 19:153–64.
Solonevich, N. G., B. A., Tikhomirov, and V. V. Ukraintseva. 1977. Preliminary results and investigation of the plant remains from the gastrointestinal tract of the Shandrin mammoth (Yakutia). In O. A. Skarlato, ed., *The anthropogene fauna and flora of the north-east Siberia.* Academy of Sciences of the USSR, Proceedings of the Zoological Institute, vol. 63.
Solonevich, N. G., and V. V. Vikhireva-Vasil'kova. 1977. Plant remains in the contents of the gastrointestinal tract of the Selerikan fossil horse. In O. A. Skarlato, ed., *The anthropogene fauna and flora of the north-east Siberia.* Academy of Sciences of the USSR, Proceedings of the Zoological Institute, vol. 63.
Soper, J. 1941. History, range and home life of the northern bison. *Ecol. Monog.* 11:347–412.
Spiess, A. E. 1979. *Reindeer and caribou hunters: An archaeological study.* New York: Academic Press.
Stuart, A. J. 1982. *Pleistocene vertebrates in the British Isles.* London: Longman.
Sutcliffe, A. J. 1970. Spotted hyena, crusher, gnawer, digester and collector of bones. *Nature* 227:1110–13.
Swiezynski, K. 1962. Pleistocene musculatural systems of the European bison, *Bison bonasus. Acta Theriol.* 6:165–217.
Tankersley, N. G. 1981. Mineral lick use by moose in central Alaska Range. M.S. thesis, University of Alaska, Fairbanks.
Telfer, E. S., and J. P. Kelsall. 1984. Adaptation of some large North American mammals for survival in snow. *Ecology* 65:1828–34.
Thorsen, R., and R. D. Guthrie. The Colorado Creek Mammoth. In preparation.

Tikhomirov, B. A. 1958. Natural conditions and vegetation in the mammoth epoch in northern Siberia. *Problems in the North* 1:168–88.

Tolmachoff, I. P. 1929. The carcasses of the mammoth and rhinoceros found in frozen ground in Siberia. *American Philosophical Society of Philadelphia* 8:1–89.

Tseytlin, S. M. 1979. *Geologiya paleolita severnay Azii* (The geology of Paleolithic of north Asia.) Moscow: Nauka.

Turner, A. 1982. Hominids and fellow travelers. *So. Afr. J. Sci.* 78:231–37.

Ukraintseva, V. V. 1981. Vegetation of warm late Pleistocene intervals and the extinction of some large herbiverous mammals. *Polar Geography and Geology* 4:189–203.

Vangengeim, E. A. 1967. The effect of the Bering Land Bridge on the Quaternary Faunas of Siberia and North America. In D. M. Hopkins, ed., *The Bering Land Bridge.* Stanford: Stanford University Press.

———. 1975. Sur la fauna periglaciare du Pleistocene. *Biuletyn Periglacjalny* 24:81–88.

Van Orsdol, K. G., J. B. Hanby, and J. D. Bygott. 1985. Ecological correlates of lion social organization (*Panthera leo*). *J. Zool. Lond.* 206:97–112.

Van Vuren, D. 1984. Summer diets of bison and cattle in southern Utah. *J. Range Mgmt.* 37:260–61.

Van Zyll de Jong, C. G. 1986. *A systematic study of recent bison, with particular consideration of the wood bison* (*Bison bison athabascae* Rhoads 1898). National Museum of Canada, Ottawa.

Vavra, M., and F. Sneva. 1978. *Seasonal diets of five ungulates grazing the cold desert biome.* Proceedings of the International Rangelands Conference.

Velikovsky, I. 1955. Earth in upheaval. New York: Doubleday.

Vereshchagin, N. K. 1959. *Mammals of the Caucasus: The history of the fauna.* Leningrad: Academy of Science.

———. 1971. The cave lion and its history in the Holarctic and in the territory of the USSR. *Trudy Zool. Inst. Akad. Nauk. SSSR* 49:123–97.

———. 1977. Some problems of the history of the formation of theriofaunas. In O. A. Skarlato, ed., *The anthropogene fauna and flora of the northeast Siberia.* Academy of Sciences of the USSR. Proceedings of the Zoological Institute, vol. 63.

Vereshchagin, N. K., and G. F. Baryshnikov. 1982. Paleoecology of the mammoth fauna in the Eurasian Arctic. In D. M. Hopkins et al., eds., *Paleoecology of Beringia,* 267–80. New York: Academic Press.

———. 1984. Quaternary mammalian extinctions in northern Eurasia. In P. S. Martin and R. G. Klein, eds., *Quaternary extinctions: A prehistoric revolution,* 483–516. Tucson: University of Arizona Press.

Vereshchagin, N. K., and I. E. Kuz'mina, eds. 1982. *The mammoth fauna of the Asiatic part of the USSR.* Akademiya Nauk, Trudy Zoologicheskologo Instituta, vol. 3.

Vereshchagin, N. K., and P. A. Lazarev. 1977. Description of soft parts and skeleton of the Selerikan horse. In O. A. Skarlato, ed., *Fauni i flori anthropogena Severo-Vostoka Sibiri,* 85–185. Akademiya Nauk, Trudy Zoologischeskologo Instituta, vol. 63.

Vereshchagin, N. K., and V. M. Mikhel'son. 1981. *The Magadan baby-mammoth, Mammuthus primigenius.* Akademiya Nauk, Trudy Zoologischeskologo Instituta.

Vereshchagin, N. K., and A. I. Nikolaev. 1982. Excavations of the Khatanga mammoth. In N. K. Vereshchagin and A. I. Nikolaev, eds., *The mammoth fauna of the Asiatic part of the USSR.* Akademiya Nauk, Trudy Zoologischeskologo Instituta, vol. 3.

Vollosovich, C. A. 1909. On the digging out of the Sanga-Yurakh mammoth, in 1908 (in Russian). *Bull. Acad. Sci.* 3:437–58.

Vrba, E. S. 1976. The fossil Bovidae of Sterkfontein, Swartkrans, and Kromdraii. *Transvaal Museum Memoirs* 21:1–166.

Waddington, J. B. C. 1969. A stratigraphic record of the pollen influx to a lake in the big woods of Minnesota. *Geol. Soc. Amer. Special Paper* 123:236–82.

Walker, D. N., and M. S. Boyce. 1984. Review of *North American bison: Their classification and evolution,* by J. N. McDonald. *Zooarchaeol. Res. News.* 3:8–11.

Walther, F. 1966. *Mit Horn und Huff.* Berlin: Parley-Verlag.

———. 1974. Some reflections on expressive behavior in combats and courtship of certain horned ungulates. In V. Geist and F. Walther, eds., *The behavior of ungulates and its relation to management,* 56–106. IUCN Publication No. 24. Morges, Switzerland.

Washburn, A. L. 1980. *Geocryology.* New York: John Wiley and Sons.

Wasilewski, W. 1967. Differences in the wear of incisors in the European bison living under natural and reserve conditions. *Acta Theriol.* 12:459–62.

Watanabe, O. 1969. On permafrost ice (in Japanese). *Seppyo* (J. Japanese Soc. of Snow and Ice) 31:53–62.

Weeks, H. P., Jr. 1978. Salt preferences and sodium drive phenology in fox squirrels and woodchucks. *J. Mammal.* 59:531–42.

Weeks, H. P., Jr., and C. M. Kirkpatrick. 1976. Adaptations of white-tailed deer to naturally occurring sodium deficiencies. *J. Wildl. Mgmt.* 40:610–25.

Wells, P. V. 1970. Postglacial vegetation on the Great Plains. *Science* 153:970–75.

West, F. H. 1981. *The archaeology of Beringia.* New York: Columbia University Press.

Wijmstra, T. A., and T. Van der Hammen. 1974. The last Interglacial-Glacial cycle: State of affairs of correlation between data obtained from the land and from the ocean. *Geol. en. Mijnb.* 53:386–92.

Wilson, M. 1975. Holocene fossil bison from Wyoming and adjacent areas. M.A. thesis, University of Wyoming, Laramie.

———. 1978. Archaeological kill site populations and the Holocene evolu-

tion of the genus Bison. In L. Davis and M. Wilson, ed., *Bison evolution and utilization*, 9–23. *Plains Anthropologist*, Memoir No. 14.
Woillard, G. 1978. Grande Pile peat bog: A continuous pollen record for the last 140,000 years. *Quart. Res.* 9:1–21.
Wroblewski, K. 1927. *Zubr Puszczy Bialowieskiej (monografia).* Nakl. Ogrodu zool. w Poznaniu. Wyd. Polskie.
Wynne-Edwards, V. C. 1962. Animal dispersion in relation to social behavior. New York: Hafner Publishing.
Wu, T. H. 1984. Soil movements of permafrost slopes near Fairbanks, Alaska. *Can. Geotech. J.* 21:699–709.
Yegorova, T. V. 1977. Carpological analysis of the plant remains of the food of the Selerikan horse. In O. A. Skarlato, ed., *The anthropogene fauna and flora of the north-east Siberia.* Academy of Sciences of the USSR, Proceedings of the Zoological Institute, vol. 63.
Young, S. B. 1976. Is steppe tundra alive and well in Alaska? *AMQUA Abstracts* 4:84–88.
———. 1982. The vegetation of land-bridge Beringia. In D. M. Hopkins, ed., *The Bering Land Bridge.* Stanford: Stanford University Press.
Yudichev, Yu. R., and A. I. Averikhin. 1982. On the macro- and micromorphology of the organs of the abdominal cavity of the Shandrin mammoth and causes of its death. In N. K. Vereshchagin and A. I. Nikolaev, eds., *The mammoth fauna of the Asiatic part of the USSR*, 35–37. Akademiya Nauk, Trudy Zoologischeskologo Instituta, vol. 3.
Yurtsev, B. A. 1974. Steppe communities in the Chukotka tundra and the Pleistocene "tundra-steppe." *Bot. Zhur.* 59:484–501.
———. 1982. Relics of the xerophytic vegetation of Beringia in northeastern Asia. In D. M. Hopkins et al., eds., *Paleoecology of Beringia*, 157–78. New York: Academic Press.
Zablocki, Von M. 1967. Territorium und Markierung beim Wisent. *Zeitschrift für Saugetierkund* 33:121–23.

INDEX